The Oblique Derivative Problem of Potential Theory

CONTEMPORARY SOVIET MATHEMATICS

Series Editor: Revaz Gamkrelidze, *Steklov Institute, Moscow, USSR*

ASYMPTOTICS OF OPERATOR AND PSEUDO-DIFFERENTIAL EQUATIONS
V. P. Maslov and V. E. Nazaikinskii

COHOMOLOGY OF INFINITE-DIMENSIONAL LIE ALGEBRAS
D. B.Fuks

DIFFERENTIAL GEOMETRY AND TOPOLOGY
A. T. Fomenko

HOMOLOGY OF ANALYTIC SHEAVES AND DUALITY THEOREMS
V. D. Golovin

LINEAR DIFFERENTIAL EQUATIONS OF PRINCIPAL TYPE
Yu. V. Egorov

THE OBLIQUE DERIVATIVE PROBLEM OF POTENTIAL THEORY
A. I. Yanushauskas

OPTIMAL CONTROL
V. M. Alekseev, V. M. Tikhomirov, and S. V. Fomin

THEORY OF SOLITONS: The Inverse Scattering Method
S. Novikov, S. V. Manakov, L. P. Pitaevskii, and V. E. Zakharov

TOPICS IN MODERN MATHEMATICS: Petrovskii Seminar No. 5
Edited by O. A. Oleinik

The Oblique Derivative Problem of Potential Theory

A. I. Yanushauskas
Mathematics Institute
Academy of Sciences of the Lithuanian SSR
Vilnius, USSR

Translated from Russian by
Norman Stein

Consultants Bureau • New York and London

Library of Congress Cataloging in Publication Data

IAnushauskas, Al'gimantas Ionasovich.
[Zadacha o naklonnoĭ proizvodnoĭ teorii potent͡siala. English]
The oblique derivative problem of potential theory / A. I. Yanushauskas.
p. cm.—(Contemporary Soviet mathematics)
Translation of: Zadacha o naklonnoĭ proizvodnoĭ teorii potent͡siala.
Bibliography: p.
Includes index.
ISBN 978-1-4684-1676-3
1. Differential equations, Elliptic. 2. Boundary value problems. 3. Harmonic functions. 4. Fredholm equations. I. Title. II. Series.
QA374.I2613 1989 89-595
515.3′53—dc19 CIP

ISBN-13: 978-1-4684-1676-3 e-ISBN-13: 978-1-4684-1674-9
DOI: 10.1007/978-1-4684-1674-9

This translation is published under an agreement with the
Copyright Agency of the USSR (VAAP)

Softcover reprint of the hardcover 1st edition 1989

A Division of Plenum Publishing Corporation
233 Spring Street, New York, N.Y. 10013

PREFACE

An important part of the theory of partial differential equations is the theory of boundary problems for elliptic equations and systems of equations. Among such problems those of greatest interest are the so-called non-Fredholm boundary problems, whose investigation reduces, as a rule, to the study of singular integral equations, where the Fredholm alternative is violated for these problems. Thanks to developments in the theory of one-dimensional singular integral equations [28, 29], boundary problems for elliptic equations with two independent variables have been completely studied at the present time [13, 29], which cannot be said about boundary problems for elliptic equations with many independent variables. A number of important questions in this area have not yet been solved, since one does not have sufficiently general methods for investigating them.

Among the boundary problems of great interest is the oblique derivative problem for harmonic functions, which can be formulated as follows: In a domain D with sufficiently smooth boundary Γ find a harmonic function $u(X)$ which, on Γ, satisfies the condition

$$\sum_{i=1}^{n} \alpha_i(X) \frac{\partial u}{\partial x_i} = f(X), \quad \sum_{i=1}^{n} [\alpha_i(X)]^2 = 1,$$

where $\alpha_1, \ldots, \alpha_n, f$ are sufficiently smooth functions defined on Γ. Obviously the left side of the boundary condition is the derivative of the function $u(X)$ in the direction of the vector $P(X)$ with components $\alpha_1(X), \ldots, \alpha_n(X)$. If the vector field $P(X)$ does not go out to the plane tangent to Γ at any point of the surface Γ, then the oblique derivative problem is Fredholm for any $n \geq 2$, while if the field goes out to the tangent plane to Γ, then this problem has been investigated conclusively only for $n = 2$ [26]. One can always reduce the oblique derivative problem equivalently to a singular integral equation on Γ; however, for $n > 2$, one is only able to investigate this equation successfully when the vector field $P(X)$ does not go out to the tangent plane to Γ at any point [52]. Such integral equations have not been studied when $P(X)$ goes out to the tangent plane to Γ.

One can solve the oblique derivative problem for harmonic functions of two independent variables either by reducing it to singular integral equations, or directly, using the properties of conjugate harmonic functions of two independent variables [29, 57]. One can also apply the second route as a method for studying the corresponding singular integral equations. Such an approach seems tempting for constructing a theory of multidimensional integral equations; however, the methods of direct investigation of the oblique derivative problem for harmonic functions of two variables which Liènard [57] applied turn out to be of little use in the multidimensional case. Thus, two problems are basic in the theory of non-Fredholm boundary problems for multidimensional elliptic equations: a) investigation of the oblique derivative problem for harmonic functions; b) construction of a general theory of multidimensional singular integral equations. These two problems are equivalent.

In this book our attention is focused mainly on the study of the measure of subdefiniteness or the measure of overdeterminedness of the oblique derivative problem for harmonic functions of many variables, and constructive methods of investigation of the given problem and some related questions. In Chapter 1 we review the basic facts of the theory of linear elliptic equations and other auxiliary information. We consider the three-dimensional case mainly and give the generalizations to the multidimensional one without detailed proofs. We study various boundary problems for second-order equations and methods for reducing them to singular integral equations.

Chapter 2 is devoted to the oblique derivative problem. In it we review the method of Bouligand–Giraud for reducing the oblique derivative problem to an equivalent Fredholm equation, when the problem itself is Fredholm. We illustrate the basic effects caused by the exit of the direction of differentiation to the tangent plane to the boundary of the domain and show that one can apply the Bouligand–Giraud method for seeking part of the solutions of the problem even when it is not Fredholm. Now the problem is not equivalent to a Fredholm equation, so the question of seeking solutions that are not obtained from a Fedholm equation requires special consideration.

In Chapter 3 we study the non-Fredholm oblique derivative problem. We give constructive methods for investigating it and establish an analog of the argument principle of the theory of analytic functions of one complex variable for fields of gradients of harmonic functions of three variables.

A number of questions closely related to the oblique derivative problem and connected with multidimensional generalizations of the Cauchy–Riemann system are considered in Chapter 4. For the simplest generalizations of the Cauchy–Riemann system we investigate the connection of the analog of the Riemann–Hilbert problem with the oblique derivative problem. In this same chapter we study second-order elliptic partial differential equations, for which the Noetherian property of the Dirichlet problem is lost. For a number of such problems we investigate the Dirichlet problem in domains of special form.

Chapter 5 is devoted to the oblique derivative problem for harmonic functions of two independent variables [57].

CONTENTS

CHAPTER 1. FOUNDATIONS OF POTENTIAL THEORY 1

1. Harmonic Functions and Potential Theory 1
2. Green's Formula and Its Consequences 8
3. Basic Boundary Problems of Potential Theory 12
4. Investigation of Boundary Problems by the Method of Integral Equations 17
5. Harmonic Functions in Axially Symmetric Domains 27
6. General Second-Order Elliptic Equations 31
7. Functions Represented by Potential-Type Integrals 45
8. Gradient Vector Fields of Functions 54

CHAPTER 2. OBLIQUE DERIVATIVE PROBLEM FOR ELLIPTIC EQUATIONS 61

1. Reduction of the Oblique Derivative Problem to Fredholm Integral Equations 62
2. Reduction of the Oblique Derivative Problem for Harmonic Functions to Fredholm Equations 81
3. Simplest Properties of the Non-Fredholm Oblique Derivative Problem. 87
4. Global Methods of Investigation of the Non-Fredholm Oblique Derivative Problem 101

CHAPTER 3. OBLIQUE DERIVATIVE PROBLEM WITH DIRECTION OF DIFFERENTIATION GOING INTO THE TANGENT PLANE 113

1. Simplest Consequences of the Maximum Principle 113
2. Generalizations of the Argument Principle 118

3. Measure of Overdeterminedness of the Oblique Derivative Problem..... 132
4. Oblique Derivative Problem with Polynomial Coefficients................ 136
5. Reduction of the Oblique Derivative Problem to a Fredholm Integrodifferential Equation.. 145
6. Boundary Problem for a System of Harmonic Functions.................. 161

CHAPTER 4. SYSTEMS OF PARTIAL DIFFERENTIAL EQUATIONS RELATED TO MULTIDIMENSIONAL GENERALIZATIONS OF THE CAUCHY–RIEMANN SYSTEM.. 169

1. Analog of the Riemann–Hilbert Problem 170
2. Generalization of a Holomorphic Vector..................................... 180
3. Second-Order Systems of Equations... 189
4. Elliptic Systems Depending on a Parameter 200

CHAPTER 5. OBLIQUE DERIVATIVE PROBLEM FOR HARMONIC FUNCTIONS OF TWO VARIABLES.......................... 213

1. Boundary Properties of Conjugate Harmonic Functions.................. 213
2. An Auxiliary Problem.. 217
3. Oblique Derivative Problem 222
4. Oblique Derivative Problem with Discontinuities in the Boundary Condition.. 230
5. Variation of Level Lines of a Harmonic Function of a Closed Contour 235
6. Multiply Connected Domains.. 238

REFERENCES... 249

Chapter 1

FOUNDATIONS OF POTENTIAL THEORY

1. HARMONIC FUNCTIONS AND POTENTIAL THEORY

In the domain D with boundary Γ in the space R^n of variables $x_1, \ldots, x_n$ we consider a function $U(x_1, \ldots, x_n) = U(X)$ of a point $X = (x_1, \ldots, x_n)$. The differential equation

$$\Delta u \equiv \sum_{k=1}^{n} \frac{\partial^2 u}{\partial x_k^2} = 0 \tag{1.1}$$

is called Laplace's equation, and its solutions are called harmonic functions. The corresponding inhomogeneous equation, which we shall write in the form

$$\Delta u = -\omega_n \mu(X), \tag{1.2}$$

where $\omega_n = 2(\sqrt{\pi})^n[\Gamma(n/2)]^{-1}$ is the area of the unit sphere of the space R^n, is the Poisson equation, where $\mu(X)$ is a given function of a point $X \in D$. If the domain D is bounded, then a solution of Laplace's equation which has continuous second derivatives in D can be considered a regular harmonic function. Analogously, if $\mu(X)$ is continuous in D, then a solution of Poisson's equation with continuous second derivatives is called regular in D.

Let $n = 3$, and $f(w, t)$ be an analytic function of the complex variable w. By direct substitution into Laplace's equation it is easy to verify that the function $u = f(z + ix \cos t + iy \sin t, t)$ satisfies this equation for any t, and it follows from this that for any fixed a and b

$$u(x, y, z) = \int_a^b f(z + ix \cos t + iy \sin t, t)\, dt \tag{1.3}$$

is a solution of Laplace's equation. If in this formula we set $f(w, t) = w^n \exp imt$, $a = -\pi$, $b = \pi$, then we get a homogeneous harmonic polynomial

$$u(x, y, z) = \int_{-\pi}^{\pi} (z + ix\cos t + iy\sin t)^n \exp imtdt,$$

which, in spherical coordinates $x = r\sin\theta\cos\varphi$, $y = r\sin\theta\sin\varphi$, $z = r\cos\theta$, assumes the form

$$n = 2r^n e^{im\varphi} \int_0^{\pi} (\cos\theta + i\sin\theta\cos t)^n$$
$$\times \cos mtdt = r^n P_n^m(\cos\theta)\exp im\varphi,$$

where $P_n^m(\cos\theta)$ are the associated Legendre functions

$$P_n^m(\cos\theta) = (-1)^m \frac{(n+m)!}{2^m(n-m)!m!}(\sin\theta)^m F(m-n,\ m+n+1;\ m+1;\ (1-\cos\theta)/2).$$

Here $F(\alpha, \beta; \gamma; s)$ is the Gauss hypergeometric function which for $m \le n$ becomes a polynomial [14].

In spherical coordinates for $n = 3$ the Laplace operator

$$\Delta u = \frac{1}{r^2\sin\theta}\left[\frac{\partial}{\partial r}(r^2 u_z \sin\theta) + \frac{\partial}{\partial\theta}(u_\theta\sin\theta) + \frac{\partial}{\partial\varphi}(u_\varphi/\sin\theta)\right].$$

Let $u(x, y, z)$ be a regular harmonic function in a bounded domain; we consider the function

$$v(x, y, z) = r^{-1}u(x/r^2, y/r^2, z/r^2),\ r^2 = x^2 + y^2 + z^2.$$

It is defined in the domain D' obtained from the domain D by inversion with respect to the unit sphere S: $\{x^2 + y^2 + z^2 = 1\}$. By direct calculation one verifies the relation

$$r^5 \frac{1}{r^2\sin\theta}\frac{\partial}{\partial r}(r^2 v_r\sin\theta) = \frac{1}{\rho^2\sin\theta}\frac{\partial}{\partial\rho}(\rho^2 u_\rho\sin\theta),$$

where $\rho = r^{-1}$. It follows from this equation and the form of the Laplace operator in spherical coordinates that the function $v(x, y, z)$ is harmonic at all finite points of the domain D'. In the space R^n, $n > 3$, along with the function $u(x_1, \ldots, x_n)$, the function

$$v(x_1, \ldots, x_n) = r^{2-n}u(x_1/r^2, \ldots, x_n/r^2) \tag{1.4}$$

is harmonic, which can be verified by substitution in the equation.

Let the harmonic function u be regular in the bounded domain D. We take an interior point of the domain D and perform inversion with respect to the unit sphere with center at this point, assuming, without loss of generality, that this point is the origin. The domain D goes into a domain D' here, lying in the exterior of the image Γ' of the boundary Γ of the domain D. We shall call the harmonic function v, which is obtained from u by (1.4), regular in the domain D'. Consequently, the regularity of a harmonic function in the domain D, which goes to infinity, is defined as follows: with the help of inversion with respect to a sphere with center at a point which is exterior to D, we carry the domain D into a bounded domain D'. The harmonic function u is said to be regular in D, if the function v corresponding to it is regular in D'. In particular, if D contains a neighborhood of infinity, and u is such that v is regular in D', then u is said to be regular at infinity. By virtue of this definition for $n \geq 3$ the harmonic function $u(x_1, \ldots, x_n) \equiv \text{const}$ is not regular at infinity.

We find solutions $\psi(r)$ of Laplace's equation, which depend only on $r = \sqrt{x_1^2 + \ldots + x_n^2}$. By direct calculation from Laplace's equation we get for $\psi(r)$

$$\psi'' + (n-1)r^{-1}\psi' = 0,$$

whose general solution has the form $\psi(r) = C_1 \ln r + C_2$, $n = 2$, $\psi(r) = C_1 r^{2-n} + C_2$, $n \geq 3$, where C_1 and C_2 are arbitrary constants. We consider the functions

$$\gamma(r) = \frac{1}{(n-2)\,\omega_n} r^{2-n}, \quad n \geqslant 3,$$

$$\gamma(r) = \frac{1}{2\pi} \ln \frac{1}{r}, \quad n = 2, \quad r^2 = \sum_{k=1}^{n} (x_k - \xi_k)^2.$$

These functions for $r = 0$ have a so-called characteristic singularity. Any solution of Laplace's equation, defined in the domain D,

$$\psi(x_1, \ldots, x_n, \xi_1, \ldots, \xi_n) = \gamma(r) + w,$$

where $\Xi = (\xi_1, \ldots, \xi_n)$ is an interior point of the domain D and w is a regular harmonic function in the domain D, is called a fundamental solution with singularity at the point Ξ.

It is also easy to construct a solution with characteristic singularity for the more general equation $\Delta u + cu = 0$, where c is a constant. For solutions of this equation of the form

$$u = \psi(r), \quad r^2 = \sum_{k=1}^{n} (x_k - \xi_k)^2,$$

we now get

$$\psi'' + \frac{n-1}{r}\psi' + c\psi = 0,$$

and as a result of the change of variables $\rho = r\sqrt{c}$, $\psi(r) = r^{-(n-2)/2}\varphi(r\sqrt{c})$ we arrive at Bessel's equation [19]

$$\varphi'' + \frac{1}{\rho}\varphi' + \left[1 - \left(\frac{n-2}{2}\right)^2 \frac{1}{\rho^2}\right]\varphi = 0.$$

The solution with characteristic singularity sought is a solution of this equation which is unbounded for $\rho = 0$ and has the form

$$\psi(r) = r^{-\frac{1}{2}(n-2)} J_{-\frac{1}{2}(n-2)}(r\sqrt{c}), \quad n = 2k+1,$$

$$\psi(r) = r^{-\frac{1}{2}(n-2)} N_{\frac{1}{2}(n-2)}(r\sqrt{c}), \quad n = 2k.$$

Here J_ν is a Bessel function and N_ν is a Neumann function:

$$J_\nu(z) = \frac{z^\nu}{2^\nu}\sum_{k=0}^{\infty}(-1)^k \frac{z^{2k}}{2^k k!\,\Gamma(\nu+k+1)},$$

$$\pi N_l(z) = 2J_l(z)\,[\ln z/2 + C]$$

$$-\sum_{k=0}^{l-1}\frac{(l-k-1)!}{k!}\left(\frac{z}{2}\right)^{2k-l} - \left(\frac{z}{2}\right)^{l}\frac{1}{l!}\sum_{k=1}^{l}\frac{1}{k}$$

$$-\sum_{k=1}^{\infty}\frac{(-1)^k}{k!\,(k+l)!}\left(\frac{z}{2}\right)^{l+2k}\left[\sum_{m=1}^{l+k}\frac{1}{m}+\sum_{m=1}^{k}\frac{1}{m}\right],$$

where $l+1$ is a natural number and C is Euler's constant.

For $n = 3$ one can take a fundamental solution of Laplace's equation in the form $1/4\pi r$. Physically the function

$$r^{-1} = [(x-\xi)^2 + (y-\eta)^2 + (z-\zeta)^2]^{-1/2}$$

is the gravitational potential which is originated at the point $P = (x, y, z)$ by a unit mass concentrated at the point $Q = (\xi, \eta, \zeta)$. Let $\mu(\xi, \eta, \zeta)$ be a function defined in the domain D. The integral

$$u(x, y, z) = \int_D \frac{\mu(\xi, \eta, \zeta)}{r}\, d\xi d\eta d\zeta \tag{1.5}$$

is called the potential of the spatial mass distribution with density μ in the domain D. In general the integral

$$u(X) = \int_D \mu(Q)\,\gamma(r)\, d\xi_1 \ldots d\xi_n, \tag{1.5a}$$

$$X = (x_1, \ldots, x_n),\ Q = (\xi_1, \ldots, \xi_n),\ r^2 = \sum_{k=1}^{n}(x_k - \xi_k)^2,$$

is also called the potential of a mass distribution in the domain D with density μ. If the point X lies in the exterior of the domain D, then the potential $u(X)$ at this point is a harmonic function. This is easy to show by differentiating under the integral sign. Now if $X \in D$ and μ has continuous derivatives, then the potential (1.5a) satisfies the Poisson equation $\Delta u = -\mu(X)$. We consider only the case $n = 3$ in more detail.

Theorem 1.1. If the density $\mu(X)$ in (1.5) is bounded and integrable in the domain D, then the potential (1.5) and its first derivatives are uniformly continuous, while one can calculate these derivatives by differentiating under the integral sign.

Proof. We consider the function

$$u_\delta(x, y, z) = \int_D \mu(\xi, \eta, \zeta) f_\delta(r)\, d\xi d\eta d\zeta,$$

$f_\delta(r)$ being an auxiliary positive function, which coincides with $1/r$ for $r \geq \delta$, i.e.,

$$f_\delta(r) = \begin{cases} \frac{1}{2\delta}(3 - r^2/\delta^2), & r \leqslant \delta. \\ 1/r, & r > \delta. \end{cases}$$

One has the inequality

$$\begin{aligned} |u_\delta - u| &= \left| \int_{\Sigma(\delta)} \mu(\xi, \eta, \zeta)[f_\delta(r) - r^{-1}]\, d\xi d\eta d\zeta \right. \\ &\leqslant M \int_{\Sigma(\delta)} [f_\delta(r) + r^{-1}]\, d\xi d\eta d\zeta \\ &= 4\pi M \int_0^\delta [f_\delta(r) + r^{-1}]\, r^2 dr = \frac{18}{5}\pi M \delta^2, \end{aligned}$$

where M is the maximum of $|\mu|$, and $\Sigma(\delta)$ is the ball $r < \delta$. It follows from this inequality that as $\delta \to 0$ the sequence u_δ converges uniformly to the potential u and u is uniformly continuous in D.

From the differentiability of the functions $f_\delta(r) = g(x - \xi, y - \eta, z - \zeta)$ it follows that u_δ is differentiable, while

$$\frac{\partial u_\delta}{\partial x} = \int_D \mu(\xi, \eta, \zeta) \frac{\partial}{\partial x} f_\delta(r)\, d\xi d\eta d\zeta.$$

We consider the convergent integral

$$w(x, y, z) = \int_D \mu(\xi, \eta, \zeta) \frac{\partial}{\partial x}\left(\frac{1}{r}\right) d\xi d\eta d\zeta, \tag{1.6}$$

which is obtained by formal differentiation of the expression under the integral sign in (1.5). We have

$$\frac{\partial}{\partial x} u_\delta - w = \int_{\Sigma(\delta)} \mu(\xi, \eta, \zeta) \left[\frac{\partial}{\partial x} f_\delta(r) - \frac{\partial}{\partial x} r^{-1}\right] d\xi d\eta d\zeta,$$

from which it follows that

$$\left|\frac{\partial}{\partial x} u_\delta - w\right| \leqslant 4\pi M \int_0^\delta \left(\left|\frac{\partial}{\partial x} f_\delta(r)\right| + r^{-2}\right) r^2 dr = 5\pi M \delta.$$

This means that the sequence $\frac{\partial}{\partial x}u_\delta$ converges uniformly to w, so $w = u_x$ and w is uniformly continuous.

Definition 1.1. We call the function μ Hölder continuous in the domain D with exponent α, $0 < \alpha \leq 1$, and with coefficient K, if for any pair of points P and Q of the domain D one has

$$|\mu(P) - \mu(Q)| \leqslant K[L(P, Q)]^\alpha,$$

where $L(P, Q)$ is the distance between the points P and Q.

The inequality in this definition is called a Hölder condition and one says about the function μ that it satisfies a Hölder condition.

Theorem 1.2. If the density $\mu(X)$ of the potential (1.5) satisfies a Hölder condition in the domain D, then this potential has continuous second derivatives and satisfies the Poisson equation $\Delta u = -4\pi\mu$.

We give a proof for the case when the density $\mu(X)$ is continuously differentiable. Under this assumption in (1.6) one can integrate by parts. We have

$$u_x = \int_D \mu(\xi, \eta, \zeta) \frac{\partial}{\partial x}\left(\frac{1}{r}\right) d\xi d\eta d\zeta = -\int_D \mu(\xi, \eta, \zeta) \frac{\partial}{\partial \xi}\left(\frac{1}{r}\right) d\xi d\eta d\zeta$$
$$= \int_D \mu_\xi \frac{1}{r} d\xi d\eta d\zeta - \int_\Gamma \mu r^{-1} \nu_1 dS,$$

where dS is the area element of the boundary Γ of the domain D, and ν_1 is the cosine of the angle between the exterior normal to Γ and the $O\xi$ axis. In the equation for u_x one can differentiate the expression under the integral sign by virtue of the same circumstances as in the proof of Theorem 1.1. After differentiating we get

$$u_{xx}(P) = -\int_\Gamma \mu \frac{\partial}{\partial x}\left(\frac{1}{r}\right) \nu_1 dS + \int_D \mu_\xi \frac{\partial}{\partial x}\left(\frac{1}{r}\right) d\xi d\eta d\zeta. \quad (1.7)$$

Since the point $P = (x, y, z)$ is independent of the point $Q = (\xi, \eta, \zeta)$, $\frac{\partial}{\partial \xi}\mu(P) = 0$ and one can write the last integral in (1.7) as follows:

$$J = \int_D \frac{\partial}{\partial \xi}[\mu(Q) - \mu(P)] \frac{\partial}{\partial x}\left(\frac{1}{r}\right) d\xi d\eta d\zeta.$$

Due to the fact that the function $h(Q) = \mu(Q) - \mu(P)$ has a zero of first order at the point P, in the preceding integral one can integrate by parts, as a result of which we find

$$J = \int_\Gamma [\mu(Q) - \mu(P)] \frac{\partial}{\partial x}\left(\frac{1}{r}\right) \nu_1 dS - \int_D [\mu(Q) - \mu(P)] \frac{\partial^2}{\partial \xi \partial x}\left(\frac{1}{r}\right) d\xi d\eta d\zeta.$$

Substituting this expression for J into (1.7) and using the equation $(\partial^2/\partial\xi\partial x)r^{-1} = -(\partial^2/\partial x^2)r^{-1}$, we reduce (1.7) to the form

$$u_{xx}(P) = -\mu(P)\int_\Gamma \frac{\partial}{\partial x}\left(\frac{1}{r}\right)\nu_1 dS + \int_D [\mu(Q) - \mu(P)]\frac{\partial^2}{\partial x^2}\left(\frac{1}{r}\right) d\xi d\eta d\zeta. \quad (1.8)$$

One also gets analogous formulas for u_{yy}, u_{zz}:

$$u_{yy} = -\mu(P)\int_\Gamma \frac{\partial}{\partial y}\left(\frac{1}{r}\right)\nu_2 dS + \int_D [\mu(Q) - \mu(P)]\frac{\partial^2}{\partial y^2}\left(\frac{1}{r}\right) d\xi d\eta d\zeta,$$
$$u_{zz}(P) = -\mu(P)\int_\Gamma \frac{\partial}{\partial z}\left(\frac{1}{r}\right)\nu_3 dS + \int_D [\mu(Q) - \mu(P)]\frac{\partial^2}{\partial z^2}\left(\frac{1}{r}\right) d\xi d\eta d\zeta, \quad (1.9)$$

where ν_2 and ν_3 are the cosines of the angles formed by the exterior normal to Γ and the axes $O\eta$ and $O\zeta$, respectively.

From (1.8) and (1.9) the continuity of the second derivatives of the function u and the equation

$$\Delta u = -\mu\int_\Gamma \left[\frac{\partial}{\partial x}\left(\frac{1}{r}\right)\nu_1 + \frac{\partial}{\partial y}\left(\frac{1}{r}\right)\nu_2 + \frac{\partial}{\partial z}\left(\frac{1}{r}\right)\nu_3\right] dS = -4\pi\mu$$

follow. With the help of the Gauss–Ostrogradskii formula [19], by virtue of the harmonicity of r^{-1} one can reduce the calculation of this integral to the calculation of an analogous integral on the sphere $L(P, Q) = \delta$, and this latter integral can be calculated explicitly.

The assertion of the theorem is proved for a differentiable density μ. However, the derivatives of the function μ do not appear in (1.8) and (1.9); the integrals in these formulas make sense even in the case when μ is only Hölder continuous. To prove the assertion of the theorem in the more general case we construct a sequence μ_n of differentiable functions, for example polynomials, which converges uniformly to μ, and then we prove that the sequence of second derivatives of the corresponding potentials u_n converges uniformly to the expressions appearing in (1.8) and (1.9). This completes the proof of the theorem. □

Besides the potential of a spatial mass distribution one considers potentials of mass distributed on manifolds of lower dimension, for example, on surfaces and lines

$$u(P) = \int_\Gamma \frac{\rho(Q)}{r} dS, \quad P = (x, y, z), Q = (\xi, \eta, \zeta), \quad (1.10)$$

$$u(P) = \int_C \frac{\tau(Q)}{r} dl, \quad (1.11)$$

where Γ is a surface and C is a line. The integral in (1.10) is called the potential of a simple layer with density $\rho(Q)$ on the surface Γ. The integral

$$u(P) = \int_\Gamma \sigma(Q) \frac{\partial}{\partial \nu}\left(\frac{1}{r}\right) dS, \tag{1.12}$$

where $\partial/\partial\nu$ is differentiation in the direction of the positive normal to the surface Γ with respect to the coordinates of the point Q, is called the potential of a double layer with density σ on the surface Γ. We study the properties of these potentials below.

2. GREEN'S FORMULA AND ITS CONSEQUENCES

An important role is played in potential theory by Green's formula, which is a consequence of the Gauss–Ostrogradskii formula [15]. It can be written in the following two forms:

$$\begin{gathered}\int_D (u_x v_x + u_y v_y + u_z v_z)\, d\omega + \int_D v \Delta u d\omega = \int_\Gamma v \frac{\partial u}{\partial n}\, d\sigma, \\ \int_D (u\Delta v - v\Delta u)\, d\omega = \int_\Gamma \left(u \frac{\partial v}{\partial n} - v \frac{\partial u}{\partial n}\right) d\sigma,\end{gathered} \tag{1.13}$$

where $d\omega = dxdydz$ is the volume element, $d\sigma$ the area element of the surface Γ, and $\partial/\partial n$ is differentiation in the direction of the outer normal to D. The second formula of (1.13) follows from the first. In the first formula it is assumed that the functions u and v are continuous in $D \cup \Gamma$, the first derivatives of v and the second derivatives of u are continuous in D, and the first derivatives of u are continuous in $D \cup \Gamma$. In the second formula it is assumed that u, v, and their first derivatives are continuous in $D \cup \Gamma$, and the second derivatives are continuous in D. One assumes that the boundary Γ of the domain D is piecewise-smooth and all integrals appearing in the expressions exist. If in the first formula of (1.13) we set $\Delta u = 0$, $v = 1$, then we get the following assertions.

Theorem 1.3. If a harmonic function is regular in the bounded domain D and continuously differentiable in the closed domain $D \cup \Gamma$, then the integral over the surface Γ of its normal derivative is equal to zero.

Theorem 1.4. Suppose given a piece of the surface Γ, bounded by a curve C, and a point P, which does not belong to Γ. Then the potential of a double layer with constant density $\sigma = 1$ of the surface Γ at the point P in absolute value is equal to the solid angle under which the curve C is visible from the point P. In particular, the potential of a double layer of the surface bounding the domain D has constant value -4π at all interior points of D, and outside D it is equal to zero.

Proof. We construct the conical surface H, formed by line segments joining the point P to points of the curve C. We consider the sphere $K(\varepsilon)$ of radius ε with center at the point P. The surfaces Γ, H, and $K(\varepsilon)$ bound a domain $\Omega(\varepsilon)$. In this domain the function r^{-1} is regular, and on the surface H one has $\frac{\partial}{\partial n}(r^{-1}) = 0$. By virtue of Theorem 1.3 we have

$$\int_{\Gamma} \frac{\partial}{\partial n}\left(\frac{1}{r}\right) dS + \int_{H} \frac{\partial}{\partial n}\left(\frac{1}{r}\right) dS + \int_{K(\varepsilon)} \frac{\partial}{\partial n}\left(\frac{1}{r}\right) dS = 0.$$

Since $\frac{\partial}{\partial n}(r^{-1}) = -\varepsilon^{-2}$ on $K(\varepsilon)$, the last integral in this formula can be calculated explicitly and is equal to the solid angle at the vertex of the constructed conical surface of Γ. In order to prove the second assertion of the theorem, we note that the sign of this solid angle is uniquely determined by defining the positive and negative sides of the surface Γ, and then we split the surface Γ in two by some curve C. Now for interior points of D the solid angles are added and for exterior points they are subtracted. □

If in the first formula of (1.13) we set $u = v$, $\Delta u = 0$, then we get the identity

$$D(u) = \int_{D} (u_x^2 + u_y^2 + u_z^2)\, d\omega = \int_{\Gamma} u \frac{\partial u}{\partial n} dS, \tag{1.14}$$

which is valid for any harmonic function u which is regular in D and continuously differentiable in $D \cup \Gamma$. The integral $D(u)$ is called the Dirichlet integral. The validity of the following assertion follows from (1.14).

Theorem 1.5. Let u be a harmonic function, which is regular in the domain D and continuously differentiable in the closed domain $D \cup \Gamma$. Then: a) if u vanishes on the boundary Γ of the domain D, then it vanishes identically in D; b) if the normal derivative $\partial u/\partial n$ vanishes on the boundary, then u is constant in D.

In the second formula of (1.13), let the domain D be the ball of radius R with center at the point P. We consider the concentric ball of radius $R_0 < R$, we set $v = r^{-1}$, and we write this formula for the spherical layer bounded by the boundary Γ of the ball D and the boundary Σ of the ball B. Then for any harmonic function u, which is regular in D, considering Theorem 1.3, we get

$$\frac{1}{4\pi R_0^2} \int_{\Sigma} u\, dS = \frac{1}{4\pi R^2} \int_{\Gamma} u\, dS.$$

Using the theorem of the mean and letting R_0 tend to zero, we get

$$u(P) = \frac{1}{4\pi R^2} \int_{\Gamma} u(Q)\, dS, \tag{1.15}$$

which expresses the so-called theorem of the mean for harmonic functions. Although in deriving (1.15) the function u was assumed to be differentiable in the closed ball $D \cup \Gamma$, it is easy to show that the continuity of u in $D \cup \Gamma$ is sufficient for the validity of the formula. With the help of (1.15) one can prove the following assertion.

Theorem 1.6. Let the harmonic function u be regular in the domain D and continuous in the closed domain $D \cup \Gamma$. Then it assumes its maximal and minimal values on the boundary Γ of the domain D; the maximum and minimum are achieved in the interior of the domain if and only if u is constant.

The assertion of the theorem is called the maximum principle for harmonic functions.

In Green's formulas (1.13) it was assumed that the functions v and u have all derivatives which appear in these formulas in the domain D. Now we take the function

$$v(\xi, \eta, \zeta) = r^{-1} + w(\xi, \eta, \zeta)$$

with characteristic singularity at the fixed point $P = (x, y, z)$. The function w is assumed to be continuous and continuously differentiable in the closed domain $D \cup \Gamma$ and twice differentiable in D. We surround the point P by a ball $\Sigma(\varepsilon)$ of sufficiently small radius $\varepsilon > 0$ with center at the point P. We throw out of the domain D the intersection $D \cap \Sigma(\varepsilon)$, and for the remaining part $D(\varepsilon)$ of the domain D we write (1.13) and then we let ε tend to zero. If the point P lies on Γ, we shall assume that Γ has a continuously varying tangent plane in a neighborhood of the point P. Taking the limit as $\varepsilon \to 0$, we get

$$\begin{gathered} \int_D (u_\xi v_\xi + u_\eta v_\eta + u_\zeta v_\zeta)\, d\omega + \int_D u \Delta v\, d\omega = \lambda u + \int_\Gamma u \frac{du}{\partial n}\, dS, \\ \int_D (u\Delta v - v\Delta u)\, d\omega = \lambda u + \int_\Gamma \left(u \frac{\partial v}{\partial n} - v \frac{\partial u}{\partial n}\right) dS, \end{gathered} \tag{1.16}$$

where $d\omega$ is the volume element and

$$\lambda = \begin{cases} 4\pi, & P \in D, \\ 2\pi, & P \in \Gamma, \\ 0, & P \notin D \cup \Gamma. \end{cases}$$

We impose the same restrictions on the functions u and w as we did on u and v in (1.13).

If in the second formula of (1.16) we set $w = 0$, then for any function u which is twice differentiable in D and continuously differentiable in $D \cup \Gamma$ we get the representation

$$u = -\frac{1}{4\pi} \int_D \frac{\Delta u}{r}\, d\omega + \frac{1}{4\pi} \int_\Gamma \frac{1}{r} \frac{\partial u}{\partial n}\, dS - \frac{1}{4\pi} \int_\Gamma u \frac{\partial}{\partial n}\left(\frac{1}{r}\right) dS \tag{1.17}$$

as the sum of three potentials. For a harmonic function we have $\Delta u = 0$, so the formula assumes the form

$$u = \frac{1}{4\pi} \int_\Gamma \frac{1}{r} \frac{\partial u}{\partial n}\, dS - \frac{1}{4\pi} \int_\Gamma u \frac{\partial}{\partial n}\left(\frac{1}{r}\right) dS, \tag{1.17a}$$

i.e., one can express a harmonic function as a sum of potentials of simple and double layers. If in the second formula of (1.16) we assume that $\Delta w = 0$, namely as v we take an arbitrary fundamental solution Ω of Laplace's equation, then we get

$$u = \frac{1}{4\pi}\int_\Gamma \Omega \frac{\partial u}{\partial n}\, dS - \frac{1}{4\pi}\int_\Gamma u \frac{\partial \Omega}{\partial n}\, dS, \tag{1.18}$$

where $\Omega = r^{-1} + w$ is a fundamental solution with singularity at the point P.

The generalization of (1.13) to the multidimensional case is the following:

$$\begin{gathered} \int_D \left(\sum_{i=1}^n \frac{\partial u}{\partial x_i}\frac{\partial v}{\partial x_i} + v\Delta u \right) d\omega = \int_\Gamma v \frac{\partial u}{\partial n}\, dS, \\ \int_D (u\Delta v - v\Delta u)\, d\omega = \int_\Gamma \left(u \frac{\partial v}{\partial n} - v \frac{\partial u}{\partial n} \right) dS, \end{gathered} \tag{1.19}$$

and the formulas of (1.16) assume the form

$$\begin{gathered} \int_D \left(\sum_{i=1}^n \frac{\partial u}{\partial \xi_i}\frac{\partial v}{\partial \xi_i} + u\Delta w \right) d\omega = \lambda u + \int_\Gamma u \frac{\partial v}{\partial n}\, dS, \\ \int_D (u\Delta w - v\Delta u)\, d\omega = \lambda u + \int_\Gamma \left(u \frac{\partial v}{\partial n} - v \frac{\partial u}{\partial n} \right) dS, \\ \lambda = \begin{cases} \omega_n, & P \in D, \\ \omega_n/2, & P \in \Gamma, \\ 0, & P \notin D \cup \Gamma, \end{cases} \\ \omega_n = \frac{2(\sqrt{\pi})^n}{\Gamma(n/2)}, \quad v = \frac{1}{n-2} r^{2-n} + w. \end{gathered} \tag{1.20}$$

With the help of Green's formula we investigate the behavior of the potentials of simple and double layers and their derivatives as the point $P = (x, y, z)$ passes through the surface Γ. Let the point P_0 lie on Γ and let the surface Γ at this point have continuous principal curvatures. With center at the point P_0 we construct a ball $\Sigma(\rho)$ of sufficiently small radius that its intersection with Γ is a connected piece of a smooth surface. We denote by D the part of the ball $\Sigma(\rho)$, for which the positive direction of the normal to Γ is the direction of the exterior normal to D. We assume the densities of the potentials of simple and double layers are Hölder continuous.

Theorem 1.7. Upon passing through the surface Γ, the value of the potential of a double layer at the point P_0 has a jump, described by the formulas

$$u^+(P_0) - u(P_0) = 2\pi\sigma(P_0), \quad u^-(P_0) - u(P_0) = -2\pi\sigma(P_0),$$

where u^+ is the limit of the potential of a double layer as the point P tends to the point P_0 from the positive side of the surface Γ and u^- is the analogous limit as P tends to P_0 from the negative side of the surface Γ.

The potential of a simple layer is continuous upon passing through the point P_0, and its normal derivative has the jump

$$\left[\frac{\partial u}{\partial n}\right] = \frac{\partial u}{\partial n^+} - \frac{\partial u}{\partial n^-} = 4\pi\rho\,(P_0),$$

where $\partial/\partial n^+$ is differentiation in the direction of the positive normal to the surface Γ at the point P_0, and $\partial/\partial n^-$ is differentiation in the direction of the negative normal.

Proof. We prove the assertion of the theorem for the case when the density σ of the potential of a double layer is twice differentiable and can be extended to a twice-differentiable function $\sigma(x, y, z)$ in the domain D. In the first formula of (1.16) we set $u = \sigma$, $w = 0$. We have

$$\lambda\sigma + \int_\Gamma \sigma \frac{\partial}{\partial \nu}\left(\frac{1}{r}\right) dS = \int_D \left[\sigma_\xi \frac{\partial}{\partial \xi}\left(\frac{1}{r}\right) + \sigma_\eta \frac{\partial}{\partial \eta}\left(\frac{1}{r}\right) + \sigma_\zeta \frac{\partial}{\partial \zeta}\left(\frac{1}{r}\right)\right] d\omega.$$

The right side of this equation is continuous as the point P passes through the surface Γ, so the left side must also be continuous, and the continuity of the expression

$$\lambda\sigma + \int_\Gamma \sigma \frac{\partial}{\partial \nu}\left(\frac{1}{r}\right) dS$$

is the first assertion of the theorem.

The second assertion of the theorem is proved analogously, but the function u is chosen so that on Γ it is identically equal to zero, and $\partial u/\partial n = \rho$, where ρ is the density of the potential of a simple layer. □

3. BASIC BOUNDARY PROBLEMS OF POTENTIAL THEORY

The study of certain physical phenomena leads to the natural formulation of the following two boundary problems of potential theory [55].

Dirichlet Problem. Find a harmonic function u, which is regular in the domain D and continuous in the closed domain $D \cup \Gamma$, which assumes preassigned values f on the boundary Γ of the domain D, i.e., $u = f$ on Γ, where f is a given continuous function.

Neumann Problem. In the domain D, whose boundary Γ has a continuously varying normal, find a harmonic function, which is regular and continuously differentiable on $D \cup \Gamma$, which satisfies the condition $\partial u/\partial n = g$ on Γ, where g is a given continuous function and $\partial/\partial n$ is the derivative with respect to the normal to Γ.

The Neumann problem is a special case of the following important and more general problem.

Oblique Derivative Problem. Suppose given, on the boundary Γ, which has a continuously varying normal, of the domain D, a field l of directions, i.e., at each point of Γ there is given a vector l of unit length. Find a harmonic function u, which is regular in D and continuously differentiable in the closed domain $D \cup \Gamma$, which satisfies the condition $\partial u/\partial l = h$ on Γ, where h is a given continuous function.

All three of the problems formulated are special cases of the following boundary problem.

Poincaré Problem. In a domain D, whose boundary Γ has a continuously varying normal, find a regular, harmonic function u, which on Γ satisfies

$$\alpha(P)\frac{\partial u}{\partial l} + \beta(P)u = f(P),$$

where α, β, and f are continuous functions defined on Γ, and the function u sought is such that in the closed domain $D \cup \Gamma$ the derivative which appears in the boundary condition exists.

We consider the Dirichlet problem. If we were able to construct a fundamental solution Ω for Laplace's equation in the domain D, such that $\Omega = 0$ on the boundary Γ of the domain D, then (1.18) would express the solution of the Dirichlet problem. By the Green's function of the Laplace operator for the domain D we shall mean the fundamental solution $G(P, Q)$ of Laplace's equation, depending on a parametric point $Q = (\xi_1, \ldots, \xi_n)$, which has the form

$$G(P, Q) = \psi(r) + w(P, Q),\ r^2 = \sum_{i=1}^{n} (x_i - \xi_i)^2,$$

and is equal to zero when the point $P = (x_1, \ldots, x_n)$ lies on the surface Γ. The summand w is continuous in the closed domain $D \cup \Gamma$. With the help of the Green's function the solution of the Dirichlet problem can be described by the formula

$$u(P) = -\int_{\Gamma} f(Q)\frac{\partial G(P, Q)}{\partial_Q n}\, dS. \tag{1.21}$$

Under the assumption of the existence of the Green's function, one gets (1.21) as a consequence of Green's formula, and for this latter to apply, the function f should satisfy certain smoothness conditions. However, it is easy to verify directly that (1.21) gives a solution of the Dirichlet problem for any continuous function f [19].

For an arbitrary domain it is hard to construct the Green's function, and the problem of constructing it is in no way simpler than the original Dirichlet problem. But in some important special cases one can write this function down explicitly. Thus, the Green's function of the ball Σ: $\{x_1^2 + \ldots + x_n^2 < R^2\}$ has the form

$$G(P, Q) = \psi(r) - \psi(\sigma r_1/R),$$

$$\psi(s) = [(n-2)\,\omega_n]^{-1} s^{2-n},$$

$$r^2 = \sum_{k=1}^{n} (x_k - \xi_k)^2,$$

$$r_1^2 = \sum_{k=1}^{n} (x_k - R^2\sigma^{-2}\xi_k)^2,$$

$$\sigma^2 = \sum_{k=1}^{n} \xi_k^2.$$

Here $(x_1, \ldots, x_n)$ is the point P, and $(\xi_1, \ldots, \xi_n)$ is the point Q. The fact that this function satisfies all the requirements imposed on the Green's function can be verified directly. By direct calculation it is also easy to verify that

$$\left.\frac{\partial G(P, Q)}{\partial n}\right|_{Q\in S} = \psi'(r)\frac{R^2-\rho^2}{rR}, \quad \rho^2 = \sum_{i=1}^{n} x_i^2,$$

where S is the sphere $x_1^2 + \ldots + x_n^2 = R^2$. Now (1.21) for the ball Σ assumes the form

$$u = -\frac{R^2-\rho^2}{R}\int\limits_S \frac{\psi'(r)}{r} f dS. \tag{1.22}$$

If in this formula we pass to spherical coordinates, then we get

$$u(X) = \frac{R^{n-2}(R^2-\rho^2)}{\omega_n}\int\limits_\sigma \frac{f_1 d\sigma}{(\rho^2 - 2R\rho\cos\theta + R^2)^{n/2}}, \tag{1.23}$$

$f_1(\eta_1, \ldots, \eta_n) = f(R\eta_1, \ldots, R\eta_n)$, $\rho^2 = x_1^2, \ldots, x_n^2$, θ is the angle between the radius-vectors of the point X and the point $Q = (R\eta_1, \ldots, R\eta_n)$,and the integration is over the n-dimensional unit sphere $\sigma : \{\eta_1^2, \ldots, \eta_n^2 = 1\}$. Formula (1.23) is called the Poisson integral formula.

Obviously in the relation

$$\frac{R^{n-2}(R^2-\rho^2)}{[\rho^2+R^2-2R\rho\cos\theta]^{n/2}} = \frac{1-\rho^2/R^2}{\left[1-2\frac{\rho}{R}\cos\theta+\rho^2/R^2\right]^{n/2}} = \sum_{n=0}^{\infty}\left(\frac{\rho}{R}\right)^n g_n(\cos\theta) \tag{1.24}$$

the series converges for all $\rho < R$, while for $\rho \leq R_1 < R$ it converges absolutely and uniformly. Since

$$\cos\theta = \frac{x_1\xi_1 + \ldots + x_n\xi_n}{\sqrt{x_1^2+\ldots+x_n^2}\cdot\sqrt{\xi_1^2+\ldots+\xi_n^2}},$$

the expressions

$$H_n(X) = \frac{1}{\omega_n}\left(\frac{\rho}{R}\right)^n \int\limits_\sigma f_1 \cdot g_n(\cos\theta)\, d\sigma$$

are homogeneous functions of integral degree n, i.e., homogeneous polynomials of degree n. Substituting (1.24) in (1.23), and interchanging the order of summation and integration, we get

$$u(X) = \sum_{n=0}^{\infty} H_n(X).$$

Thus we have

Theorem 1.8. One can decompose any harmonic function which is regular in the ball Σ into a series of homogeneous polynomials, where this series converges absolutely and uniformly in any strictly interior subdomain of the ball Σ.

It follows from this theorem that any harmonic function is analytic inside its domain of regularity.

For the half-space H : $\{x_n > 0\}$ one can also write down the analog of Poisson's formula (1.23). This analog has the form

$$u(X) = \frac{2x_n}{\omega_n} \int_{-\infty}^{+\infty} \cdots \int_{-\infty}^{\infty} \frac{f(\xi_1, \ldots, \xi_{n-1})\, d\xi_1 \ldots d\xi_{n-1}}{[(x_1 - \xi_1)^2 + \ldots + (x_{n-1} - \xi_{n-1})^2 + x_n^2]^{n/2}}.$$

From Poisson's formula one can also get a number of useful consequences [19]. There are also other methods for solving the Dirichlet problem [19, 26], but we shall not dwell on this here, since our basic goal is the study of the oblique derivative problem. We just consider some properties of a Green's function.

The Green's function $G(P, Q)$ of the domain D vanishes on the boundary Γ of the domain D, and in a sufficiently small neighborhood of the point P it is positive, since $\psi(r)$ goes to $+\infty$ at this point. By the maximum principle we get that $G(P, Q) > 0$ everywhere in the domain D. It also follows from the maximum principle that a solution of the Dirichlet problem is unique, so setting $f \equiv 1$ in (1.21) we have $u \equiv 1$; hence

$$\int_\Gamma \frac{\partial G(P, Q)}{\partial_Q n}\, dS = 1.$$

If the domain D is bounded then for any point P of it one can find a number R such that the ball Σ of radius R with center at the given point contains the domain D in its interior. The Green's function of this ball has the form

$$G_1(P, Q) = \psi(r) - \psi(R) = [(n-2)\omega_n]^{-1}(r^{2-n} - R^{2-n}).$$

Let $G(P, Q)$ be the Green's function of the domain D with characteristic singularity at the same point P. The harmonic function $h(P, Q) = G_1(P, Q) - G(P, Q)$ is regular in the domain D and negative on the boundary Γ of the domain D. By the maximum principle, everywhere in the domain D we have

$$G(P, Q) \leqslant [(n-2)\omega_n]^{-1}(r^{2-n} - R^{2-n}). \tag{1.25}$$

The validity of the assertion follows from this estimate.

Lemma 1.1. Let $G(P, Q)$ be the Green's function of the domain D, and let B be a subdomain of it with diameter less than h. Then

$$\int_B G(P, Q)\, d_Q\omega < \varepsilon(h),$$

where $\varepsilon(h)$ depends only on h, and not on the form of the subdomain B, and tends to zero with h.

With the help of this lemma it is easy to show that for any function $f(X)$, which is Hölder continuous in the domain D, the expression

$$v(X) = \int_D G(X, Q) f(Q)\, d_Q\omega$$

gives a solution of the Poisson equation $\Delta v = -f(X)$, which is continuous in $D \cup \Gamma$ and equal to zero on the boundary Γ.

The fact that v is continuous in $D \cup \Gamma$ and satisfies Poisson's equation follows from Theorem 1.2. It remains to show that $v = 0$ on Γ. Let P be a point of Γ. We construct the ball of radius h with center P and we denote by B the intersection of the domain D with this ball. We have

$$v(X) = \int_{D'} G(X, Q) f(Q)\, d_Q\omega + \int_B G(X, Q) f(Q)\, d_Q\omega,$$

where D' is the part of the domain D which remains after throwing B out of D. If the point X tends to the point P, then the integral taken over D' tends to zero uniformly, and by virtue of Lemma 1.1 we have

$$\left| \int_B G(X, Q) f(Q)\, d_Q\omega \right| < M\varepsilon(h),$$

where M is the maximum of $|f|$ on the subdomain B. Since one can take h arbitrarily small, $\lim_{X \to P} v(X) = 0$, $P \in \Gamma$.

In exactly the same way as the second formula of (1.20) was derived, for the Green's function one can derive the formula

$$G(X, P) - G(P, X) = \int_\Gamma \left[G(Q, P) \frac{\partial G(Q, X)}{\partial n} - G(Q, X) \frac{\partial G(Q, P)}{\partial n} \right] dS,$$

but since the Green's function vanishes on the boundary Γ of the domain D, from this we get the symmetry property of the Green's function $G(X, P) = G(P, X)$.

We try to construct an analog of the Green's function for the Neumann problem. This function should be a fundamental solution of Laplace's equation $\Omega = r^{-1} + w$, which on Γ satisfies the equation $\partial\Omega/\partial n = 0$. Consequently, the harmonic function w, which is regular in the domain D, should, on Γ, satisfy the equation

$$\frac{\partial w}{\partial n} = -\frac{\partial}{\partial n}(r^{-1}). \tag{1.26}$$

By Theorem 1.4, the integral over Γ of the right side of this equation is equal to 4π, and it now follows from Theorem 1.3 that a harmonic function, which is regular in the domain D and satisfies (1.26), does not exist. The analogous situation holds for $n > 3$ also.

In order to bypass this difficulty, we fix a point A in the domain and we shall seek an analog of the Green's function for the Neumann problem in the form $N(X, Q; A) = \psi(r) - \psi(\rho) + w(x, Q; A)$, where r is the distance $L(X, Q)$ from the point

X to the point Q, and ρ is the distance $L(X, A)$ from the point X to A. The function w is regular in the domain D, and on the boundary Γ it satisfies the condition

$$\frac{\partial w}{\partial n} = \frac{\partial}{\partial n} \psi(\rho) - \frac{\partial}{\partial n} \psi(r).$$

In exactly the same way that we got (1.20), we can derive the relation

$$u(X) - u(A) = \int_\Gamma N(X, Q; A) \frac{\partial u}{\partial n} d_Q S,$$

by virtue of which, for a solution of the Neumann problem we get the formula

$$u(X) = C + \int_\Gamma N(X, Q; A) g(Q) d_Q S, \tag{1.27}$$

where $u(A) = C$ is an arbitrary constant, while the given function g satisfies the condition of Theorem 1.3

$$\int_\Gamma g(Q) dS = 0.$$

The function N which appears in (1.27) is called the Neumann function of the domain D.

We have not yet touched on the problem of existence of the Green and Neumann functions. The solution of this question is equivalent to the investigation of the original Dirichlet and Neumann problems. We solve both problems by reducing them to Fredholm integral equations for the case when the boundary of the domain is a sufficiently smooth surface. Let us assume that the boundary of the domain has a continuously varying tangent plane, and in some cases we shall require the continuity of the principal curvatures of the boundary.

4. INVESTIGATION OF BOUNDARY PROBLEMS BY THE METHOD OF INTEGRAL EQUATIONS

Before investigating the boundary problems we study the behavior of the potentials of simple and double layers in a neighborhood of a surface on which the densities are given in detail. Let P be a fixed point of S, T be the tangent plane to S at point P. At point P we calculate the tangential derivatives of the potential of a simple layer

$$u = \int_S r^{-1} \rho dS.$$

We take point P as the origin, we direct the Oz axis along the normal to S at point P, and we take the xOy plane to be the plane T. On T one can find a sufficiently small disk ω such that the part S' of surface S, which lies over this disk, is defined by the equation $S' : \{\zeta = \varphi(\xi, \eta), \varphi_\xi(0) = 0, \varphi_\eta(0) = 0\}$. We represent the potential u as a sum

$$u = \int_{S'} r^{-1}\rho dS' + \int_{S-S'} r^{-1}\rho dS.$$

In this sum, as the variable point approaches point P, the second summand remains analytic, so it suffices to study the behavior of the derivatives of the first summand. For points $Q = (0, 0, z)$, which lie on the normal to the surface S, we consider the derivatives with respect to x at P of the first summand

$$u_x = \int_{S'} \frac{\xi\rho(\xi, \eta, \zeta)\, dS'}{[\xi^2 + \eta^2 + (\zeta - z)^2]^{3/2}} = \int_{\omega} \frac{\xi\rho(\xi, \eta, \varphi)\, \lambda d\omega}{[\xi^2 + \eta^2 + (\zeta - z)^2]^{3/2}},$$

$$\lambda = \sqrt{1 + \varphi_\xi^2 + \varphi_\eta^2}.$$

We take a disk ω' with center at the origin, contained in ω, and we write

$$u_x = I + J, \quad I = \int_{\omega'} \xi\lambda\rho r^{-3} d\omega', \quad J = \int_{\omega-\omega'} \xi\lambda\rho r^{-3} d\omega.$$

For fixed ω' the integral J is continuous. We estimate the integral I. We have $I = I_1 + I_2$, where

$$I_1 = \rho(P) \int_{\omega'} \xi r^{-3} d\omega', \quad I_2 = \int_{\omega'} [\rho(A)\lambda(A) - \rho(P)]\, \xi r^{-3} d\omega'.$$

We set $\gamma^2 = \xi^2 + \eta^2 + z^2$. Obviously

$$\int_{\omega'} \xi\gamma^{-3} d\omega' = 0,$$

since the integrand assumes values of opposite signs at the points (ξ, η) and $(-\xi, \eta)$. Consequently,

$$I_1 = \rho(P) \int_{\omega'} \xi(r^{-3} - \gamma^{-3})\, d\omega' = \rho(P) = \int_{\omega'} \frac{\xi\zeta(2z - \zeta)}{r\gamma(r + \gamma)} [r^{-2} + r^{-1}\gamma^{-1} + \gamma^{-2}]\, d\omega'.$$

Let $\mu^2 = \xi^2 + \eta^2$. We have $r \geq \mu$, $\gamma \geq \mu$, $|\xi| \leq \mu$. If the principal curvatures of surface S are continuous, the function φ in the equation $\zeta = \varphi(\xi, \eta)$ of this surface has second derivatives and one has $|\zeta| < M\mu^2$, where M is a constant, depending on surface S. Further, we have $|2z + \zeta| \leq 2(r + \mu)$, if $\mu < M^{-1}$. Consequently,

$$|I_1| \leqslant |\rho(P)| \int_{\omega'} \mu^{-1} d\omega' = 2\pi\delta |\rho(P)|,$$

$\delta < M^{-1}$ being the radius of the disk ω'.

By virtue of the smoothness of surface S, the function $\lambda(A)$ satisfies a Lipschitz condition and $\lambda(P) = 1$, so the function $\rho(A)\lambda(A)$ satisfies the Hölder condition with the same exponent as $\rho(A)$. From this one gets the estimate

$$|I_2| \leqslant C \int_{\omega'} \mu^{\alpha-2} d\omega' = 2\pi\alpha^{-1} C \delta^{\alpha},$$

where C is a constant which depends only on the density of the potential and the surface S, and α is the Hölder exponent of the density ρ. Thus, for all z such that $0 < |z| < \delta_1$, one can choose the radius of the disk ω' sufficiently small that $|I| < \varepsilon/3$, where ε and δ_1 are previously given numbers. Fixing ε and δ_1, we choose a corresponding δ. By virtue of the continuity of the integral J, for previously given $\varepsilon > 0$ one can find a δ_2 such that for any z_1 and z_2, $0 < |z_1| < \delta_2, 0 < |z_2| < \delta_2$, one will have $|J(z_2) - J(z_1)| < \varepsilon/3$, from which we get

$$\left| \left(\frac{\partial u}{\partial x}\right)_{Q_2} - \left(\frac{\partial u}{\partial x}\right)_{Q_1} \right| < \varepsilon,$$

where $\delta \to 0$ together with ε. Since ε can be arbitrarily small, it follows from this inequality that the following limit exists:

$$\lim_{Q \to P} u_x, \quad Q \in D, \quad P \in S. \tag{1.28}$$

Let $Q = (x, y, z)$, $P = (\xi, \eta, \zeta)$, $r^2 = (x - \xi)^2 + (y - \eta)^2 + (z - \zeta)^2$. Despite the existence of the limit of (1.28), the integral

$$u_x = -\int_S \rho(P)(x - \xi) r^{-3} dS \tag{1.29}$$

diverges when both points Q and P lie on the surface S. A method follows from the course of the proof of the existence of the limit (1.28), with the help of which one can give specific meaning to this integral. This method is the following. On the tangent plane T to S at point Q we construct a disk ω' with center at point Q, of radius δ. From the cylinder with generator parallel to the normal to S at point Q and base ω' on surface S we excise a piece σ' and we consider the integral

$$J_\delta(Q) = \int_{S-\sigma'} \rho(P) \frac{x - \xi}{r^3} dS.$$

We have shown that $u_x = \lim_{\delta \to 0} J_\delta(Q)$, while the limit which appears on the right side exists. We set

$$\int_S \rho(P) \frac{x - \xi}{r^3} dS = \lim_{\delta \to 0} \int_{S-\sigma'} \rho(P) \frac{x - \xi}{r^3} dS.$$

We shall call this limit the integral in the sense of the principal value. Now one can say that (1.29) holds, where the integral is understood in the sense of the principal value.

Remark. In proving the existence of the limit (1.28) we assumed the continuity of the principal curvatures of surface S. However, it follows from the course of the proof that it is sufficient to require the Hölder continuity of the direction cosines of the normal to S. In this case the inequality $|\zeta| \le M\mu^2$ is replaced by the inequality $|\zeta| \le M\mu^{1+\beta}$, $0 < \beta < 1$.

We investigate the normal derivative of the potential of a simple layer. We keep the same notation as in calculating the limit of the tangential derivative. We have

$$u_z = \int_{S'} \frac{(\zeta - z)\,\rho(\xi, \eta, \zeta)\,dS'}{[\xi^2 + \eta^2 + (\zeta - z)^2]^{3/2}} = \int_{\omega} \frac{(\zeta - z)\,\lambda\rho(\xi, \eta, \zeta)\,d\omega}{[\xi^2 + \eta^2 + (\zeta - z)^2]^{3/2}},$$

$$\zeta = \varphi(\xi, \eta), \quad \lambda = \sqrt{1 + \varphi_\xi^2 + \varphi_\eta^2}.$$

We consider the potential

$$u_z' = -\int_{\omega} z\lambda\rho(\xi, \eta, \zeta)\,\gamma^{-3}\,d\omega, \quad \gamma^2 = \xi^2 + \eta^2 + z^2.$$

We introduce the notation

$$I_1 = -\int_{\omega'} \rho\lambda z\gamma^{-3} d\omega', \quad I_2 = -\int_{\omega - \omega'} \rho\lambda z\gamma^{-3} d\omega.$$

For fixed ω' the integral I_2 is uniformly continuous as $z \to 0$ and vanishes for $z = 0$. In calculating the integral I_1 we use the mean value theorem and Theorem 1.4:

$$I_1 = -\overline{\rho\lambda}\int_{\omega'} z\gamma^{-3} d\omega = -\overline{\rho\lambda}\,\Omega,$$

where $\overline{\rho\lambda}$ is some intermediate value of $\rho\lambda$, and Ω is the angle through which the disk ω' is visible from the point $Q = (0, 0, z)$, $z > 0$. Obviously, as $z \to 0$ this angle tends to 2π. When the disk ω' shrinks to point P, the value of $\overline{\rho\lambda}$ tends to $\rho(P)$, so shrinking ω' to a point, we get $u_z' = -2\pi\rho(P)$, as $Q \to P$ along the positive half-axis Oz. Further, we have

$$u_z - u_z' = \int_{\omega} \lambda\rho\,[(\xi - z)\,r^{-3} + z\gamma^{-3}]\,d\omega$$
$$= \int_{\omega} \lambda\rho\,[\zeta r^{-3} - z\,(r^{-3} - \gamma^{-3})]\,d\omega = \int_{\omega} \lambda\rho\zeta r^{-3} d\omega$$
$$- \int_{\omega} \lambda\rho z\,(r^{-3} - \gamma^{-3})\,d\omega, \quad r^2 = \xi^2 + \eta^2 + (\zeta - z)^2.$$

Both integrals on the right side of the equation converge, and the last integral tends to zero as $z \to 0$, so in the limit we get

$$u_z + 2\pi\rho(P) = \int_\omega \lambda\rho\zeta r^{-3} d\omega = \int_{S'} \rho\zeta r^{-3} dS' = \int_{S'} \rho(Q) \left[\frac{\partial}{\partial z} r^{-1}\right]_P dS'.$$

From this it follows that

$$\frac{\partial u}{\partial n^+} = -2\pi\rho(P) + \int_S \rho(Q) \frac{\partial}{\partial n}(r^{-1})\, dS,$$
$$\frac{\partial u}{\partial n^-} = 2\pi\rho(P) + \int_S \rho(Q) \frac{\partial}{\partial n}(r^{-1})\, dS, \tag{1.30}$$

where n^+ is the positive normal and n^- the negative one, and the integral which appears on the right side converges.

Suppose given on surface S a field of directions $l = (l_1, l_2, l_3)$, i.e., with each point of the surface there is associated a unit vector; also let $\nu(P)$ be the cosine of the angle between the vectors $l(P)$ and $n^+(P)$, and $m(P)$ be the projection of l to the tangent plane to S at point P. We calculate the derivative $\partial u/\partial l$ of the potential u with respect to the coordinates of a variable point Q, considering the field l to be extended to a neighborhood of surface S, and then we let point Q tend to surface S. As a result we have

$$\frac{\partial u}{\partial l} = -2\pi\nu(Q)\rho(Q) + \int_S \rho(P) \frac{\partial}{\partial l}(r^{-1})\, dS, \tag{1.31}$$

where the derivative $\partial/\partial l$ is taken with respect to the coordinates of point Q, and the integral is understood in the sense of the principal value.

Analogously, passing to the limit, for the potential of a double layer we get

$$u^+(P) = 2\pi\sigma(P) + \int_S \sigma(Q) \frac{\partial}{\partial n}(r^{-1})\, dS,$$
$$u^-(P) = -2\pi\sigma(P) + \int_S \sigma(Q) \frac{\partial}{\partial n}(r^{-1})\, dS, \tag{1.32}$$

where the integral which appears on the right sides of the formulas converges in the usual sense, and the derivative in the integrand is taken with respect to the coordinates of the point Q.

We consider a bounded domain D with boundary Γ, and we denote by D' the complement of $D \cup \Gamma$ in the whole space. Let us assume that the surface Γ can be represented, in a neighborhood of each of its points P, in a system of coordinates with origin at this point and with Oz axis directed along the normal to Γ at point P, by an equation $z = \varphi(x, y)$, where the function φ has continuous derivatives of the first two orders. We shall seek a solution of the Dirichlet problem in domain D in the form of the potential of a double layer

$$u(X) = \frac{1}{2\pi} \int_\Gamma \sigma(Q) \frac{\partial}{\partial n}(r^{-1})\, dS,$$

where the derivative in the integrand is taken with respect to the coordinates of point Q. Substituting this expression into the boundary condition, and using (1.32), we get for $\sigma(Q)$ a Fredholm equation of the second kind

$$f(P) = -\sigma(P) + \frac{1}{2\pi}\int_{\Gamma} \sigma(Q)\frac{\partial}{\partial n}(r^{-1})\, dS.$$

Analogously, in domain D' the Dirichlet problem reduces to the equation

$$f(P) = \sigma(P) + \frac{1}{2\pi}\int_{\Gamma} \sigma(Q)\frac{\partial}{\partial n}(r^{-1})\, dS.$$

We write these two equations in the form of one

$$F(P) = \varphi(P) - \lambda\int_{\Gamma} K(P, Q)\,\varphi(Q)\, dS, \tag{1.33}$$

where we have introduced the notation

$$\varphi(P) = \sigma(P),$$
$$K(P, Q) = \frac{1}{2\pi}\frac{\partial}{\partial_Q n}\left(\frac{1}{L(P, Q)}\right),$$

and $L(P, Q)$ is the distance between the points P and Q. For $\lambda = 1$, $F(P) = -f(P)$, (1.33) corresponds to the Dirichlet problem in domain D, and for $\lambda = -1$, $F(P) = f(P)$, that in domain D'.

We shall seek a solution of the Neumann problem in the form of the potential of a simple layer

$$u(P) = \frac{1}{2\pi}\int_{\Gamma} \rho(Q)\frac{1}{r}\, dS.$$

Substituting this expression into the boundary condition for the Neumann problem and using (1.30), for the Neumann problem in domain D we get the equation

$$g(P) = \rho(P) + \frac{1}{2\pi}\int_{\Gamma} \rho(Q)\frac{\partial}{\partial n}\left(\frac{1}{r}\right) dS,$$

where the derivative in the integrand is taken with respect to the coordinates of point P. Analogously, the Neumann problem in domain D' reduces to the equation

$$g(P) = -\rho(P) + \frac{1}{2\pi}\int_{\Gamma} \rho(Q)\frac{\partial}{\partial n}\left(\frac{1}{r}\right) dS.$$

One can also write these equations in the form of one equation

$$F(P) = \psi(P) - \lambda\int_{\Gamma} K(Q, P)\,\psi(Q)\, dS, \tag{1.34}$$

which for $\lambda = -1$, $F(P) = g(P)$ corresponds to the Neumann problem in the domain D, and for $\lambda = 1$, $F(P) = -g(P)$, that in domain D'. The kernel of (1.32) is the transpose of the kernel of (1.33), i.e., these equations are mutually adjoint.

We investigate the kernel of (1.33). We surround point P by a ball Σ with center at this point of radius so small that part σ of surface Γ which lies in Σ admits a representation $z = \varphi(x, y)$ in a coordinate system with origin at P and Oz axis directed along the normal to Γ at point P. When Q varies on $\Gamma' = \Gamma - \sigma$, the kernel K is continuous and bounded, $|K(P, Q)| \leq B$, while when $Q \in \sigma$, then as was established in investigating the derivatives of the potential of a simple layer,

$$|K(P,Q)| = \left|\frac{\partial}{\partial n}(r^{-1})\right| \leqslant M\gamma^2 r^{-3},$$

where r is the distance between P and Q, and γ is the distance between the projections P' and Q' of points P and Q to the tangent plane to Γ at point P, so $\gamma \leq r$. If now we take $M \geq BR$, where R is the length of the largest chord of the surface Γ, then for K we get the estimate $|K(P, Q)| \leq Mr^{-1}$. Consequently, this kernel is Fredholm [30].

We investigate the iterated kernels K_1 and K_2 of the kernel K. Let the distance between P and Q satisfy $L(P, Q) \geq \delta$. We have

$$K_1(P,Q) = \int_\Gamma K(P,X)\,K(X,Q)\,d_X S = \int_{\Gamma'} K(P,X)\,K(X,Q)\,d_X S + \int_{\sigma_1 \cup \sigma_2} K(P,X)\,K(X,Q)\,d_X S.$$

where σ_1 and σ_2 are the pieces of the surface Γ, lying in balls of sufficiently small radius $\alpha < \delta$ with centers at P and Q, respectively, and Γ' is the rest of Γ. The first summand I_1 of this sum is bounded and continuous for fixed α. We take $\alpha < \delta/3$, so for $X \in \sigma_1$ we have

$$|K(P,X)| \leqslant M[L(P,X)]^{-1}, \quad |K(X,Q)| \leqslant \frac{M}{\delta - 2\alpha} \leqslant 3M/\delta.$$

For the second summand I_2 we now find the estimate

$$|I_2| \leqslant 2\int_\sigma \frac{3M}{\delta} M[L(P,X)]^{-1}\,d\sigma \leqslant M_1\alpha,$$

where M_1 is a new constant which depends only on Γ. The last estimate is obtained here by the usual method with the help of passage to a coordinate system with origin at point P and Oz axis directed along the normal to Γ at point P. The boundedness $K_1(P, Q)$ for $L(P, Q) \geq \delta$ follows from this. Now let $L(P, Q) < \delta$, and σ be the part of the surface Γ, lying in the ball of radius 2δ with center at point Q, which we shall assume fixed. The integral

$$I_1 = \int_{\Gamma - \sigma} K(P,X)\,K(X,Q)\,d_X S$$

does not exceed $M^2\delta^{-2}$ in absolute value, so it is bounded. Further, in the usual way we get the estimate

$$\left|\int_{\sigma} K(P,X)\,K(X,Q)\,d_XS\right| \leqslant M^2M_1\int_{\omega}\frac{d\omega}{\rho\rho_1},$$

where M_1 is a new constant depending on the surface Γ, ρ is the length of the projection to the tangent plane to Γ at point Q of the segment joining Q with X, ρ_1 is the length of the projection of the segment joining P with X, and ω is a small disk containing the projection of σ to the tangent plane to Γ at point Q. Obviously

$$\int_{\omega}\frac{d\omega}{\rho\rho_1} = \int_0^{2\pi}\int_0^{a}\frac{\rho d\rho d\theta}{\rho\rho_1} \leqslant 4\pi C\ln\frac{a-\eta}{\eta},$$

η being the distance from Q to P, and C being a constant. Finally, for K_1 we get the estimate

$$|K_1(P,\ Q)| \leqslant B_1\ln\left[M/L(P,\ Q)\right], \tag{1.35}$$

where B_1 is a new constant.

Using (1.35), we prove analogously that the kernel

$$K_2(P,Q) = \int_{\Gamma} K_1(P,X)\,K(X,Q)\,d_XS$$

is continuous for all P and Q lying on Γ.

We consider the potentials of double and simple layers

$$W(P) = \frac{1}{2\pi}\int_{\Gamma}\sigma(Q)\frac{\partial}{\partial n}\left(\frac{1}{r}\right)d_QS,\quad V(P) = \frac{1}{2\pi}\int_{\Gamma}\rho(Q)\frac{1}{r}\,d_QS.$$

For these potentials one has the limit relations

$$\begin{aligned} W^- &= -\sigma(P) + \int_{\Gamma}\sigma(Q)\,K(P,Q)\,d_QS,\\ W^+ &= \sigma(P) + \int_{\Gamma}\sigma(Q)\,K(P,Q)\,d_QS; \end{aligned} \tag{1.36}$$

$$\begin{aligned} \frac{\partial V}{\partial n^-} &= \rho(P) + \int_{\Gamma}\rho(Q)\,K(Q,P)\,d_QS,\\ \frac{\partial V}{\partial n^+} &= -\rho(P) + \int_{\Gamma}\rho(Q)\,K(Q,P)\,d_QS. \end{aligned} \tag{1.37}$$

From (1.36) for any λ one gets

$$\frac{1-\lambda}{2}W^+ - \frac{1+\lambda}{2}W^- = \sigma(P) - \lambda\int_\Gamma \sigma(Q)\,K(P,Q)\,d_QS, \tag{1.38}$$

and from (1.37) we find

$$\frac{1-\lambda}{2}\frac{\partial V}{\partial n^-} - \frac{1+\lambda}{2}\frac{\partial V}{\partial n^+} = \rho(P) - \lambda\int_\Gamma \rho(Q)\,K(Q,P)\,d_QS. \tag{1.39}$$

We show that all eigenvalues of the kernel $K(P, Q)$ are real. Let us assume that $\lambda = \alpha + i\beta$ is a complex eigenvalue, to which corresponds the eigenfunction $\sigma_1(P) + i\sigma_2(P)$. The corresponding potential $V_1(P) + iV_2(P)$ exists and has the necessary derivatives up to Γ. Separating real and imaginary parts in (1.39), we find

$$\begin{aligned}(1-\alpha)\frac{\partial V_1}{\partial n^-} - (1+\alpha)\frac{\partial V_1}{\partial n^+} + \beta\frac{\partial V_2}{\partial n^-} + \beta\frac{\partial V_2}{\partial n^+} &= 0,\\ (1-\alpha)\frac{\partial V_2}{\partial n^-} - (1+\alpha)\frac{\partial V_2}{\partial n^+} - \beta\frac{\partial V_1}{\partial n^-} - \beta\frac{\partial V_1}{\partial n^+} &= 0.\end{aligned} \tag{1.40}$$

We multiply the first equation of (1.40) by V_2, the second by V_1, and integrate the difference of the two results over Γ. We make use of the equation

$$\int_\Gamma \left(\frac{\partial V_1}{\partial n}V_2 - V_1\frac{\partial V_2}{\partial n}\right)dS = 0$$

and introduce the notation

$$\begin{aligned} J_1 &= \int_\Gamma V_1\frac{\partial V_1}{\partial n^-}\,dS, & J_1' &= -\int_\Gamma V_1\frac{\partial V_1}{\partial n^+}\,dS,\\ J_2 &= \int_\Gamma V_2\frac{\partial V_2}{\partial n^-}\,dS, & J_2' &= -\int_\Gamma V_2\frac{\partial V_2}{\partial n^+}\,dS,\end{aligned}$$

so that after the integration we get

$$\beta(J_1 + J_2) - \beta(J_1' + J_2') = 0. \tag{1.41}$$

Analogously, adding the results of the multiplication and integrating over Γ, we find

$$(1-\alpha)(J_1 + J_2) + (1+\alpha)(J_1' + J_2') = 0. \tag{1.42}$$

Equations (1.41) and (1.42) are a system of two linear equations for $J_1 + J_2$ and $J_1' + J_2'$ with determinant 2β, so at least one of these quantities can differ from zero only for $\beta = 0$, and for $\beta \neq 0$ we have $J_1 + J_2 = 0$, $J_1' + J_2' = 0$, but these quantities are non-negative and vanish only for $\partial V_i/\partial n^\pm \equiv 0$, $i = 1, 2$. Consequently, $\beta = 0$ and $V_2 \equiv 0$. Now $J_2' = J_2 = 0$ and it follows from (1.42) that

$$(1-\alpha) J_1 + (1+\alpha) J_1' = 0, \quad \alpha = (J_1 + J_1')/(J_1 - J_1'),$$

i.e., all the eigenvalues of the kernel $K(P, Q)$ are real and no less than one in absolute value.

Let us assume that the eigenvalue λ_0 of kernel K corresponds to the eigenfunction $\psi_1(P)$ and the adjoint function $\psi_2(P)$. These functions satisfy the relations

$$\psi_1(P) = \lambda_0 \int_\Gamma \psi_1(Q) K(Q, P) \, d_Q S,$$

$$\psi_2(P) = \int_\Gamma \psi_1(Q) K(Q, P) \, d_Q S + \lambda_0 \int_\Gamma \psi_2(Q) K(P, Q) \, d_Q S,$$

and the potentials V_1 and V_2 corresponding to them satisfy the boundary conditions

$$(1-\lambda_0) \frac{\partial V_1}{\partial n^-} + (1+\lambda_0) \frac{\partial V_1}{\partial n^+} = 0,$$

$$(1-\lambda_0) \frac{\partial V_2}{\partial n^-} - (1+\lambda_0) \frac{\partial V_2}{\partial n^+} = \frac{\partial V_1}{\partial n^-} + \frac{\partial V_1}{\partial n^+}.$$

If we multiply the first of these equations by V_2, the second by V_1, add the results and integrate over Γ, then we get $J_1 - J_1' = 0$. Now if we multiply the first equation by V_1 and integrate over Γ, then we find $(1 - \lambda_0)J_1 + (1 + \lambda_0)J_1' = 0$. It follows from the last two equations that $J_1 = J_1' = 0$, by virtue of which $V_1 = 0$, and hence, $\varphi_1(\rho) = 0$ also. This contradicts the fact that an eigenfunction of the kernel $K(P, Q)$ is not identically zero. Thus, to each eigenvalue λ_0 of the kernel K there correspond only eigenfunctions and not adjoint functions.

Let us assume that $\lambda = 1$ is an eigenvalue of kernel K; then to it corresponds an eigenfunction $\psi(P) \not\equiv 0$, which defines a potential V, which coincides in D' with a regular, harmonic function, whose normal derivative is equal to zero on Γ, by virtue of the second equation of (1.37). Consequently, $V \equiv 0$ in D', but the potential of a simple layer is continuous, so the limit values of V on Γ from within D are also equal to zero, and $V \equiv 0$ in D, but this is possible only for $\psi(P) \equiv 0$. The contradiction found shows that $\lambda = 1$ is not an eigenvalue of kernel K. Now let $\lambda = -1$; then the potential of a double layer W with density $\sigma(P) \equiv 1$ on Γ is equal to zero in D' by Theorem 1.4, and the homogeneous equation corresponding to the second equation of (1.36) has a nontrivial solution $\sigma(P) \equiv 1$. It is assumed here that the surface Γ is connected. Thus, $\lambda = -1$ is an eigenvalue of kernel K. Conversely, let W be the potential of a double layer with density $\sigma(P)$, satisfying the homogeneous equation corresponding to the second equation of (1.36); then $W \equiv 0$ in D'. The normal derivative of the potential of a double layer with continuous density is continuous on Γ [19], so $\partial W^-/\partial n = 0$ on Γ; consequently W is constant in D, due to which its density is also constant by Theorem 1.7. To the eigenvalue $\lambda = -1$ corresponds a linearly independent eigenfunction $\sigma(P) = 1$.

Along with the second relation of (1.36) the homogeneous equation corresponding to the first equation of (1.37) also has one linearly independent solution $\rho(P)$ [30], so if in the second equation of (1.36) we replace $W^{+}(P)$ by

$$W^{+}(P)-c, \quad c=\int_{\Gamma} W^{+}(Q)\,\rho(Q)\,d_Q S,$$

then it will always be solvable. Let $\sigma(P)$ be a solution of it. We consider the potential of a double layer

$$W(P)=\frac{1}{2\pi}\int_{\Gamma}\sigma(Q)\frac{\partial}{\partial n}(r^{-1})\,d_Q S. \tag{1.43}$$

This potential on Γ has limit from without $W^{+}-c$, where $W^{+}=f$ is a given function and the constant c, by virtue of the homogeneous equation corresponding to the first equation of (1.37), defines the potential of a simple layer

$$V_c=\frac{c}{2\pi}\int_{\Gamma} r^{-1}d_Q S,$$

which is a regular harmonic function in D'. Now the solution of the Dirichlet problem in D' can be represented as a sum $W+V_c$. We note that in (1.43) W is independent of which solution of the second equation of (1.36) is taken as $\sigma(P)$, since two solutions of this equation differ by a constant. We formulate the results found as a theorem.

Theorem 1.9. The Dirichlet problem in the bounded domain D is always solvable and its solution can be represented by the potential of a double layer with density distributed on the boundary Γ of domain D. In an unbounded domain D' the Dirichlet problem is always solvable, but its solution can be represented as the sum of the potential of a double layer and the potential of a simple layer with constant density. In an unbounded domain D' the Neumann problem is always solvable and its solution can be represented as the potential of a simple layer with density distributed on the boundary Γ of domain D'. The Neumann problem in a bounded domain D is solvable provided the function given on Γ is orthogonal to a constant, and the solution can be represented as the sum of the potential of a simple layer and an arbitrary constant.

5. HARMONIC FUNCTIONS IN AXIALLY SYMMETRIC DOMAINS

For $n=3$, (1.24) assumes the form [19]

$$\left(1-\frac{\rho^2}{R^2}\right)\left(1-2\frac{\rho}{R}\cos\omega+\frac{\rho^2}{R^2}\right)^{-3/2}=\sum_{n=0}^{\infty}(2n+1)\left(\frac{\rho}{R}\right)^n P_n(\cos\omega), \tag{1.44}$$

where $P_n(t)$ are the Legendre polynomials [14]. We set $z = \rho\cos\theta$, $y = \rho\sin\theta\sin\varphi$, $x = \rho\sin\theta\cos\varphi$, $x_0 = R\sin\beta\cos\psi$, $y_0 = R\sin\beta\sin\psi$, $z_0 = R\cos\beta$. Obviously, $\cos\omega = \rho^{-1}R^{-1}(xx_0 + yy_0 + zz_0) = \cos\theta\cos\beta + \sin\theta\sin\beta\cos(\varphi - \psi)$. The following equation holds:

$$(2n+1)\,P_n(\cos\omega) = P_n(\cos\theta)\,P_n(\cos\beta)$$
$$+\,2\sum_{l=1}^{n}\frac{(n-l)!}{(n+l)!}P_n^l(\cos\theta)\,P_n^l(\cos\beta)\cdot\cos l(\varphi-\psi).$$

Here $P_n^l(t)$ are the adjoint Legendre functions [14]

$$P_n^l(t) = \frac{(-1)^l}{2^n n!}(1-t^2)^{l/2}\frac{d^{n+l}}{dt^{n+l}}(t^2-1)^n.$$

Now the decomposition of a harmonic function as a series of homogeneous polynomials assumes the form

$$u(x, y, z) = \sum_{n=0}^{\infty}\sum_{l=0}^{n} r^n P_n^l(\cos\theta)\,[a_{nl}\cos l\varphi + b_{nl}\sin l\varphi],$$
$$a_{nl} = 2\frac{(n-l)!}{(n+l)!}\cdot\frac{1}{4\pi R^n}\int_\sigma P_n^l(\cos\beta)\cos l\psi f_1(\beta, \psi)\,d\sigma,$$
$$b_{nl} = 2\frac{(n-l)!}{(n+l)!}\cdot\frac{1}{4\pi R^n}\int_\sigma P_n^l(\cos\beta)\sin l\psi f_1(\beta,\psi)\,d\sigma, \tag{1.45}$$
$$a_{n0} = \frac{2n+1}{4\pi R^n}\int_\sigma P_n(\cos\beta)\,f_1(\beta, \psi)\,d\sigma,$$
$$b_{n0} = 0,\quad n = 0, 1, 2, \ldots,\quad l = 1, \ldots, n,$$

where the integrals are taken over the unit sphere σ: $\{0 \le \beta \le \pi, 0 \le \psi \le 2\pi\}$.

We transform the function $r^n P_n^l(\cos\theta)\exp il\varphi$ as follows:

$$r^n P_n^l(\cos\theta)\exp il\varphi = r^{n-l}C_n^l(\cos\theta)\,r^l\sin^l\theta\exp il\varphi = r^{n-l}\;C_n^l(\cos\theta)\,(x+iy)^l,$$

where $C_n^l(\cos\theta)$ is a polynomial of degree no higher than $n - l$, as follows from the representation of the adjoint Legendre functions. Obviously the function $r^{n-l}C_n^l(\cos\theta)$ depends only on the variables z and $x^2 + y^2$. Since the series (1.45) converges uniformly in the ball Σ: $\{x^2 + y^2 + z^2 < R^2\}$, its terms can be regrouped arbitrarily. After regrouping terms, we transform (1.45) into the series

$$u(x, y, z) = \sum_{l=0}^{\infty}\{g_l(x^2+y^2, z)\,p_l(x, y) + h_l(x^2+y^2, z)\,q_l(x, y)\}, \tag{1.46}$$

where $p_l(x, y) = \operatorname{Re}(x + iy)^l$, $q_l(x, y) = \operatorname{Im}(x + iy)^l$, and

$$g_l(x^2+y^2, z) = \sum_{n=l}^{\infty} a_{nl}r^{n-l}C_n^l(\cos\theta),\; h_l(x^2+y^2, z)$$
$$= \sum_{n=l}^{\infty} b_{nl}r^{n-l}C_n^l(\cos\theta).$$

The terms of the series (1.46) are harmonic functions, since they are sums of absolutely and uniformly convergent series of harmonic functions. It follows from this that $g_l(\sigma, z)$ and $h_l(\sigma, z)$ satisfy the equation

$$4\sigma \frac{\partial^2 \omega_l}{\partial \sigma^2} + \frac{\partial^2 w_l}{\partial z^2} + 4(l+1)\frac{\partial w_l}{\partial \sigma} = 0, \tag{1.47}$$

where g_l and h_l are solutions of this equation which are holomorphic for $\sigma = 0$. A solution of (1.47), which is holomorphic for $\sigma = 0$, is given by the series [46]

$$w_l(\sigma, z) = \sum_{k=0}^{\infty} \frac{(-1)^k l!}{k!\,(k+l)!} \cdot \frac{\sigma^k}{4^k} \cdot \frac{d^{2k}\omega_l(z)}{dz^{2k}}, \tag{1.48}$$

where $\omega_l(z)$ is an analytic function of the variable z, which one can represent by the Cauchy integral formula [32]

$$\omega_l(z) = \frac{1}{2\pi i} \int_\Gamma \frac{\omega_l(t)}{t-z}\, dt. \tag{1.49}$$

Here Γ is the closed curve bounding a domain in which ω_l is analytic and continuous in the closed domain. Substituting (1.49) into (1.46) and interchanging the order of summation and integration, we get

$$w_l(\sigma, z) = \frac{1}{2\pi i} \int_\Gamma \sum_{k=0}^{\infty} \frac{(-1)^k l!\,(2k)!}{k!\,(k+l)!\,4^k} \cdot \frac{\sigma^k \omega_l(t)}{(t-z)^{2k+1}}\, dt$$

$$= \frac{1}{2\pi i} \int_\Gamma \frac{\omega_l(t)}{t-z} \sum_{k=0}^{\infty} \frac{\Gamma(k+1/2)\,k!\,l!}{\Gamma(1/2)\,k!\,(k+l)!} \left[-\frac{\sigma}{(t-z)^2}\right]^k dt$$

$$= \frac{1}{2\pi i} \int_\Gamma F\left(1, 1/2; l; -\frac{\sigma}{(t-z)^2}\right) \frac{\omega_l(t)}{t-z}\, dt,$$

where $F(\alpha, \beta; l; s)$ is the Gauss hypergeometric function [46]. Making use of the familiar formula

$$F(\alpha, \beta; l; s) = (1-s)^{-\alpha} F\left(\alpha, l-\beta; l; \frac{s}{s-1}\right),$$

we finally get the integral representation

$$w_l(\sigma, z)$$
$$= \frac{1}{2\pi i} \int_\Gamma F\left(l-1, 1/2; l; \frac{\sigma}{\sigma + (t-z)^2}\right) \frac{\omega_l(t)\, dt}{\sqrt{\sigma + (t-z)^2}}. \tag{1.50}$$

Let the variables σ and z be real. With the help of (1.50) it is easy to construct the domain of regularity of the function w_l in the real space of the variables σ and z. For real σ and z and any t, the hypergeometric function is analytic, since its argument does not land in the piece of the real positive half-axis included between one and infinity. If $t = \tau + i\delta$, then the singularities which arise due to the root lie on the line $\sigma + (\tau - z)^2 = \delta^2$, $2(\tau - z)\delta = 0$, i.e., $z = \tau$, $\sigma = \delta^2$. Thus, to each point $t = \tau + i\delta$ of the curve Γ in the space of the variables x, y, z there corresponds a circle $K(t)$:

$\{x^2 + y^2 = \delta^2, z = \tau\}$. If the functions w_l are real for real values of the independent variables, then the functions $\omega_l(z) = w_l(0, z)$ are also real for real z, so by the Schwarz symmetry principle [32] their domain of analyticity E is symmetric with respect to the real axis. Now we have the following simple method of constructing the domain of regularity of the harmonic function $u_l(x, y, z) = g_l(x^2 + y^2, z)p_l(x, y)$ from the domain of analyticity of the function $G_l(z) = g_l(0, z)$ of the complex variable z. We imbed the domain of analyticity E of the function $G_l(z)$ in the plane $y = 0$ of the space of variables x, y, z so that the origins coincide and the real axis of the plane of the complex variable z lies on the Oz axis of the space R^3; rotating the image E' of the domain E after this imbedding about the Oz axis, we get the domain of regularity of the function u_l. In particular, if u_l is regular in the unit ball, then G_l must be analytic in the unit disk.

Now let the harmonic function $u(x, y, z)$ be regular in a domain B, which is symmetric with respect to the Oz axis and contains a connected piece L of this axis. The function u is analytic in any axially symmetric subdomain B' of the domain B, so in cylindrical coordinates it can be expanded in a Fourier series [3]:

$$u(x, y, z) = \sum_{l=0}^{\infty} [E_l(\rho, z)\cos l\varphi + D_l(\rho, z)\sin l\varphi] = \sum_{l=0}^{\infty} \{[\rho^{-l}E_l(\rho, z)] \times p_l(x, y) + [\rho^{-l}D_l(\rho, z)]\, q_l(x, y)\},$$

which converges absolutely in B'. Obviously in a neighborhood of points of the axis $\rho = 0$ one must have

$$\rho^{-l}E_l(\rho, z) = a_l(x^2 + y^2, z), \quad \rho^{-l}D_l(\rho, z) = b_l(x^2 + y^2, z),$$

where $a_l(\sigma, z)$ and $b_l(\sigma, z)$ are solutions of (1.47). Consequently, a function which is regular in an axially symmetric domain B, which intersects the Oz axis in a connected segment, can be represented by a series (1.46), where the functions g_l and h_l are uniquely determined by the analytic functions $G_l(z) = g_l(0, z)$, $H_l(z) = h_l(0, z)$ of the complex variable z. These functions extend analytically to a domain which coincides with a meridional section of the domain B, if the plane of the complex variable z is matched with the plane $y = 0$ so that the origins coincide and the real axis of the complex variable z lies on the Oz axis.

One can represent a harmonic function which is regular in the unit ball by a series (1.45). The function

$$v(x, y, z) = r^{-1}u(x/r^2, y/r^2, z/r^2) = \sum_{n=0}^{\infty}\sum_{l=0}^{n} r^{-n-1}P_n^l(\cos\theta)\,[a_{nl}\cos l\varphi + b_{nl}\sin l\varphi]$$

is regular in the exterior of the unit ball, while the series converges in the exterior of the ball. The analog of the series (1.46) for the function v has the form

$$v(x, y, z) = \sum_{l=0}^{\infty} [g_l(x^2 + y^2, z)\, p_l(x, y) + h_l(x^2 + y^2, z)\, q_l(x, y)],$$

$$g_l(x^2+y^2,z)=\sum_{n=l}^{\infty} a_{nl}r^{-n-l-1}C_n^l(\cos\theta),$$

$$h_l(x^2+y^2,z)=\sum_{n=l}^{\infty} b_{nl}r^{-n-l-1}C_n^l(\cos\theta).$$

Obviously as $z\to\infty$ the functions $g_l(0, z)$ and $h_l(0, z)$ decrease no slower than z^{-2l-1}. Consequently, when we consider in (1.50) functions w_l which correspond to harmonic functions which are regular in a neighborhood of infinity, then the $\omega_l(z)$ of the complex variable z must not only be holomorphic in a neighborhood of infinity, but also decrease at infinity no slower than z^{-2l-1}.

6. GENERAL SECOND-ORDER ELLIPTIC EQUATIONS

The basic facts concerning harmonic functions carry over to more general elliptic equations. We consider the operator

$$M=\sum_{i,j=1}^{n} a_{ij}(x)\frac{\partial^2}{\partial x_i \partial x_j}+\sum_{i=1}^{n} b_i(X)\frac{\partial}{\partial x_i}+c(X), \tag{1.51}$$

whose coefficients we shall assume to be twice continuously differentiable in a domain E. The operator M is called elliptic at the point X, if the quadratic form $\chi(\Lambda)=\sum_{i,j=1}^{n} a_{ij}(X)\lambda_i\lambda_j$ in the variables $\lambda_1, \ldots, \lambda_n$ is positive or negative definite. If this form is definite and has the same sign at all points of the domain E, then the operator M is called elliptic in the given domain. Everywhere below we shall assume that the form $\chi(\Lambda)$ is positive definite. This means that its discriminant $\Delta(X)$ is everywhere positive in the domain E. Let $f(X)$ be a given continuous function in the domain E, and let M be an elliptic operator. The partial differential equation

$$M(u)=f \tag{1.52}$$

is elliptic in the domain E. A function $u(X)$, which is twice continuously differentiable in domain E and which satisfies (1.52), will be called a regular solution of this equation.

Theorem 1.10. If in the domain $D\subset E$ the conditions $c\le 0, f<0$ $[f>0]$ or $c<0, f\le 0$ $[f\ge 0]$ hold, then a regular solution of (1.52) does not have negative relative minimum points [positive relative maximum] in D.

Proof. Suppose, for example, a solution u of (1.52) has a negative minimum at the point $X\in D$. At point X, by familiar theorems of analysis one has

$$\frac{\partial u}{\partial x_i}=0, \quad i=1,\ldots,n, \quad \sum_{i,j=1}^{n}\frac{\partial^2 u}{\partial x_i \partial x_j}\lambda_i\lambda_j\geqslant 0. \tag{1.53}$$

The quadratic form $\chi(\Lambda)$ at point X can be represented as a sum of squares of linear forms [20]

$$\chi(\Lambda)=\sum_{i,j=1}^{n} a_{ij}(X)\,\lambda_i\lambda_j=\sum_{k=1}^{n}\left[\sum_{l=1}^{n} g_{kl}(X)\lambda_l\right]^2.$$

One also has the obvious equation

$$\sum_{i=1}^{n}\sum_{k,j=1}^{n} g_{ij}(X)\frac{\partial^2 u}{\partial x_i \partial x_k} g_{ik}(X)=\sum_{i,j=1}^{n} a_{ij}(X)\frac{\partial^2 u}{\partial x_i \partial x_j},$$

so by (1.53), at point X we have $M(u)-c(X)u \geq 0$. If $u<0$ at point X, then it follows from the hypotheses of the theorem that $-cu \leq 0$ at this point. Now again by the hypotheses of the theorem, at point X, $M(u)-cu=f-cu<0$. The contradiction found proves the assertion of the theorem for a minimum. The proof for a maximum is analogous.

A more general assertion is valid [26], which we give without proof.

Theorem 1.11. Let the coefficients of (1.52) be bounded in $D \subset E$ and the discriminant $\Delta(X)$ have a positive lower bound in D. If $c \leq 0, f \leq 0$ [$f \geq 0$] in D, then no regular solution u of (1.52) can have a negative relative minimum [positive relative maximum] at the point $X_0 \in D$ under the condition that it is not constant in any domain D_0, containing X_0, in which $u(X) \geq u(X_0)$ [$u(X) \leq u(X_0)$].

Theorem 1.12 (*Zaremba–Giraud Principle*). Let $D \subset E$ be a bounded domain, whose boundary Γ has a normal with direction cosines satisfying a Hölder condition at each point, $u(X)$ be a solution of (1.52), which is regular in D and continuous in $D \cup \Gamma$ and not identically constant. If $c \leq 0, f \leq 0$ [$f \geq 0$], then at any point Y of the boundary Γ, at which u achieves a negative minimum [positive maximum], for each ray l issuing from point Y and which forms an obtuse angle with the direction of the exterior normal to Γ at point Y, there exists a positive constant C such that for $X \in l$ and sufficiently close to Y one has

$$u(X)-u(Y)>CL(X,\ Y)\quad [u(X)-u(Y)<-CL(X,\ Y)], \tag{1.54}$$

where $L(X, Y)$ is the distance between points X and Y.

Proof. We prove the assertion of the theorem for the case when the surface Γ, in a coordinate system in which the Ox_n axis coincides with the normal at point X and the plane Γ coincides with the tangent plane T to $x_n=0$ at point X, in a neighborhood of each point X admits a representation of the form $x_n=\omega(x_1, \ldots, x_{n-1})$, where the function ω has continuous second derivatives. Now on the interior normal at point X there exists a point X_1 such that the ball of radius $L(X, X_1)$ with center at point X_1 is completely contained in domain D, and the sphere bounding it has only the one point X in common with boundary Γ of domain D. We construct such a ball $\Sigma(X_1, \rho)$ for a minimum point Y of function $u(X)$. We consider the ball $\Sigma(Y, \rho_1)$ with center at point Y, of radius $\rho_1<\rho$. We denote the intersection of these balls by B. By Theorem 1.11, everywhere in B we have $u(X)-u(Y)>0$. In the first ball we construct the function $v(X)=\exp(-k\rho^2)-\exp(-kr^2)$, where $r=L(X_1, X)$ is the distance between the points X_1 and X, and k is chosen sufficiently large that one has $M(v)<0$ for all X from B. One can always do this since $M(v)\times \exp(kr^2)$ is a quadratic trinomial with respect to k with negative coefficient of k^2

and with the other coefficients bounded. The function v vanishes on the part of the boundary of the domain B which coincides with the boundary of the ball $\Sigma(X_1, \rho)$. Obviously for sufficiently small $\lambda > 0$ one has

$$\lambda v(X) < u(X) - u(Y) \tag{1.55}$$

for all X lying on the boundary of the domain B except possibly $X = Y$. Since for sufficiently large k, $M(v) < 0$, we have

$$M[u(X) - u(Y) + \lambda v(X)] = f(X) - c(X)u(Y) + \lambda M(v) < 0.$$

By Theorem 1.11, (1.55) holds everywhere in B. Now the assertion of the theorem follows from the inequality

$$-\lambda \left.\frac{dv}{dl}\right|_{X=Y} = -2\lambda k \exp(-k\rho^2) \cos(l, n) > 0,$$

since for an obtuse angle the cosine is negative. One gets the second assertion of the theorem from the first by replacing $u(X)$ by $-u(X)$. □

We have assumed that the coefficients of the operator M are twice continuously differentiable, and for such coefficients one can transform $M(u)$ to the form

$$M(u) = \sum_{i,j=1}^{n} \frac{\partial}{\partial x_j}\left[a_{ij}(X)\frac{\partial u}{\partial x_i}\right] + \sum_{i=1}^{n} e_i(X)\frac{\partial u}{\partial x_i} + cu,$$

$$e_i(X) = b_i(X) - \sum_{k=1}^{n} \frac{\partial}{\partial x_k} a_{ik}(X).$$

The operator N, which acts as

$$N(v) = \sum_{i,j=1}^{n} \frac{\partial}{\partial x_i}\left[a_{ij}(X)\frac{\partial v}{\partial x_i}\right] - \sum_{i=1}^{n} \frac{\partial}{\partial x_i}[e_i(X)v] + cV,$$

is called the adjoint of the operator M. Obviously the adjoint of the operator N is the operator M. It is easy to verify the following identity, which is valid for any twice differentiable functions $u(X)$ and $v(X)$:

$$vM(u) - uN(v) = \sum_{i,j=1}^{n} \frac{\partial}{\partial x_j}\left[a_{ij}(X)\left(v\frac{\partial u}{\partial x_i} - u\frac{\partial v}{\partial x_i}\right)\right] + \sum_{i=1}^{n} \frac{\partial}{\partial x_i}[e_i(X)uv].$$

We take a domain D, whose boundary Γ has a continuously varying tangent plane; assuming that the functions u and v are continuously differentiable in the closed domain $D \cup \Gamma$, we integrate this identity over the domain and make use of the Gauss–Ostrogradskii formula. We get

$$\int_D [vM(u) - uN(v)]\, d\omega = \int_\Gamma \left[a(X)\left(v\frac{du}{d\nu} - u\frac{dv}{d\nu}\right) + b(X)uv\right] d\sigma, \qquad (1.56)$$

where ν is the conormal, i.e., the direction with direction cosines

$$Y_i = \frac{1}{a}\sum_{j=1}^{n} a_{ij}(X)\, n_j(X), \quad a = \left[\sum_{i=1}^{n}\left(\sum_{j=1}^{n} a_{ij} n_j\right)^2\right]^{1/2},$$

and $n_j = \cos(x_j, n)$ are the direction cosines of the exterior normal to Γ at point X, and

$$b(X) = \sum_{i=1}^{n} e_i(X)\cos(X_i, n).$$

Equation (1.56) generalizes the second formula of (1.13), and it will also be called Green's formula. Just as (1.13) does in potential theory, Green's formula (1.56) plays an important role in the theory of elliptic equations.

We consider n^2, $n > 2$, functions, $a_{kl}(Y)$ of a variable $Y = (y_1, \dots, y_n)$. We denote by $A(Y)$ the determinant of the matrix $\|a_{kl}\|$, and by $A_{kl}(Y)$ the cofactor of the element a_{kl} of this matrix, divided by the determinant A of the given matrix. We construct the function

$$H(X, Y) = \frac{1}{(n-2)\, w_n \sqrt{A(Y)}}\left[\sum_{k,l=1}^{n} A_{kl}(Y)(x_k - y_k)(x_l - y_l)\right]^{(2-n)/2}.$$

Since the form $\Sigma_{i,j=1}^{n} a_{ij}(Y)\lambda_i\lambda_j = \chi(\Lambda)$ is positive definite, the form $\Psi(\Lambda) = \Sigma_{i,j=1}^{n} \times A_{ij}(Y)\lambda_i\lambda_j$ is also positive definite [27], so $\Psi(\Lambda) = O(\lambda_1^2 + \dots + \lambda_n^2)$, where the symbol O here denotes the existence of two constants $c \geq 0$ and $C > 0$, such that for all $\lambda_1, \dots, \lambda_n$ one has

$$c(\lambda_1^2 + \dots + \lambda_n^2) \leqslant \Psi(\Lambda) \leqslant C(\lambda_1^2 + \dots + \lambda_n^2).$$

It follows from this that $H(X, Y) = O(r^{2-n})$, where r is the distance between the points X and Y. By direct calculation one verifies the relations

$$M_Y(H) \equiv \sum_{i,j=1}^{n} a_{ij}(Y)\frac{\partial^2 H}{\partial x_i \partial x_j} = 0, \qquad (1.57)$$

$$H = O(r^{2-n}), \quad \frac{\partial H}{\partial x_i} = O(r^{1-n}), \quad \frac{\partial^2 H}{\partial x_i \partial x_j} = O(r^{-n}). \qquad (1.58)$$

where the functions a_{ij} as usual are assumed to be twice continuously differentiable (although the continuity of the functions suffices).

We shall call a function $\Lambda(X, Y)$, which depends on a pair of points X and Y and a domain D continuously along with its derivatives of the first and second orders with respect to the coordinates of the point X, when $X \neq Y$, Levi function, if in each closed subdomain of domain D it satisfies the following estimates uniformly:

$$\Lambda - H = O\left(r^{\lambda+2-n}\right), \quad \frac{\partial}{\partial x_i}(\Lambda - H) = O\left(r^{\lambda+1-n}\right),$$
$$\frac{\partial^2 (\Lambda - H)}{\partial x_i \partial x_j} = O\left(r^{\lambda-n}\right), \quad r = L(X, Y), \tag{1.59}$$

with some positive λ. Obviously the functions $H(X, Y)$ and $H(Y, X)$ are Levi functions with $\lambda = 1$. If the coefficients of the operator M are Hölder continuous with exponent $\lambda > 0$, then by (1.57) we have

$$\sum_{i,j=1}^{n} [a_{ij}(X) - a_{ij}(Y)] \frac{\partial^2 H}{\partial x_i \partial x_j} = O\left(r^{\lambda-n}\right),$$

so $M(H) = O(r^{\lambda-n})$, and it follows from (1.59) that any Levi function satisfies the relation

$$M[\Lambda(X, Y)] = O(r^{\lambda-n}), \quad r = L(X, Y). \tag{1.60}$$

We take a domain D entirely contained in domain E in which the operator M is defined and in which its coefficients are twice continuously differentiable. Let Y be an interior point of domain D; by $I(Y, \rho)$ we denote the neighborhood of point Y defined by the inequality

$$\sum_{i,j=1}^{n} a_{ij}(Y)(x_i - y_i)(x_j - y_j) \leqslant \rho^2,$$

and we choose ρ sufficiently small that this neighborhood is entirely contained in domain D. By $D(\rho)$ we denote the domain which is obtained from D by removing the neighborhood $I(Y, \rho)$, and by $\Gamma(Y, \rho)$ the boundary of $I(Y, \rho)$. We write (1.56) for the domain $D(\rho)$, setting $v = \Lambda$,

$$\int_{D(\rho)} [\Lambda M(u) - uN(\Lambda)]\, d\omega = \int_{\Gamma} \left[a(X)\left(u \frac{d\Lambda}{d\nu} - \Lambda \frac{du}{d\nu}\right) - b(X) u\Lambda\right] d\sigma$$
$$- \int_{\Gamma(Y,\rho)} \left[a(X)\left(u \frac{d\Lambda}{d\nu} - \Lambda \frac{du}{d\nu}\right) - b(X) u\Lambda\right] d\sigma. \tag{1.61}$$

Obviously, by virtue of the properties of Λ we have

$$a(X) u \frac{d\Lambda}{d\nu} = \frac{u(Y)}{\rho^n \omega_n \sqrt{A(Y)}} \sum_{k=1}^{n} (x_k - y_k) n_k(X) + O\left(\rho^{\lambda+1-n}\right),$$

$$a(X)\Lambda\frac{du}{d\nu}=O\left(\rho^{2-n}\right),\quad b(X)u\Lambda=O\left(\rho^{2-n}\right),$$

so the integral over $\Gamma(Y,\rho)$ can be transformed as follows:

$$\int\limits_{\Gamma(Y,\rho)}\left[a(X)\left(u\frac{d\Lambda}{d\nu}-\Lambda\frac{du}{d\nu}\right)d\sigma=\frac{u(Y)}{\rho^n\omega_n\sqrt{A(Y)}}\int\limits_{\Gamma(Y,\rho)}\sum_{k=1}^{n}(x_k\right.$$
$$\left.-y_k)n_k(X)\,d\sigma+\Sigma\cdot O\left(\rho^{\lambda+1-n}\right)=u(Y)\frac{n\Omega}{\rho^n\omega_n\sqrt{A(Y)}}+\Sigma\cdot O\left(\rho^{\lambda+1-n}\right),$$

where Σ is the area of the surface $\Gamma(Y,\rho)$, and Ω is the volume of the domain $I(Y,\rho)$. One calculates the volume, and the area of the boundary is easily estimated. We have

$$\Omega=n^{-1}\rho^n\omega_n\sqrt{A(Y)},\quad \Sigma=O\left(\rho^{n-1}\right),$$

so letting ρ tend to zero, we find

$$\lim_{\rho\to0}\int\limits_{\Gamma(Y,\rho)}\left[a(X)\left(u\frac{d\Lambda}{d\nu}-\Lambda\frac{du}{d\nu}\right)\right]d\sigma=u(Y),$$

and letting ρ tend to zero in (1.61), we get

$$u(Y)=\int\limits_{D}[\Lambda M(u)-uN(\Lambda)]\,d\omega+\int\limits_{\Gamma}\left[a(X)\left(\Lambda\frac{du}{d\nu}-u\frac{d\Lambda}{d\nu}\right)+b(X)\Lambda u\right]d\sigma, \tag{1.62}$$

the derivative $d/d\nu$ here is taken with respect to point X. Equation (1.62) is the analog of the second equation of (1.20) and is called Stokes' formula.

A Levi function $\Lambda(X, Y)$ which is a solution of the equation $M_x(\Lambda) = 0$ [$N_x(\Lambda) = 0$] is called a fundamental or elementary solution of the equation $M(u) = 0$ [$N(u) = 0$]. By the Green's function of a boundary problem is meant a fundamental solution $\Omega(X, Y)$ of the adjoint equation with respect to the variable Y, which satisfies the boundary condition of the adjoint homogeneous problem.

With the help of a Levi function one can introduce so-called generalized potentials of double and simple layers and solid ones

$$W(X)=\int\limits_{\Gamma}\frac{d\Lambda(X,Y)}{d_Y\nu}\sigma(Y)\,dS, \tag{1.63}$$

$$V(X)=\int\limits_{\Gamma}\Lambda(X,Y)\rho(Y)\,dS, \tag{1.64}$$

$$u(X)=\int\limits_{D}\Lambda(X,Y)\omega(Y)\,d\Omega \tag{1.65}$$

respectively, where D is the domain and Γ is the boundary of the domain. Many properties of ordinary potentials generalize to these potentials. Below, following

[40], we introduce, in a somewhat different form, potentials similar to (1.63)–(1.65), which are even more similar to the ordinary potentials.

First of all we recall some concepts of Riemannian geometry [43, 53]. In a Riemannian space suppose the line element has the form

$$ds^2 = \sum_{i,k=1}^{n} g_{ik}dx_i dx_k, \quad g_{ik} = g_{ki}, \tag{1.66}$$

then the volume element in this space is given by the formula $dV = \sqrt{g}\, dx_1 \ldots dx_n$, where g is the determinant of the matrix $\|g_{ik}\|$. The form (1.66) is positive definite, so $g > 0$. A submanifold M of dimension $n - 1$ in a Riemannian space is defined by an equation $F(x_1, \ldots, x_n) = \sigma = \text{const}$. Instead of this one can define the manifold M with the help of $n - 1$ independent parameters $\lambda_1, \ldots, \lambda_{n-1}$ in the following way:

$$x_i = x_i(\lambda_1, \ldots, \lambda_{n-1}), \quad i = 1, \ldots, n. \tag{1.67}$$

Substituting the expressions for x_i and dx_i from these formulas into (1.68), we get a quadratic form with respect to $d\lambda_i$, which will represent a line element on the manifold M. One gets the volume element of M analogously.

We express the volume element of manifold M directly in terms of the g_{ik} and the quantities of (1.67). We denote by g^{ik} the cofactor of the elements g_{ik} of the matrix $\|g_{ik}\|$. Further, we set

$$D_i = (-1)^{i-1} \frac{\partial (x_1, \ldots, x_{i-1}, x_{i+1}, \ldots, x_n)}{\partial (\lambda_1, \ldots, \lambda_{n-1})},$$

$$i = 1, \ldots, n.$$

We show that the volume element of manifold M is given by the formula

$$d\sigma = \sqrt{g} \sqrt{\sum_{k,l=1}^{n} g^{lk} D_l D_k}\, d\lambda_1 \ldots d\lambda_{n-1}. \tag{1.68}$$

For this it suffices to prove that $d\sigma$ is independent of the choice of parameters $\lambda_1, \ldots, \lambda_{n-1}$ and of the choice of coordinates $x_1, \ldots, x_n$ in the original space, and that for some special choice of the parameters and coordinates the volume actually has the form (1.68).

The quantities g^{lk} and g are independent of the parameters $\lambda_1, \ldots, \lambda_{n-1}$ and depend only on the original space. The determinants D_i under change of the parameters λ_i are multiplied by the Jacobian of the substitution, so the form $d\sigma$ is invariant with respect to transformation of parameters. It remains to show that the expression

$$\nu = \left(g \sum_{k,l=1}^{n} g^{lk} D_l D_k \right)^{1/2}$$

is independent of the coordinate system in the original space. We substitute the expressions (1.67) into the equation of the manifold M: $\{\sigma = F(x_1, \ldots, x_n)\}$ and we differentiate the identity found with respect to $\lambda_1, \ldots, \lambda_{n-1}$:

$$\sum_{j=1}^{n} F_j \frac{\partial x_j}{\partial \lambda_i} = 0, \quad i = 1, \ldots, n-1, \quad F_j = \frac{\partial F}{\partial x_j}.$$

From this system of $n - 1$ equations for the quantities F_j we find $F_j = \mu D_j$, $j = 1, \ldots, n$, where μ is a fixed quantity. Consequently,

$$F_j \left(\sum_{i,k=1}^{n} g^{ik} F_i F_k \right)^{-1/2} = D_j \left(\sum_{i,k=1}^{n} g^{ik} D_i D_k \right)^{-1/2}.$$

The coefficient of F_j on the left side of this equation is invariant. We consider $\sigma = F(x_1, \ldots, x_n)$, $\lambda_1, \ldots, \lambda_{n-1}$ as new independent variables. We have

$$D = \frac{\partial (x, \ldots, x_n)}{\partial (\sigma, \lambda_1, \ldots, \lambda_{n-1})} = \sum_{j=1}^{n} \frac{\partial x_j}{\partial \sigma} D_j = \sum_{j=1}^{n} \frac{\partial x_j}{\partial \sigma} F_j$$

$$\times \left[\frac{\sum_{k,l=1}^{n} g^{lk} D_l D_k}{\sum_{k,l=1}^{n} g^{lk} F_l F_k} \right]^{1/2} = \left[\frac{\sum_{k,l=1}^{n} g^{lk} D_l D_k}{\sum_{k,l=1}^{n} g^{lk} F_l F_k} \right]^{1/2},$$

so one has

$$\left(\sum_{k,l=1}^{n} g^{lk} D_l D_k \right)^{1/2} = D \left(\sum_{k,l=1}^{n} g^{lk} F_l F_k \right)^{1/2}.$$

Thus the expression which appears on the left side of this equation changes like the Jacobian of a change of variables, so the quantity ν remains invariant under all transformations of the coordinate system.

We choose a coordinate system so that in a sufficiently small neighborhood of the point $A \in M$ the manifold M is defined by the equation $x_n = \text{const}$, and we set $\lambda_i = x_i$, $i = 1, \ldots, n - 1$. For the given choice of coordinate system the volume element of M has the form $\sqrt{\gamma}\, dx_1 \ldots dx_{n-1}$, where γ is the determinant of the matrix $\|g_{\alpha\beta}\|$, $\alpha, \beta = 1, \ldots, n - 1$. Obviously, $\gamma = g g^{nn}$, and $D_j = 0$, $j = 1, \ldots, n - 1$, so (1.68) assumes the form $d\sigma = \sqrt{g g^{nn}}\, dx_1 \ldots dx_{n-1}$. Consequently, for such a special choice of coordinates, (1.68) gives the volume element of manifold M. This also proves that $d\sigma$ is the volume element of manifold M.

Now we consider a self-adjoint elliptic operator with sufficiently smooth coefficients

$$N(u) = \sum_{i,k=1}^{n} \frac{\partial}{\partial x_k} \left(A^{ik} \frac{\partial u}{\partial x_i} \right), \quad n \geqslant 3.$$

We introduce the notation

$$g^{ik} = A^{1/(n-2)}A^{ik}, \quad A = \det \|A^{ik}\|, \quad i, k = 1, \ldots, n,$$

and rewrite this operator in the form of the second differential parameter corresponding to the line element (1.66):

$$\frac{1}{\sqrt{g}} N(u) \equiv \Delta_2 u = \frac{1}{\sqrt{g}} \sum_{i,k=1}^{n} \frac{\partial}{\partial x_k}\left(g^{ik}\sqrt{g}\,\frac{\partial u}{\partial x_i}\right), \tag{1.69}$$
$$g = \det \|g_{ik}\| = [\det \|g^{ik}\|]^{-1}, \quad i, k = 1, \ldots, n.$$

For any domain G with sufficiently smooth boundary Γ such that the closed domain $G \cup \Gamma$ lies entirely in the Riemannian space corresponding to the line element (1.66), defined by the operator (1.69), we prove Green's formula

$$\int_G (v\Delta_2 u - u\Delta_2 v)\, d\omega = \int_\Gamma \left(v\frac{\partial u}{\partial n} - u\frac{\partial v}{\partial n}\right) d\sigma, \tag{1.70}$$

where $d\omega = \sqrt{g}\, dx_1 \ldots dx_n$ is the volume element, $d\sigma$ is the volume element of the boundary Γ (1.68), and

$$\frac{\partial}{\partial n} = \sum_{l,k=1}^{n} g^{lk} \frac{D_k}{\sqrt{\sum_{i,j=1}^{n} g^{ij} D_i D_j}} \frac{\partial}{\partial x_l}$$

is normalized differentiation in the direction of the outer normal to the surface Γ in the sense of the corresponding Riemannian metric. This definition of the normal derivative differs from the derivative in the direction of the conormal only by the factor $(g\Sigma_{i,j=1}^{n} g^{ij}D_iD_j)^{-1/2}$. Here, as usual, it is assumed that Γ can be represented in the form of a union of a finite number of generally intersecting pieces of manifolds Δ_k, $k = 1, \ldots, N$, such that each Δ_k is parametrized by (1.67), and on the intersection of Δ_k with Δ_l one can pass from one parameter to the others by a change of variables with positive Jacobian. We take an infinitely small displacement $dx_1, \ldots, dx_n$ from a point of the manifold, and we consider the expression $\delta = D_1\, dx_1 + \ldots + D_n dx_n$, where D_i are the same as in (1.68). On Δ_k we shall take only parametrizations for which $\delta > 0$ for any displacement from a point of the manifold Γ into the exterior of domain G. This convention fixes an orientation of the parametrizations of Δ_k and, consequently, also of Γ. Obviously for any collection of pieces of manifolds Δ_k covering Γ, one can construct a collection of manifolds, whose closures cover Γ, and these manifolds do not have common interior points. We take Δ_1 and we remove all Δ_k contained in Δ_1, and from the remaining Δ_k we remove all points which are common with Δ_1; as a result we get a new collection $\Delta_1, \Delta_2', \ldots, \Delta_p', p < \infty$. Further, with $\Delta_2', \ldots, \Delta_p'$ we make the analogous construction. After a finite number of steps we get the required collection $\Delta_1, \Delta_2', \ldots, \Delta_s', s < \infty$. One can understand the integral on the right side of (1.70) as the sum of integrals

..., Δ_s', $s < \infty$. One can understand the integral on the right side of (1.70) as the sum of integrals

$$\int_{E_j} \left(v \frac{\partial u}{\partial n} - u \frac{\partial v}{\partial n} \right) d\sigma,$$

taken over the domain of the change of parameters (1.67), corresponding to Δ_j'. With this understanding one can write the multidimensional Gauss formula in the form

$$\int_G \sum_{i=1}^{n} \frac{\partial A_i}{\partial x_i} dx_1 \ldots dx_n = \int_\Gamma \sum_{i=1}^{n} A_i D_i d\lambda_1 \ldots d\lambda_{n-1}, \tag{1.71}$$

where the integrals are taken in Euclidean space and the sum of integrals appears on the right. In (1.71), letting

$$A_k = \sum_{l=1}^{n} g^{lk} \left(v \frac{\partial u}{\partial x_l} - u \frac{\partial v}{\partial x_l} \right),$$

by simple identity transformations we get (1.70). Thus Green's formula is written in exactly the same way as for the Laplace equation. In particular, if G is a ball in the corresponding Riemannian metric, then $\partial/\partial n$ coincides with the derivatives with respect to r, which one can get from the differential equations of geodesic lines [43], or by introducing the geodesic distance r as an independent variable in the line element.

An arbitrary non-self-adjoint elliptic operator can be written as follows:

$$L(u) = \Delta_2 u + \sum_{i=1}^{n} b_i \frac{\partial u}{\partial x_i} + cu = \frac{1}{\sqrt{g}} \sum_{i,k=1}^{n} \frac{\partial}{\partial x_k} \left(g^{ik} \sqrt{g} \frac{\partial u}{\partial x_i} \right) + \sum_{i=1}^{n} b_i \frac{\partial u}{\partial x_i} + cu,$$

and the operator adjoint to it has the form

$$M(v) = \Delta_2(v) - \frac{1}{\sqrt{g}} \sum_{i=1}^{n} \frac{\partial}{\partial x_i} (b_i \sqrt{g}\, v) + cv.$$

In exactly the same way that (1.70) was derived one can prove the equation

$$\int_G [vL(u) - uM(v)]\, d\omega = \int_\Gamma \left[v \frac{\partial u}{\partial n} - u \frac{\partial v}{\partial n} + Nuv \right] d\sigma, \tag{1.72}$$

$$N = \sum_{i=1}^{n} b_i D_i \left(\sum_{l,k=1}^{n} g^{lk} D_l D_k \right)^{-1/2} = \sum_{i=1}^{n} b_i \cos(n, x_i).$$

Thus we have again arrived at (1.56). One can use this formula just as in potential theory; it is only necessary to construct a fundamental solution, which was already considered.

Before considering the fundamental solution, we construct a Levi function of special form, which will be a large part of this fundamental solution. Let X and Y be two points of the domain G; we denote by $s(X, Y)$ the geodesic distance between these points in the Riemannian metric corresponding to the operator Δ_2. We show that

$$\Delta_2[s(X, Y)]^{2-n} = O(s^{2-n}). \tag{1.73}$$

For this we introduce canonical Riemannian coordinates $\eta_1, \ldots, \eta_n$ [43], which are chosen so that the equations of geodesic lines issuing from the point $Y = (\xi_1, \ldots, \xi_n)$ have the form

$$\eta_j = c_j s, \quad c_j = \text{const} = \left.\frac{d\eta_j}{ds}\right|_{s=0},$$

where as parameter we take the arc length of the geodesic. If the line element (1.66) is put in these coordinates, then we have

$$\sum_{i,k=1}^{n} g_{ik} c_i c_k = 1.$$

We denote by $g_{ik}{}^0$ the value of g_{ik} at point Y, from which we get

$$\sum_{i,k=1}^{n} g_{ik}^0 c_i c_k = 1 \quad \text{or} \quad s^2 = \sum_{i,k=1}^{n} g_{ik}^0 c_i c_k S^2 = \sum_{i,k=1}^{n} g_{ik}^0 \eta_i \eta_k,$$

i.e., in canonical coordinates the distance can be expressed by a quadratic form with constant coefficients. Differentiating the last equation with respect to η_i and considering that $\eta_j = c_j s$, we find

$$2s \frac{\partial s}{\partial \eta_j} = 2 \sum_{k=1}^{n} g_{ik}^0 \eta_k,$$

$$\frac{\partial s}{\partial \eta_j} = \frac{1}{s} \sum_{k=1}^{n} g_{ik}^0 \eta_k = \sum_{k=1}^{n} g_{ik}^0 c_k,$$

but we have the identities

$$\sum_{k=1}^{n} g_{ik} c_k = \sum_{k=1}^{n} g_{ik}^0 c_k, \quad k = 1, \ldots, n, \tag{1.74}$$

by virtue of which we get

$$\frac{\partial s}{\partial \eta_j} = \sum_{k=1}^{n} g_{kj} c_k = \frac{1}{s} \sum_{k=1}^{n} g_{kj} \eta_k. \tag{1.75}$$

In order to justify (1.74), we note that the equations of geodesic lines in any coordinates can be written as follows:

$$\frac{d}{ds} \sum_{k=1}^{n} g_{ik} \frac{d\eta_k}{ds} - \frac{1}{2} \sum_{k,j=1}^{n} \frac{\partial g_{kj}}{\partial \eta_i} \frac{d\eta_k}{\partial s} \frac{d\eta_j}{ds} = 0.$$

Since in Riemannian coordinates along geodesics we have $\eta_j = c_j s$, we get

$$\frac{d}{ds} \sum_{k=1}^{n} g_{ik} c_k - \frac{1}{2} \frac{\partial}{\partial \eta_i} \sum_{k,j=1}^{n} g_{kj} c_k c_j = 0,$$

but the following relation holds:

$$\sum_{k,j=1}^{n} g_{kj} c_k c_j = \left(\sum_{k,j=1}^{n} g_{kj} d\eta_k d\eta_j \right) \Big/ ds^2 = 1.$$

so along a geodesic $\sum_{k=1}^{n} g_{ik} c_k = \text{const}$, and this is (1.74).

With the help of (1.75) we calculate further

$$\Delta_2 s^{2-n} = \frac{1}{\sqrt{g}} \sum_{i,k=1}^{n} \frac{\partial}{\partial \eta_k} \left[g^{ik} \sqrt{g} \frac{\partial}{\partial \eta_i} s^{2-n} \right] = \frac{2-n}{\sqrt{g}} \sum_{i,k=1}^{n} \frac{\partial}{\partial \eta_k} \left[g^{ik} \sqrt{g} \sum_{j=1}^{n} g_{ji} \eta_j s^{-n} \right].$$

Using the identities

$$\sum_{i=1}^{n} g^{ik} g_{ji} = \begin{cases} 1, & j = k, \\ 0, & j \neq k, \end{cases}$$

we continue the transformations of the expression

$$\Delta_2 s^{2-n} = \frac{2-n}{\sqrt{g}} \sum_{k=1}^{n} \frac{\partial}{\partial \eta_k} [\sqrt{g}\, \eta_k s^{-n}]$$

$$= \frac{2-n}{\sqrt{g}} \sum_{k=1}^{n} \left[\frac{\partial \sqrt{g}}{\partial \eta_k} \eta_k s^{-n} + s^{-n} \sqrt{g} + n \eta_k \sqrt{g} \sum_{j=1}^{n} g_{jk} \eta_j s^{-1} s^{-n-1} \right]$$

$$= \frac{2-n}{\sqrt{g}} \left[\sum_{k=1}^{n} c_k \frac{\partial \sqrt{g}}{\partial \eta_k} s^{-n+1} + n s^{-n} \sqrt{g} \right.$$

$$\left. - n \sqrt{g} \sum_{i,k=1}^{n} g_{jk} \eta_j \eta_k s^{-n-2} \right],$$

but the last two summands in the brackets annihilate each other by virtue of the identity

$$\sum_{j,k=1}^{n} g_{jk}\eta_j\eta_k = s^2 \sum_{j,k=1}^{n} g_{jk}c_jc_k = s^2.$$

Since along geodesics issuing from point Y we have $\eta_j = c_js$, the operator $\Sigma_{k=1}^{n} \times c_k\, \partial/\partial\eta_k$ is the operator of differentiation with respect to s along these geodesics. Thus, we arrive at the equation

$$\Delta_2 s^{2-n} = (2-n)\, s^{1-n} \frac{\partial \ln \sqrt{g}}{\partial s}.$$

In canonical coordinates all the derivatives $\partial g_{ik}/\partial\eta_j$ vanish at point Y, which is taken as the origin, so $\ln\sqrt{g} = g_0 + O(s^2)$ in a neighborhood of point Y, so $(\partial/\partial s)\ln\sqrt{g} = O(s)$, from which the validity of (1.73) follows.

With the help of the Levi function $[s(X, Y)]^{2-n}$ one can construct the potentials (1.63)–(1.65) and a fundamental solution by the same method as for a fundamental solution of a hyperbolic equation [1]. This construction is analogous to Giraud's [26, 51] and Levi's [21] construction, but some of the estimates along the way are found more simply. We seek a fundamental solution $\Gamma(X, Y)$ in the form [1]

$$\Gamma(X, Y) = [s(X, Y)]^{2-n} + [s(X, Y)]^{4-n}U(X, Y), \quad n = 2k+1,$$
$$\Gamma(X, Y) = [s(X, Y)]^{2-n} + U_0(X, Y)\ln[s(X, Y)] + [s(X, Y)]^{4-n}U(X, Y), \quad n = 2k,$$
$$U(X, Y) = \sum_{l=0}^{\infty} [s(X, Y)]^l A_l(X, Y).$$

Without loss of generality one can assume the fundamental solution positive in any bounded domain, by adding a suitably chosen constant if necessary. We introduce the notation

$$\rho(X, Y) = [\Gamma(X, Y)]^{1/(2-n)}.$$

Obviously one has

$$\lim_{X\to Y} s/\rho = \lim_{s\to 0} s/\rho = 1,$$

from which it follows that for sufficiently small constant $\gamma > 0$ the surfaces $\rho = \gamma$ are closed and the equation $\rho = 0$ defines only point Y. We consider the family of hypersurfaces ρ = const, and on the fixed hypersurface $\rho = \rho_1$ we introduce parameters $\varphi_1, \ldots, \varphi_{n-1}$. Through the points of the hypersurface $\rho = \rho_1$ we draw the orthogonal trajectories of the family of hypersurfaces ρ = const. The orthogonal trajectories tend to point Y, intersecting all hypersurfaces ρ = const $< \rho_1$. Along these trajectories we transfer the values of $\varphi_1, \ldots, \varphi_{n-1}$ from the initial hypersurface to all the other hypersurfaces ρ = const $< \rho_1$. Thus, to each point of the domain $\rho \le \rho_1$ there is associated a system of parameters $\varphi_1, \ldots, \varphi_{n-1}, \rho$, which one can take as coordinates. By the orthogonality of the lines $\varphi_j = c_j$, $j = 1, \ldots, n-1$ to the hypersurfaces ρ = const the line element (1.66) in these coordinates assumes the form

$$ds^2 = a_0 d\rho^2 + \sum_{k,l=1}^{n} a_{kl} d\varphi_k d\varphi_l.$$

The volume element of the hypersurface $\rho = \text{const}$ can be expressed as follows: $\sqrt{a}\, d\varphi_1 \ldots \varphi_{n-1}$, $a = \det \|a_{kl}\|$, $k, l = 1, \ldots, n-1$, and the volume element of the space $-dV = \sqrt{a_0 a}\, d\rho d\varphi_1 \ldots \varphi_{n-1}$. Differentiation in the direction of the normal to the hypersurface $\rho = \text{const}$ reduces to $a_0^{-1/2}\partial/\partial\rho$ and the operator Δ_2 assumes the form

$$\Delta_2 u = \frac{1}{\sqrt{a_0 a}} \frac{\partial}{\partial \rho}\left[\frac{\partial}{\partial \rho}\left(u \sqrt{\frac{a}{a_0}}\right)\right] + \sum_{k,l=1}^{n} \frac{\partial}{\partial \varphi_l}\left(a^{kl} \frac{\partial u}{\partial \varphi_k} \sqrt{a_0 a}\right),$$

where the a^{kl} can be calculated in terms of the elements a_{kl} of the $(n-1)$-rowed matrix $\|a_{kl}\|$. From the fact that $u = \rho^{2-n}$ is a solution of the equation $\Delta_2 u = 0$, we get

$$\frac{\partial}{\partial \rho}\left[(2-n)\,\rho^{1-n}\sqrt{\frac{a}{a_0}}\right] = 0, \quad \sqrt{\frac{a}{a_0}} = \Phi(\varphi_1, \ldots, \varphi_{n-1})\,\rho^{n-1}.$$

It follows from this that the following integral, taken over the surface Σ: $\{\rho = \text{const}\}$,

$$\int_{\Sigma} \rho^{1-n} \sqrt{\frac{a}{a_0}}\, d\varphi_1 \ldots d\varphi_{n-1} = E(Y)$$

is a positive quantity, independent of ρ. Instead of the fundamental solution $\Gamma(X, Y)$ it is sometimes more convenient to consider the so-called normalized fundamental solution $K(X, Y) = \Gamma(X, Y)/E(Y)$, which is symmetric with respect to the points X and Y and satisfies the equation $\Delta_2 u = 0$ with respect to the coordinates of the point $Y = (\xi_1, \ldots, \xi_n)$ also.

We consider the potentials analogous to (1.63)–(1.65)

$$U(Y) = \int_G \mu(X)\, K(X, Y)\, dV, \tag{1.76}$$

$$V(Y) = \int_\Gamma \sigma(X)\, K(X, Y)\, d\Gamma, \tag{1.77}$$

$$W(Y) = \int_\Gamma \tau(X) \frac{\partial}{\partial x^n}[K(X, Y)]\, d\Gamma, \tag{1.78}$$

where μ, σ, and τ are sufficiently smooth functions. All the properties of the ordinary potentials of the theory of harmonic functions generalize to these potentials. For example, if v is a function which is twice continuously differentiable in domain G, then just as for Laplace's equation, removing a small neighborhood of point Y, by passage to the limit we get the formula

$$\int_G \Delta_2 v K(X, Y)\, d_X G = v(Y) + \int_\Gamma \left(v \frac{\partial K}{\partial n} - K \frac{\partial v}{\partial n}\right) d\Gamma.$$

7. FUNCTIONS REPRESENTED BY POTENTIAL-TYPE INTEGRALS

In the theory of elliptic functions, an important role is played by functions which can be expressed by integrals of the form

$$F(X) = \int_G K(X, Y)\, \sigma(Y)\, d\omega, \tag{1.79}$$

$$Q(X) = \int_\Gamma \acute{M}(X, Y)\, \tau(Y)\, d\Gamma, \tag{1.80}$$

where G is a domain with boundary Γ, and K and M are kernels which admit singularities for $X = Y$. We study the properties of such functions in more detail [26]. Let G be a domain with Lyapunov boundary Γ, $K(X, Y)$ be a function which is defined for any $X, Y \in G$ and continuous for $X \neq Y$. We shall call the function $K(X, Y)$ a kernel of class $N^{(\alpha)}$, $\alpha < n$, if the condition $K = O[L^{\alpha-n}(X, Y)]$ holds uniformly in G, where $L(X, Y)$ is the distance between the points X and Y; now if the condition $K = O[\ln 2R/L(X, Y)]$ holds uniformly in G, where R is the diameter of the domain G, then we call K a kernel of class $N^{(n)}$. If the function $K(X, Y)$ belongs to the class $N^{(\alpha)}$ and there exists a number h, $0 < h < \min(1, \alpha)$, such that the relation

$$K(X, Y) - K(Z, Y) = O[L^h(X, Z) l^{\alpha-h-n}], \tag{1.81}$$

holds uniformly for all X, Y, and Z from the domain G, where $l = l(X, Y, Z)$ is the shortest distance from point Y to the line segment joining points X and Z, then we shall say that K is a kernel of class $N^{(\alpha,h)}$.

If the kernel $K(X, Y)$ belongs to the class $N^{(\alpha,h)}$, then it also belongs to any class $N^{(\alpha,\lambda)}$, $\lambda < h$. In order to establish this it suffices to show that along with (1.81) one has

$$K(X, Y) - K(Z, Y) = O[L^\lambda(X, Z) l^{\alpha-\lambda-n}], \quad \lambda < h.$$

We take a fixed value of g, $0 < g < 1$. If $L(X, Z) \geq gL(X, Y)$, then obviously

$$K(X, Y) = O[L^\lambda(X, Z) L^{\alpha-\lambda-n}(X, Y)] = O(L^\lambda(X, Z) l^{\alpha-\lambda-n}],$$

since $l \leq L(X, Y)$. An analogous estimate also holds for $K(Z, Y)$, so the required relation holds for any X, Y, and Z. Now let $L(X, Z) < gL(X, Y)$; then

$$L(X, Y) \geqslant l \geqslant L(X, Y) - L(X, Z) > (1-g) L(X, Y),$$

and since $0 < \lambda < h$, we find that

$$K(X, Y) - K(Z, Y) = O(L^h(X, Z)L^{\alpha-h-n}(X, Y)]$$
$$= O[L^\lambda(X, Z)L^{\alpha-\lambda-n}(X, Y)] = O[L^\lambda(X, Z)l^{\alpha-\lambda-n}].$$

Thus, for any $\lambda < h$ we have $N^{(\alpha,\lambda)} \supset N^{(\alpha,h)}$.

As an example of kernels of class $N^{(\alpha,h)}$ one has the kernel of the form $K(X, Y) = [H(X) - H(Y)]K_1(X, Y)$, where H is a function which is Hölder continuous with exponent α, and K_1 is a function which is continuously differentiable for $X \neq Y$, satisfying the relations

$$K_1 = O(L^{-n}), \quad \frac{\partial K_1}{\partial x_i} = O\left(L^{-1-n}\right).$$

Theorem 1.13. Let the function $H(X, Y)$ of two points X and Y of the domain D of the space of $n > 1$ independent variables be continuous for $X \neq Y$ and satisfy the relation

$$H(X, Y) = O[L^{\mu-n}(X, Y)], \ 0 < \mu < n.$$

Suppose further that $G(X, Y)$ is a function which is continuous for $X \neq Y$ in D, satisfying the conditions

$$G(X, Y) = O[L^{\alpha-n}(X, Y)], \quad 0 < \alpha < n,$$
$$G(X, Y) - G(Z, Y) = O[L^h(X, Z)l^{\alpha-h-n}],$$
$$0 < h < \min(1, \alpha).$$

We consider the function

$$F(X, Y) = \int\limits_E G(X, A)\, H(A, Y)\, d\Omega,$$

where E is a bounded closed subset of the domain; then for $\mu + \alpha < n + h$ we have

$$F(X, Y) - F(Z, Y) = O[L^\lambda(X, Z)l^{\alpha+\mu-\lambda-n}], \tag{1.82}$$

where λ is any positive number not exceeding h, less than α and greater than $\alpha + \mu - n$, if $\alpha + \mu > n$, i.e., the function $F(X, Y)$ satisfies a Hölder condition with respect to X for any Y, with exponent λ, not exceeding h and less than $\alpha + \mu - n$.

Proof. We fix points X, Y, and Z, and we consider the case $4L(X, Z) \leq L(X, Y)$. For the part of the integral expressing $F(X, Y)$, taken over the domain satisfying the inequality $L(X, A) < 2L(X, Z)$, one has an estimate of the form $O[L^\alpha(X, Z)L^{\mu-n}(X, Y)]$, from which there follows an estimate of the form $O[L^\lambda(X, Z)l^{\alpha+\mu-\lambda-n}]$. For the domain $2L(X, Z) < L(X, A) < L(X, Y)/2$ one has the relations

$$\begin{gathered} G(X, A) - G(Z, A) = O[L^\lambda(X, Z)L^{\alpha-\lambda-n}(X, A)], \\ H(A, Z) = O[L^{\mu-n}(X, Z)], \end{gathered} \tag{1.83}$$

so for the part F_1 of the integral representing F, taken over this domain, we have

$$F_1(X, Y) - F_1(Z, Y) = \int O[L^\lambda(X, Z)L^{\alpha-\lambda-n}(X, A)]O[L^{\mu-n}(X, Z)]d\Omega$$
$$= O[L^{\lambda+\mu-n}(X, Z)] \times \int O[L^{\alpha-\lambda-n}(X, A)]d\Omega = O[L^\lambda(X, Z)l^{\alpha+\mu-\lambda-n}].$$

Obviously here the condition $\lambda < \alpha$ is needed since for the integral to converge one must have $\alpha - \lambda - n > 0$. For the part F_2 of the integral expressing F, taken over the domain satisfying the inequality $L(Y, A) < L(X, Y)/2$, the assertion of the theorem is proved analogously with the help of (1.83). On the rest of the set E one has the inequalities $L(Y, A) < L(X, Y)/2, L(X, A) > L(X, Y)/2$ simultaneously, from which it follows that $1/3 < L(X, A)[L(Y, A)]^{-1} < 3$, so the integrand for the difference $F_3(X, Y) - F_3(Z, Y)$ corresponding to the integral, taken over this subset, has the estimate $O[L^\lambda(X, Z)\, L^{\alpha+\mu-\lambda-2n}(X, A)]$. If $\alpha + \mu < n + \lambda$, then this estimate, after integration, gives the estimate $O[L^\lambda(X, Z)\, l^{\alpha+\mu-\lambda-n}]$; now if $\alpha + \mu > n$, then one can write the estimate of the integrand in the form

$$O[L^\gamma(X, Z)L^{\alpha+\mu-\gamma-2n}(X, A)], \quad \gamma \leqslant \alpha,$$
$$\gamma < \alpha + \mu - n,$$

from which, after integration for the corresponding part F_3 of the function F, one gets

$$F(X, Y) - F(Z, Y) = O[L^\gamma(X, Z)].$$

Now we consider the part of set E satisfying the inequality $4L(X, Z) > L(X, Y)$. One can assume that $4L(X, Z) > L(Z, Y)$, because, if not, the proof goes in exactly the same way as the preceding case. Let us assume $\alpha + \mu \geq n$, because, if not, one has $F(X, Y) = O[L^{\alpha+\mu-n}(X, Y)]$, from which one gets the assertion of the theorem exactly as we proved the inclusion $N^{(\alpha,h)} \subset N^{(\alpha,\lambda)}$, $\lambda < h$. The parts of $F(X, Y)$ corresponding to the domains $L(X, A) < L(X, Y)/2$ and $L(Y, A) < L(X, Y)/2$ satisfy the estimate

$$O\left[L^{\alpha+\mu-n}(X, Y)\right] = \begin{cases} O\left[L^\lambda(X, Z)\, l^{\alpha+\mu-n-\lambda}\right], & \alpha+\mu < n+\lambda, \\ O\left[L^\lambda(X, Z)\right], & \alpha+\mu \geqslant n+\lambda. \end{cases}$$

The part of $F(X, Y)$ corresponding to the domain $L(Y, A) < 8L(X, Z)$ is obtained by integration of a quantity which is $O[L^{\alpha+\mu-\lambda-2n}(Y, A)]$, so for it one has the estimates

$$O[L^{\alpha+\mu-n}(X, Z)], \quad \alpha+\mu > n,$$
$$O\left[\ln \frac{16L(X, Z)}{L(X, Y)}\right], \quad \alpha+\mu = n.$$

By virtue of the obvious inequality $x \geq x^\lambda/\lambda e$ for $x > 1$ in both cases we get the same estimate. In the domain $L(Y, A) > 8L(X, Z)$ we take the relation $G(X, A) - G(Z, A) = O[L^h(X, Z)L^{\alpha-h-n}(X, A)]$, since $L(X, A) \geq L(Y, A) - L(X, Y) > 4L(X, Z)$; further one can replace $L(X, A)$ by $L(Y, A)$, since $L(Y, A) > 2L(X, Y)$. Consequently, the integrand for F under these restrictions has the estimate

$O[L^{\alpha+\mu-\lambda-2n}(Y, A)]\cdot L^{\lambda}(X, Z)$, and the integration is over a domain contained in the set $L_0 > L(Y, A) > 8L(X, Z)$, where $L_0 = \max_{Y,A\in E} L(Y, A)$. This completes the proof of the theorem. □

If in Theorem 1.13 $H(X, Y)$ is independent of Y, then one can take $\mu = n$ and from this theorem we get that the integrals (1.79) and (1.80) with kernels satisfying the same conditions as the kernel G in Theorem 1.13 are Hölder continuous for any integrable densities.

Theorem 1.14. If kernels K and K_1 of potentials of the form (1.79) belong to the classes $N^{(\alpha)}$ and $N^{(\beta)}$, respectively, and $\alpha, \beta > 0$, then the kernel

$$E(X, Y) = \int_G K(X, A)\cdot K_1(A, Y)\, d_A\omega$$

belongs to the class $N^{(\alpha+\beta)}$. Now if kernels M and M_1 of potentials of the form (1.80) belong to the classes $N^{(\alpha)}$ and $N^{(\beta)}$, respectively, and $\alpha, \beta > 0$, then the kernel

$$F(X, Y) = \int_\Gamma M(X, A)\, M_1(A, Y)\, d_A\Gamma$$

belongs to the class $N^{(\alpha+\beta-1)}$.

In (1.31) we have already met an integral in the sense of the principal value. Now we consider such integrals. Let E be a bounded closed set in Euclidean space R^n, in which there is contained a countable sequence of measurable sets $E(j)$, $j = 1, 2, \ldots$, such that $E(j + 1) \subset E(j)$ and the measure of $E(j)$ tends to zero as $j \to \infty$. Let $F(A)$ be a function which is not summable on E, but is summable on the complement $E - E(j)$ of each $E(j)$ in E. If

$$\lim_{j\to\infty} \int_{E-E(j)} F(A)\, d\Omega,$$

exists, then this limit is called the integral in the sense of the principal values. The set of excluded subsets $E(j)$ is not necessarily countable, but one can also allow dependence of this set on a continuous parameter. Suppose, for example, the origin is contained in the set E, and $F(A)$ is a homogeneous function with respect to the coordinates of point A of degree n, which is continuous everywhere away from point O. Let us assume that there exists a positive definite quadratic form

$$\chi(A) = \sum_{i,j=1}^{n} c_{ij} a_i a_j$$

such that the integral of the function $F(A)$, taken over the domain which satisfies the inequalities $\eta^2 < \chi(A) < \zeta^2$, $0 < \eta < \zeta$, is equal to zero for any η and ζ. Now the integral in the sense of the principal value is the limit to which the integral of $F(A)$, taken over the set E, with the set of point $\chi(A) < \eta^2$ removed, tends as $\eta \to 0$. Obviously for this limit to exist it is necessary and sufficient that one have

$$\int_{\eta^2<\chi(A)<\zeta^2} F(A)\, d\Omega = 0 \tag{1.84}$$

for any $0 < \eta < \zeta$.

Theorem 1.15. If F is a function which is homogeneous of degree n, which is continuous everywhere except the origin, and satisfies (1.84), then in the definition of the integral in the sense of the principal value, as the set to be removed one can take the sets described by the relations

$$\sum_{i,j=1}^{n} c_{ij}a_i a_j < \eta^2 + \eta^2 f(A),$$

where f is a function of the point A, which satisfies the condition $f(A) = O[L^h(O, A)]$, $h > 0$, where the limit as $\eta \to 0$ is independent of the choice of f.

Proof. $L(O, A)$ has order $O(\eta)$, and F is bounded. The integral of the absolute value of F over the domain

$$\eta^2 - k\eta^{2+h} < \sum_{i,j=1}^{n} c_{ij}a_i a_j < \eta^2 + k\eta^{2+h}$$

for sufficiently small η has order

$$O[\ln((1 + k\eta^h)/(1 - k\eta^h))] = O(\eta^h).$$

This means that the integral cited tends to zero as $\eta \to 0$ for any finite k, which implies the validity of the assertion of the theorem. □

Sometimes it is necessary to represent integrals in the sense of principal values of functions defined on a manifold V. Let us assume that on the manifold V there is defined a symmetric matrix $\|A_{ij}\|$ whose elements are functions of a point $X \in V$. For each point $X \in V$ the quadratic form

$$\chi(Z) = \sum_{i,j=1}^{n} A_{ij}(X) z_i z_j$$

is positive definite. If the A_{ij} are expressed in terms of the coordinate $x_1, \ldots, x_n$, varying in a domain W_1 of Euclidean space, and the A_{ij} are the same things, but expressed in terms of coordinates $t_1, \ldots, t_n$, varying in a domain W_2 of Euclidean space, then at points $Y = (x_1, \ldots, x_n)$ and $T = (t_1, \ldots, t_n)$ corresponding to the same point X of manifold V, one has

$$A'_{ij} = \sum_{k,l=1}^{n} A_{kl} \frac{\partial x_k}{\partial t_i} \cdot \frac{\partial x_l}{\partial t_j}.$$

Thus, on V there is defined a symmetric tensor which does not necessarily coincide with the metric tensor of the manifold. Suppose further that we have a function $G(X, Y)$ of two points $X, Y \in V$, which is continuous when points X and Y are different, and satisfies a Hölder condition with exponent h with respect to X when X and Y do not belong to the same domain $W_j \subset R^n$ of variation of the parameters, i.e., X and Y do not belong to the same chart W_j of manifold V. If X is an interior point of a domain of variation of the parameters W_1, then obviously for sufficiently small $\eta > 0$ the domain satisfying the inequalities

$$\sum_{i,j=1}^{n} A_{ij}(X)(x_i - y_i)(x_j - y_j) < \eta^2, \quad \eta > 0, \tag{1.85}$$

belongs entirely to W_1. For a compact manifold V there exists a positive constant η_0, independent of point $X \in V$, such that at least one of the coordinate domains W_j containing point X lies entirely in the domain satisfying (1.85), when $\eta \le \eta_0$. With each point $X \in V$ we associate such a coordinate domain W_j. When $X, Y \in W_j$, let us assume that function $G(X, Y)$ can be represented in the form $G(X, Y) = G_1(X, Y) + G_2(X, Y)$, where $G_2(X, Y)$ is an integrable function of class $N^{(a,\lambda)}$, and G_1 can be written in the form

$$G_1(X, Y) = G_1^*(x_1 - y_1, \ldots, x_n - y_n; v_1(X), \ldots, v_q(X)),$$

where the functions v_j, $j = 1, \ldots, q$ satisfy a Hölder condition with exponent h, and the functions

$$G_1^*(\omega_1, \ldots, \omega_n, v_1, \ldots, v_q), \quad \partial G_1^*/\partial v_j, \quad j = 1, \ldots, q,$$

are homogeneous of degree n with respect to the variables $\omega_1, \ldots, \omega_n$, and are continuous everywhere except the point $\omega_j = 0$, $j = 1, \ldots, n$. The derivatives of the function $G_1{}^*$ with respect to all the variables ω_j exist and are continuous everywhere except $\omega_j = 0$, $j = 1, \ldots, n$.. These derivatives are homogeneous functions of degree $n - 1$ with respect to the variables ω_j. Finally, the integral

$$\int_{\Delta} G_1(X, A)\, d_A\Omega,$$

taken over the domain

$$\Delta : \eta^2 < \sum_{i,j=1}^{n} A_{ij}(X)(x_i - a_i)(x_j - a_j) < \zeta^2, \quad 0 < \eta < \zeta,$$

is equal to zero. Now the integral

$$F(X) = \int_{V} G(X, A)\, \rho(A)\, d_A V$$

is understood in the sense of the principal value, in the definition of which one excludes the domains defined by (1.85) from the manifold V. As in the proof of Theorem 1.15, one can show that the value of the integral in the sense of the principal value is unchanged if one excludes from V more general domains of this type as in Theorem 1.15.

On the part of V corresponding to two coordinate subdomains W_1 and W_2 two coordinate systems $x_1, \ldots, x_n$ and $t_1, \ldots, t_n$ are defined. The function $G_1{}^{*\prime}$, analogous to the function $G_1{}^*$ in the variables t_j, can be written as follows:

$$G_1^*\left(\sum_{j=1}^{n}(t_j-u_j)\frac{\partial x_x}{\partial t_j}, \ldots, \sum_{j=1}^{n}(t_j-u_j)\frac{\partial x_n}{\partial t_j}, \quad v_1(X), \ldots, v_q(X)\right),$$

where u_j are the coordinates of point Y, and $v_j(X)$ can also be expressed in the variables t_j. This function is homogeneous of degree n with respect to $t_j - u_j$ since in the homogeneous function G_1 one makes a linear homogeneous change of independent variables.

We show that one has

$$\begin{aligned} &G_1^*(x_1-y_1, \ldots, x_n-y_n, v_1(X), \ldots, v_q(X)) \\ &-G_1^*\left(\sum_{j=1}^{n}(t_j-u_j)\frac{\partial x_1}{\partial t_j}, \ldots, \sum_{j=1}^{n}(t_j-u_j)\frac{\partial x_n}{\partial t_j},\right. \\ &\quad v_1(X), \ \ldots, \ v_q(X)) = O[L^{h-n}(X, \ Y)]. \end{aligned} \tag{1.86}$$

It is sufficient to prove this relation for sufficiently small distance $L(X, Y)$ and under the condition that one has

$$x_j - y_j - \sum_{k=1}^{n}(t_k-u_k)\frac{\partial x_j}{\partial t_k} = O\left[L^{1+h}(X, Y)\right].$$

We define the coordinates z_j of point Z by the relations

$$z_j - y_j = \sum_{k=1}^{n}(t_k-u_k)\frac{\partial x_j}{\partial t_k}.$$

Now we have $L(X, Z) = O[L^{1+h}(X, Y)]$, and it follows from this that if $L(X, Y)$ is sufficiently small, then $2L(X, Z) < L(X, Y)$, so the shortest distance l from point Y to the segment XZ is greater than $L(X, Y)/2$. The difference (1.86) now behaves like $O[L(X, Z)l^{-1-n}] = O[L^{h-n}(X, Y)]$, and this is (1.86). Thus, the definition of the integral in the sense of the principal value on a manifold is independent of the coordinate system.

We consider some properties of the potentials (1.61)–(1.63). We start with the solid potential

$$u(X) = \int_D \Lambda(X, Y)\,\omega(Y)\,dV. \tag{1.87}$$

Theorem 1.16. If ω is bounded, then u has Hölder continuous first derivatives in the entire domain of definition of the Levi function Λ and

$$\frac{\partial u}{\partial x_i} = \int_D \omega(Y) \frac{\partial}{\partial x_i} \Lambda(X, Y)\, dV; \tag{1.88}$$

if ω is Hölder continuous with exponent α, then u has second derivatives in the open domain D and for X lying in the open domain D, one has

$$\frac{\partial^2 u}{\partial x_i \partial x_k} = -\frac{1}{n} A_{ik}(X)\, \omega(X) + \int_D \frac{\partial^2 \Lambda(X, Y)}{\partial x_i \partial x_k} \times \omega(Y)\, dV, \tag{1.89}$$

where the integral is understood in the sense of the principal value, and A_{ik} are the quantities which occur in the expression for the function $H(X, Y)$, which appears in (1.55), (1.56).

Proof. By definition of the Levi function $\Lambda(X, Y)$ we have $\Lambda(X, Y) = H(X, Y) + K(X, Y)$, where by (1.56) and (1.57),

$$\frac{\partial \Lambda}{\partial x_i} \in N^{(1)}, \quad K \in N^{(2)}, \quad \frac{\partial K}{\partial x_i} \in N^{(1+\lambda)}, \quad \frac{\partial^2 K}{\partial x_i \partial x_j} \in N^{(\lambda)}, \tag{1.90}$$

so (1.88) is a direct consequence of the theorem on differentiation under the integral sign [15] and Theorem 1.13. To prove (1.89) we set

$$u_1(X) = \int_D H(X, Y)\, \omega(Y)\, dV, \quad u_2(X) = \int_D K(X, Y)\, \omega(Y)\, dV.$$

By the arguments used in justifying (1.88), the function u_2 has continuous second derivatives, which one can calculate by differentiation under the integral sign.

We denote by $I(X, \rho)$ the same ball as in Section 6 of this chapter, and by $\Gamma(X, \rho)$ its boundary, as in (1.54). We consider the function

$$\varphi_i(X, \rho) = \int_{D-I(X,\rho)} \frac{\partial H(X, Y)}{\partial X_i} \omega(Y)\, dV.$$

Differentiating this integral with respect to x_k, considering that the domain of integration depends on a parameter, we find [15]

$$\frac{\partial \varphi_i(X)}{\partial x_k} = \int_{D-I(X,\rho)} \frac{\partial^2 H(X, Y)}{\partial x_i \partial x_k} \omega(Y)\, dV - \int_{\Gamma(X,\rho)} \frac{\partial H(X, Y)}{\partial x_i} \omega(Y)\, n_k(Y)\, d\sigma, \tag{1.91}$$

where $n_k(Y)$ are the direction cosines of the exterior normal to the sphere $\Gamma(X, \rho)$ at point Y. The last summand on the right side of this formula can be rewritten as follows:

$$\int_{\Gamma(X,\rho)} \left[\frac{\partial H(X, Y)}{\partial x_i} \omega(Y) - \frac{\partial H(Y, X)}{\partial y_i} \omega(X)\right] n_k(Y)\, d\sigma + \omega(X) \int_{\Gamma(X,\rho)} \frac{\partial H(Y, X)}{\partial y_i} n_k(Y)\, d\sigma.$$

The first summand of this expression tends to zero with ρ uniformly with respect to X, since the integrand behaves like $O(\rho^{\lambda+1-n})$, and the second integral can be calculated and is equal to $n^{-1}\omega(X)A_{ik}(X)$. Obviously, $\lim_{\rho\to 0}\varphi_i(X) = \partial u/\partial x_i$, so to justify (1.89) it suffices to prove that the limit of the first summand on the right side of (1.91) exists as $\rho \to 0$. We transform this summand:

$$\int\limits_{D-I(X,\rho)} \frac{\partial^2 H(X,Y)}{\partial x_i \partial x_k}\omega(Y)\,dV = \int\limits_{D-I(X,\rho)} \left[\frac{\partial^2 H(X,Y)}{\partial x_i \partial x_k}\omega(Y) - \frac{\partial^2 H(Y,X)}{\partial y_i \partial y_k}\omega(X)\right] dV + \int\limits_{D-I(X,\rho)} \frac{\partial^2 H(Y,X)}{\partial y_i \partial y_k}\omega(X)\,dV$$

$$= \omega(X)\int\limits_{\Gamma} \frac{\partial H(Y,X)}{\partial y_i} n_k(Y)\,d\sigma + \int\limits_{D-I(X,\rho)} \left[\frac{\partial^2 H(X,Y)}{\partial x_i \partial x_k}\omega(Y) - \frac{\partial^2 H(Y,X)}{\partial y_i \partial y_k}\omega(X)\right] dV + n^{-1}A_{ik}(X)\,\omega(X).$$

The integrand in the last integral behaves like $O[L^{\lambda-n}(X, Y)]$, so the limit of this integral exists as $\rho \to 0$. □

Let the second-order operator

$$M(u) = \sum_{i,j=1}^{n} \frac{\partial}{\partial x_i}\left(a_{ij}(X)\frac{\partial u}{\partial x_j}\right) + \sum_{i=1}^{n} e_i(X)\frac{\partial u}{\partial x_i} + c(X)\,u$$

be defined in a domain E, and let the density $\omega(Y)$ of the potential (1.87) be distributed over the domain $D \subset E$. Obviously at any point of the complement of D in E one can differentiate twice under the integral sign in (1.87), and in the complement one has

$$\frac{\partial^2 u}{\partial x_i \partial x_k} = \int\limits_{D} \frac{\partial^2 \Lambda(X,Y)}{\partial x_i \partial x_k}\omega(Y)\,dV.$$

From this formula and (1.89) we get the following generalization of Poisson's formula for the equation $M(u) = 0$ in domain D with boundary Γ:

$$M(u) = \begin{cases} \int\limits_{D} M_X[\Lambda(X,Y)]\,\omega(Y)\,dV, & X \in E - \overline{D}, \\ -\omega(X) + \int\limits_{D} M_X[\Lambda(X,Y)]\,\omega(Y)\,dV, & X \in D. \end{cases} \tag{1.92}$$

We consider the generalized potential of a simple layer

$$V(X) = \int\limits_{\Gamma} \Lambda(X,Y)\,\rho(Y)\,dS. \tag{1.93}$$

By Theorem 1.13 this potential is an everywhere continuous function in E for any integrable density $\rho(X)$. Obviously $V(X)$ has continuous second derivatives at all points of the complement of Γ in E. Let X_0 be a point of the boundary Γ of domain D; through this point we draw a line l such that $\cos(l, n) \neq 0$, where n is the normal to Γ, i.e., line l is not tangent to the surface Γ at point X_0 (does not lie in the tangent plane to Γ at point X_0). We calculate the derivatives of the function $V(X)$ at point X_0 once, when X tends to X_0 along l from within D, another time when X tends to X_0 along l from without D, i.e., from within $E - D$. We denote these limits by $(dV/dl)^-$ and $(dV/dl)^+$, respectively. Without proof we cite a theorem about the behavior of these derivatives.

Theorem 1.17. If $\rho(X)$ is Hölder continuous, then for any point $X_0 \in \Gamma$ we have

$$\left(\frac{dV}{dl}\right)^{\pm} = \mp \frac{\cos(n, l)}{2a(X_0)\cos(n, \nu)} \rho(X_0) + \int_\Gamma \frac{d\Lambda(X_0, Y)}{dl} \rho(Y)\, dS,$$

where the integral is understood in the sense of the principal value, $\cos(n, \nu)$ is the cosine of the angle made by the normal to Γ and the conormal, $\cos(n, l)$ is the cosine of the acute angle made by line l and the normal to Γ, and

$$a(X) = \left[\sum_{i=1}^{n} \left(\sum_{j=1}^{n} a_{ij}(X)\, n_j(X)\right)^2\right]^{1/2},$$

where the outer normal with respect to D is taken here.

This theorem can be applied for reducing the oblique derivative problem to singular integral equations. Below we shall not apply multidimensional singular integral equations to investigate the oblique derivative problem, so we shall not use Theorem 1.17.

All the properties of the potentials (1.61)–(1.63) remain valid for the potentials (1.76)–(1.78) also, as one can prove directly. Such direct proofs are even somewhat simpler technically.

8. GRADIENT VECTOR FIELDS OF FUNCTIONS

In Chapter 3 we shall formulate the analog of the argument principle for harmonic functions of three independent variables. We shall recount here some information about vector fields which is needed for this [18]. Most of the assertions of the present section are given without proof, since the proofs are not related to the theory of partial differential equations.

Suppose given, at each point X of the planar set Ω, a vector $\Phi(X)$ which lies in this plane. Then we shall say that a vector field Ω is given on $\Phi(X)$. If a coordinate system is introduced in the plane, then giving a vector field Φ is equivalent to giving a pair of real functions $[\Phi(X), \psi(X)]$ of a point $X \in \Omega$. If the functions $\varphi(X)$ and $\psi(X)$ are continuous, then we shall call the field Φ continuous. We consider a continuous vector field $\Phi(X)$, defined on a Jordan curve Γ for which there exists a tangent at each point, and which is represented in parametric form $x = x(t)$, $y = y(t)$, $a \leq t \leq b$. Defining a vector field $\Phi(x, y) = \{\varphi(x, y), \psi(x, y)\}$ on Γ is equivalent to defining a vector function $\Phi(t) = \{\varphi(x(t), y(t)), \psi(x(t), y(t))\}$ on the segment $[a, b]$. If the vector-function $\Phi(t)$ does not vanish at any point of this segment, then for each t we define the angle between the vectors $\Phi(t)$ and $\Phi(a)$, computed from $\Phi(a)$ counterclockwise, which is a multivalued function of t. A continuous branch $\theta(t)$ of this function, which vanishes for $t = a$, will be called the angle function of the field Φ on curve Γ. The increment of the angle function $\theta(t)$ on the whole segment $[a, b]$, expressed in units of full revolution, will be called the rotation of the field Φ on curve Γ. The rotation of a field $\gamma(\Phi, \Gamma)$ is given by the formula

$$\gamma(\Phi, \Gamma) = \frac{1}{2\pi}[\theta(b) - \theta(a)] = \frac{1}{2\pi}\theta(a).$$

The rotation of a field depends on the orientation of curve Γ. Upon passing to the opposite orientation the absolute value of the rotation is preserved, but the sign changes to the opposite.

A vector field can also be defined on a closed curve. Such a curve can consist of ν nonintersecting one-dimensional manifolds. The curve Γ bounds a ν-connected domain Ω. By the positive direction of traversing the curve Γ is meant the direction for which the domain Ω remains on the left. The rotation of the vector field $\Phi(X)$ on the boundary Γ of the ν-connected domain Ω is defined as the sum of the rotations over all the components Γ_j of the curve Γ, where each component is taken with positive orientation. In order to define the concept of rotation for a component, we split it by two points M_1 and M_2 into two parts, for each of which the concept of rotation is already defined above. By the rotation of the field Φ on the connected closed curve Γ_j we shall mean the sum of the rotations of the field Φ on the curves obtained as a result of splitting Γ_j by two points M_1 and M_2, where the positive direction on each piece coincides with the positive directions on Γ_j. It is easy to show that the rotation of a field is independent of the method of splitting the curve Γ_j and is an integer. The following theorem holds.

Theorem 1.18. The rotation of the field of tangents to a connected smooth curve is equal to one. If the vector field Φ, defined in the closed domain $\overline{\Omega}$, has no zero vectors, then its rotation $\gamma(\Phi, \Gamma)$ on the boundary Γ of the domain Ω is equal to zero.

Points of the domain Ω at which the field vanishes, is discontinuous, or undefined are called singular points of the field Φ. A singular point M_0 will be called isolated if in a neighborhood of it there are no other singular points. A field Φ on any circle of sufficiently small radius with center at an isolated singular point M_0 has no zero vectors and is continuous, so the rotation of this field is defined on all such circles, and it is identical for all the given circles. This common rotation for all such circles is called the index of the singular point. The sum $\gamma_1 + \ldots + \gamma_k$ of the indices $\gamma_1, \ldots, \gamma_k$ of the singular points $M_1, \ldots, M_k$ is called the algebraic number of singular points.

Theorem 1.19. The algebraic number of isolated singular points of the vector field Φ lying in the domain Ω is equal to the rotation of the field Φ on boundary Γ of domain Ω.

We consider a family of vector fields $\Phi(X, \lambda)$, which depend on a parameter λ, $0 \leq \lambda \leq 1$, and are given on a closed set Γ. Let the vector-function $\Phi(X, \lambda)$ be continuous in the collection of variables and not have zero vectors. We shall say that the family $\Phi(X, \lambda)$ connects the fields $\Phi_0(X) = \Phi(X, 0)$ and $\Phi_1(X) = \Phi(X, 1)$ homotopically. We shall call two fields Φ and Ψ homotopic if they can be homotopically joined by some family. Each continuous field $\Phi(X)$ without zero vectors is homotopic to the field $\Phi_2(X)$, obtained from the field $\Phi(X)$ by rotating all vectors of this field by the constant angle $\pi\lambda$.

Theorem 1.20. Let Φ and Ψ be continuous vector fields, defined on a smooth curve Γ, which bounds a ν-connected domain. If the fields Φ and Ψ are homotopic, then their rotations coincide; if the fields Φ and Ψ have the same rotation, then these fields are homotopic.

Theorem 1.21. If the vector fields Φ and Ψ are not opposite at any point, then these fields are homotopic.

Let f be a smooth real function, defined on manifold M. We choose, in a neighborhood U of point p of manifold M, a coordinate system $x_1, \ldots, x_n$. We shall call the point $p \in M$ a critical point of function f, if at this point

$$\frac{\partial f}{\partial x_1} = \ldots = \frac{\partial f}{\partial x_n} = 0.$$

We call the number $f(p)$ a critical value of function f. It is easy to show that the critical points of a function do not depend on the choice of coordinate system. With each smooth function f one associates the gradient vector field grad$f$, defined on manifold M. The field grad$f$ is defined as follows: in a neighborhood U of a point $p \in M$ we introduce a coordinate system $x_1, \ldots, x_n$ and at each point $X \in M$ we define the vector

$$\operatorname{grad} f = \left(\frac{\partial f}{\partial x_1}, \ldots, \frac{\partial f}{\partial x_n}\right).$$

The singular points of the vector field grad$f$ are the critical points of function f. We shall denote by M^a the set of all points $X \in M$ at which $f(X) \leq a$. If a is not a critical value of f, then M^a is a smooth manifold with boundary, the level set $N(a)$: $\{f = a\}$ [25].

Theorem 1.22. Let f be a smooth real function on manifold M and let $a < b$. Let us assume that the set consisting of all points $p \in M$, such that $a \leq f(p) \leq b$, is compact and contains no critical points of f. Then M^a is diffeomorphic to M^b.

With each smooth function f one can associate a first-order system of differential equations

$$\frac{dx_1}{fx_1} = \ldots = \frac{dx_n}{fx_n} = dt. \tag{1.94}$$

The integral curves of this system are the orthogonal trajectories to the level surfaces of function f. In order to construct a diffeomorphism of the level surface $N(a)$: $\{f = a\}$ to the level surface $N(b)$: $\{f = b\}$, we proceed as follows. Through each point X of the surface $N(a)$ we draw the integral curve of (1.94) and we extent it to its intersection with the surface $N(b)$ at some point Y. We define the diffeomorphism τ required by $\tau(X) = Y$. One can make this construction by virtue of the existence and uniqueness theorem for solutions of the Cauchy problem for (1.94).

We consider a real-analytic function $f(x, y)$ of two variables x and y in the domain Ω. Now (1.94) assumes the form

$$\frac{dx}{dt} = f_x(x, y), \quad \frac{dy}{dt} = f_y(x, y). \tag{1.95}$$

Since the vectors of the field grad$f$ are orthogonal to the level lines of function f, the field grad$f$ is homotopic to the field of tangent vectors to the level lines of the function f. Consequently, the singular points of these fields and the indices of the singular points coincide. The isolated singular points of the field of tangent vectors to the level lines of the function f are extreme and saddle points of the function f and cusps of level lines of the function f. It follows from Theorem 1.18 that the index of an isolated extreme point of the function f is equal to +1. It is also obvious that the index of an isolated cusp of a level line of the function f is equal to zero. Now the index of a saddle point depends on the behavior of the level line of function f which passes through this point. The following theorem holds [47].

Theorem 1.23. Let $f(x, y)$ be a real function which is analytic in the domain Ω, and $f(0, 0) = 0$. If the domain Ω is a disk of sufficiently small radius σ with center at the point $0 = (0, 0)$, then the curve $f(x, y) = 0$ has no more than a finite number of real branches in Ω, which are of one of the following types:

$$x = 0 \quad \text{or} \quad y = \alpha_k x^{k/n} + \alpha_{k+1} x^{(k+1)/n} + \ldots,$$

where k and n are natural numbers, and all the coefficients α_j are real. For sufficiently small σ the real branches of a level line intersect in Ω only at the origin.

The index of the saddle point O, which is an isolated critical point of function f, is equal to $1 - s/2$, where s is the number of sectors into which the level line of function f passing through point O divides a disk Ω of sufficiently small radius with center O. Thus, the index of a saddle point is always negative.

We consider a real function $f(x, y)$ which is analytic in domain D with smooth boundary Γ, whose gradient is different from zero everywhere on Γ. The field grad$f$ on the curve Γ is homotopic to the normalized field

$$P\colon \{f_x(f_x^2 + f_y^2)^{-1/2}, \quad f_y(f_x^2 + f_y^2)^{-1/2}\},$$

so the rotations of these vector fields on Γ coincide. The isolated singular points of the fields grad$f$ and P, which lie in D, also coincide. The field grad$f$ can vanish on entire level curves of function f. Obviously if a curve l, on which grad$f$, in a neighborhood of a point $X \in l$, is a one-dimensional manifold, and on l the function f does not achieve an extremum, then all points of curve l in a neighborhood of point X are nonsingular points of the normalized field P, because upon passing through l the field P remains continuous, although its component, tangent to l, is equal to zero on l. Thus, the singular points of the normalized field P are isolated singular points of the field grad$f$, one-dimensional lines of extremum of the function f, and points of self-intersection of lines on which grad$f = 0$, but f does not achieve an extremum on these curves. The singular points of the field P of the first and third types are isolated singular points.

Let the function f achieve an extremum on ν-connected curves l_i, and let the extremum on the curve l_i be equal to c_i, $i = 1, \ldots, \nu$. For sufficiently small $\varepsilon > 0$ the level curves L_i^-: $\{f = c_i - \varepsilon\}$ and L_i^+: $\{f = c_i + \varepsilon\}$ bound a domain of finite connectivity μ_i, where $\varepsilon > 0$ can be chosen so that in the closure $\overline{H_i(\varepsilon)}$ of the domain $H_i(\varepsilon)$, bounded by these lines, there are no critical points of the function which are not points of the line l_i. The boundary $\Gamma_i(\varepsilon)$ of the domain $H_i(\varepsilon)$ consists of μ_i closed smooth Jordan curves $\gamma_{ij}(\varepsilon)$, $j = 0, \ldots, \mu_i - 1$, where these curves are indexed so that $\gamma_{i0}(\varepsilon)$ bounds a simply connected domain $D_i(\varepsilon)$, in which all the other components of the boundary $\Gamma_i(\varepsilon)$ of the domain $H_i(\varepsilon)$ lie. We choose ϵ so small that the $H_i(\varepsilon)$ do not intersect one another, and we remove from the domain D all the domains $D_i(\varepsilon)$, $i = 1, \ldots, \nu$. As a result we get a domain $D(\varepsilon)$ of finite connectivity $\lambda \leq \nu$. Let the rotation of the field grad f on the curve Γ be equal to n; then by Theorems 1.18 and 1.20 the algebraic sum of all singular points of the field P which lie in the domain $D(\varepsilon)$ is equal to $n - \lambda$.

We take a fixed domain $D_i(\varepsilon)$ containing the domain $H_i(\varepsilon)$, in which the curve l_i lies. The rotation of the field P on boundary $\Gamma_i(\varepsilon)$ of domain $H_i(\varepsilon)$ is equal to $\mu_i + 1$. Since curve l_i is connected, and the rotation of field P on each curve $\gamma_{ij}(\varepsilon)$, $j = 1, \ldots, \mu_i - 1$ is equal to -1 and is independent of ε, the Euler characteristic χ of the set bounded by curve l_i is equal to one. If l_i is not a manifold, the number χ is determined as follows: let m_i be the number of points of self-intersection of the curve l_i, n_i be the number of connected components into which l_i splits after removing the points of self-intersection, p_i be the number of connected components of the open set bounded by curve l_i; then $\chi = m_i - n_i + p_i$. Extended over all points of self-intersection of the curve l_i, the sum of the number of sectors into which l_i divides a neighborhood of a point of self-intersection of l_i is equal to n_i, and $p_i = \mu_i - 1$. Suppose that at a point of self-intersection Y_j of curve l_i there occur β_j connected components of l_i; then the number of sectors into which l_i divides a neighborhood of Y_j is also equal to β_j. We consider the sum

$$\sum_{Y_j} (1 - \beta_j/2) = \chi - p_i = 1 + 1 - \mu_i = m_i - n_i,$$

but $1 = \chi = m_i - n_i + p_i$, so the sum which occurs on the left side of the equation is equal to μ_i. It follows from this that if we connect $D(\varepsilon)$ to $H_i(\varepsilon)$, then the rotation of field P on the boundary of the domain obtained will be equal to the algebraic sum of all isolated singular points of the field grad f, contained in this domain, and points of self-intersection of the curve l_i, considered as saddle points of function f of the corresponding multiplicity. When l_i has no points of self-intersection, i.e., is a smooth closed manifold, $H_i(\varepsilon)$ is homeomorphic to a circular ring and the rotation of P on the boundary of $H_i(\varepsilon)$ is equal to zero. After jointing $H_i(\varepsilon)$ to $D(\varepsilon)$, neither the number of isolated singular points contained in the domain nor the rotation of the field P changes.

After a finite number of joinings of domains $H_i(\varepsilon)$ we get a domain $E(\varepsilon)$ of finite connectivity m, where the rotation of field P on the boundary of $E(\varepsilon)$ is equal to $n - m$ and, on the other hand, it is equal to the algebraic sum of the number of isolated extrema and the number of saddle points of function f, lying in $E(\varepsilon)$. Now the complement of $E(\varepsilon)$ in D contains only isolated singular points of field P, the algebraic sum of which is equal to m. Consequently, the following theorem holds.

Theorem 1.24. The algebraic sum of the number of saddle points and isolated extreme points of function f which lie in domain D is equal to the rotation of the field grad f on boundary Γ of domain D.

By saddle point in this theorem we mean a point X for which the level line of the function passing through it splits, after removing point X, in a sufficiently small disk with center at point X, into more than two connected components; i.e., a saddle point is a point of self-intersection of a level line.

Chapter 2

OBLIQUE DERIVATIVE PROBLEM FOR ELLIPTIC EQUATIONS

In a domain Ω of Euclidean space R^m of independent variables $x_1, \ldots, x_m$ we consider the homogeneous elliptic equation

$$\sum_{i,j=1}^{m} A_{ij}(X) \frac{\partial^2 u}{\partial x_i \partial x_j} + \sum_{i=1}^{m} B_i(X) \frac{\partial u}{\partial x_i} + C(X) u = 0, \tag{2.1}$$

$$X = (x_1, \ldots, x_m),$$

with sufficiently smooth coefficients. Suppose given a domain $D \subset \Omega$ with a sufficiently smooth boundary Γ, and at each point X of the boundary Γ suppose given a vector $P(X) = (a_1(X), \ldots, a_m(X))$. Thus, on the surface Γ there is given a vector field $P(X)$. This field will also be assumed sufficiently smooth. The oblique derivative problem for Eq. (2.1) is formulated as follows: find a solution of (2.1) which is regular in domain D, continuously differentiable in closed domain $D \cup \Gamma$, and which on boundary Γ of domain D satisfies the condition

$$(\operatorname{grad} u, P) \equiv \sum_{i=1}^{m} a_i(X) \frac{\partial u}{\partial x_i} = f(X), \tag{2.2}$$

where f is a given function which is continuous on Γ.

The oblique derivative problem is a special case of the Poincaré problem, in which the boundary condition (2.2) is replaced by the condition

$$\sum_{i=1}^{m} a_i(X) \frac{\partial u}{\partial x_i} + b(X) u = f(X) \tag{2.3}$$

on Γ, which becomes (2.2) for $b \equiv 0$. Like the Poincaré problem, the oblique derivative problem reduces to the investigation of singular integral equations. When the vector field P is not in the tangent plane to Γ at any point of Γ, i.e., when throughout Γ one has

$$(\vec{n}(X), P(X)) \neq 0, \tag{2.4}$$

where $\vec{n}(X)$ is the unit vector normal to Γ at point X, for these singular integral equations the Fredholm alternative holds. Now if (2.4) fails at least at one point of Γ, the corresponding singular integral equations do not yield to investigation by contemporary methods [7].

1. REDUCTION OF THE OBLIQUE DERIVATIVE PROBLEM TO FREDHOLM INTEGRAL EQUATIONS

Since, when (2.4) holds, the oblique derivative problem reduces to a singular integral equation for which the Fredholm alternative is valid, it is natural to raise the question of whether it is possible to reduce this problem to a Fredholm equation. Such a reduction is possible [50] and we shall perform it here. For this it is necessary to consider a number of auxiliary problems. We start with the following one: In the half-space $x_m > 0$ find a regular solution of the equation

$$\Delta u - g^2 u = f(X), \quad g \equiv \text{const}, \tag{2.5}$$

which is equal to zero at infinity, whose derivative $T(u)$ in the direction $(\sin\theta, 0, \ldots, 0, -\cos\theta)$ exists at all points of the hypersurface $x_m = 0$ and satisfies the condition

$$T(u) \equiv \frac{\partial u}{\partial x_1} \sin\theta - \frac{\partial u}{\partial x_m} \cos\theta = \varphi(X) \cos\theta. \tag{2.6}$$

First we consider the case $\varphi(X) \equiv 0$, $f(X)$ being a continuously differentiable function. Since θ is a constant, any solution of the problem (2.5), (2.6) satisfies the equation

$$\Delta\left(\frac{\partial u}{\partial x_1} \tan\theta - \frac{\partial u}{\partial x_m}\right) - g^2\left(\frac{\partial u}{\partial x_1} \tan\theta - \frac{\partial u}{\partial x_m}\right) = \frac{\partial f}{\partial x_1} \tan\theta - \frac{\partial f}{\partial x_m},$$

and for $x_m = 0$ we have

$$\frac{\partial u}{\partial x_1} \tan\theta - \frac{\partial u}{\partial x_m} = 0.$$

Consequently, the function $v = (\partial u/\partial x_1) \tan\theta - (\partial u/\partial x_m)$ is a solution of the following problem:

$$\begin{gathered} \Delta v - g^2 v = \frac{\partial f}{\partial x_1} \tan\theta - \frac{\partial f}{\partial x_m} \equiv g(X), \\ v\,|_{x_m} = 0. \end{gathered} \tag{2.7}$$

We construct the Green's function for this problem. Let L be the distance between two points $X = (x_1, \ldots, x_m)$ and $A = (a_1, \ldots, a_m)$. A principal fundamental solution $\Psi(L)$ of (2.5) is given by the formula [51]

$$\Psi(L) = \frac{\exp(-gL)}{2g\Gamma\left(\frac{m-1}{2}\right)}\left(\frac{g}{2\pi L}\right)^{\frac{m-1}{2}} \int_0^{\infty} t^{\frac{m-3}{2}} \left(1 - \frac{t}{2gL}\right)^{\frac{m-3}{2}} e^{-t} dt.$$

This fundamental solution and its derivatives of order p with respect to L satisfy

$$\Psi^{(p)}(L) = \begin{cases} O(L^{2-m-p}), & m + p > 2, \ gL \leqslant Q, \\ O(\ln 2Q/gL), & m + p = 2, \ gL \leqslant Q. \end{cases}$$

Now if gL is bounded below by a positive quantity independent of g, then we have

$$\Psi^{(p)}(L) = O\left(g^{\frac{m-3}{2}+p}\right) L^{\frac{1-m}{2}} \exp(-gL).$$

We denote by $A' = (a_1, \ldots, a_{m-1}, -a_m)$ the point symmetric to point A with respect to the plane $a_m = 0$. Obviously the Green's function for problem (2.7) is given by the formula $\Psi(L(X, A)) - \Psi(L(X, A'))$. Thus, we find

$$\frac{\partial u}{\partial x_1} \tan\theta - \frac{\partial u}{\partial x_m} = -\int_Q \{\Psi(L(X, A)) - \Psi(L(X, A'))\} \left(\frac{\partial f}{\partial a_1} \tan\theta - \frac{\partial f}{\partial a_m}\right) dV_A, \tag{2.8}$$

where Q is the half-space $a_m > 0$, and dV_A is the volume element of this half-space.

In (2.8) we introduce new variables with the help of the relations

$$x_1 = y_1 \cos\theta - y_m \sin\theta, \ x_m = y_1 \sin\theta + y_m \cos\theta,$$
$$x_j = y_j, \ j = 2, \ldots, m-1;$$
$$a_1 = b_1 \cos\theta - b_m \sin\theta, \ a_m = b_1 \sin\theta + b_m \cos\theta,$$
$$a_j = b_j, \ j = 2, \ldots, m-1;$$
$$a_1 = b_1' \cos\theta - b_m' \sin\theta, \quad -a_m = b_1' \sin\theta + b_m' \cos\theta,$$
$$a_j = b_j', \quad j = 2, \ldots, m-1,$$

from which we find

$$y_1 = x_1 \cos\theta + x_m \sin\theta, \ y_m = -x_1 \sin\theta + x_m \cos\theta,$$
$$y_j = x_j, \ j = 2, \ldots, m-1;$$
$$b_1 = a_1 \cos\theta + a_m \sin\theta, \ b_m = -a_1 \sin\theta + a_m \cos\theta,$$
$$b_j = a_j, \ j = 2, \ldots, m-1;$$
$$b_1' = a_1 \cos\theta - a_m \sin\theta, \quad b_m' = -a_1 \sin\theta - a_m \cos\theta,$$
$$b_j' = a_j, \quad j = 2, \ldots, m-1.$$

From the last two groups of formulas we get

$$b_1' = b_1 \cos 2\theta - b_m \sin 2\theta, \ b_m' = -b_1 \sin 2\theta - b_m \cos 2\theta,$$
$$b_j' = b_j, \ j = 2, \ldots, m-1.$$

After introduction of the new variables and integration by parts, (2.8) assumes the form

$$\frac{\partial u}{\partial y_m} = -\int_Q \{\Psi(L(Y,B)) - \Psi(L(Y,B'))\} \frac{\partial f}{\partial b_m} dV_B$$
$$= -\int_Q \Big\{ \Psi'(L(Y,B)) \frac{y_m - b_m}{L(Y,B)}$$
$$+ \Psi'(L(Y,B')) \frac{y_1 \sin 2\theta + y_m \cos 2\theta + b_m}{L(Y,B')} \Big\} f dV_B. \tag{2.9}$$

To determine u it is necessary to integrate this formula with respect to y_m so that the result tends to zero at infinity. Obviously

$$\frac{\partial}{\partial y_m} \Psi(L(Y,B)) = \Psi'(L(Y,B)) \frac{y_m - b_m}{L(Y,B)},$$
$$\frac{\partial}{\partial y_m} \Psi(L(Y,B')) = \Psi'(L(Y,B')) \frac{y_m + b_m}{L(Y,B')}.$$

We introduce the angle Ω and function $H_1(X, A)$ as follows:

$$L(X,A') \cos \Omega = y_m - b_m' = (x_m + a_m) \cos \theta - (x_1 - a_1) \sin \theta, \tag{2.10}$$
$$H_1(X,A) = \Psi(L(X,A)) + \Psi(L(X,A')) \cos 2\theta$$
$$- \sin 2\theta \, [(x_1 - a_1) \cos \theta + (x_m + a_m) \sin \theta]$$
$$\times \int_{\cos\Omega}^{\infty} \Psi'(L(X,A') \sqrt{\sin^2 \Omega + t^2}) / \sqrt{\sin^2 \Omega + t^2}\, dt. \tag{2.11}$$

The first summand in the expression for H_1 is obtained by integration of the first summand in (2.9) with respect to y_m and then passing to the old variables. The second summand under the integral sign in (2.9) transforms as follows:

$$\Psi'(L(Y,B'))(y' \sin 2\theta + y_m \cos 2\theta + b_m)[L(Y,B')]^{-1}$$
$$= \cos 2\theta \frac{\partial}{\partial y_m} \{\Psi(L(Y,B'))\} + \Psi'(L(Y,B')) \frac{y_1 - b_1'}{L(Y,B')} \sin 2\theta\}$$
$$= \cos 2\theta \frac{\partial}{\partial y_m} \{\Psi(L(Y,B'))\} + \sin 2\theta \, [(x_1 - a_1)$$
$$\times \cos \theta + (x_m + a_m) \sin \theta] \, \Psi'(L(Y,B'))[L(Y,B')]^{-1}.$$

Interchanging signs on both sides of the equation and integrating from $y_m - b_m'$ to ∞, we get

$$\int_{y_m - b'_m}^{\infty} \Psi'(L(Y, B'))\,[y_1 \sin 2\theta + y_m \cos 2\theta + b_m] \frac{d(y_m - b'_m)}{L(Y, B')} = \cos 2\theta\, \Psi(L(Y, B'))$$

$$- \sin 2\theta\,[(x_1 - a_1)\cos\theta + (x_m + a_m)\sin\theta]$$

$$\times \int_{y_m - b'_m}^{\infty} \Psi'(L(Y, B')) \frac{d(y_m - b'_m)}{L(Y, B')} = \cos 2\theta \Psi(L(X, A'))$$

$$- \sin 2\theta\,[(x_1 - a_1)\cos\theta + (x_m + a_m)\sin\theta] \int_{\cos\Omega}^{\infty} \frac{\Psi'\left(L(X, A')\sqrt{\sin^2\Omega + t^2}\right)}{\sqrt{\sin^2\Omega + t^2}}\, dt.$$

Here, under the integral, we introduce the change of variables

$$y_m - b'_m = tL(X, A'), \quad [L(Y, B)]^2 - (y_m - b'_m)^2 = [L(X, A')]^2 \sin^2\Omega,$$

and the lower limit of integration is recalculated considering (2.10).

We show that the formula

$$u(X) = - \int_Q H_1(X, A)\, f(A)\, dV_A, \qquad (2.12)$$

where Q is the half-space $a_m > 0$, gives a solution of the problem (2.5), (2.6) if $\varphi(X) \equiv 0$. First we investigate the properties of the function H_1, which we shall call the Green's function of the problem considered. This function, as a function of X, is analytic everywhere except the point A and manifold $\cos\Omega = -1$ or $\Omega = \pi$, since for $\Omega = \pi$ the integral which occurs in the expression for H_1 assumes the form

$$\int_{-1}^{\infty} |t|^{-1}\, \Psi'(L(X, A')\,|t|)\, dt,$$

and this latter integral diverges. The manifold $\Omega = \pi$ can be defined as follows:

$$[(x_1 - a_1)^2 + \ldots + (x_{m-1} - a_{m-1})^2 + (x_m + a_m)^2]^{1/2} = (x_1 - a_1)\sin\theta - (x_m + a_m)\cos\theta.$$

If we introduce the notation $\xi_1 = (x_m + a_m)\cos\theta - (x_1 - a_1)\sin\theta$, $\xi_2 = (x_m + a_m)\sin\theta + (x_1 - a_1)\cos\theta$, then it follows from the preceding equation that the manifold $\Omega = \pi$ can be defined by the equations $\xi_2 = 0$, $x_2 = a_2$, ..., $x_{m-1} = a_{m-1}$, $\xi_1 \geq 0$, and these equations define a ray of the line issuing from point A' in the direction of the vector $(\sin\theta, 0, \ldots, 0, -\cos\theta)$. Thus, in the half-space $x_m > 0$ there is contained only one singular point A of function H_1, and it is obvious that the function $H_1(X, A) - \Psi(L(X, A))$ is analytic at point A. Consequently, H_1 has the same singularity at point A as the fundamental solution.

We investigate the behavior of H_1 in a neighborhood of a ray of the line $\Omega = \pi$. For this we transform the integral $J(L)$:

$$J(L) \equiv \int_{\cos\Omega}^{\infty} \frac{\Psi'\left(L\sqrt{\sin^2\Omega + t^2}\right)}{\sqrt{\sin^2\Omega + t^2}}\, dt = \int_{\cot\Omega}^{\infty} \frac{\Psi'\left(L\sin\Omega\sqrt{1 + t^2}\right)}{\sqrt{1 + t^2}}\, dt.$$

It was noted above that

$$\Psi^{(p)}(L) = O(L^{2-m-p}), \quad m+p>2.$$

In the expression for $J(L)$ we have $p = 1$, so

$$J(L) = O\left[(L\sin\Omega)^{1-m}\right] \int_{-\infty}^{+\infty} (1+t^2)^{-m/2}\,dt = O\left[(L\sin\Omega)^{1-m}\right].$$

The obvious inequality $|(x_1 - a_1)\cos\theta + (x_m + a_m)\sin\theta| \le L(X, A')\sin\Omega$ holds, so by (2.10) we get that the last summand in (2.11) behaves like $O(\sin^{2-m}\Omega)L^{2-m}(X, A')$. We note that for $m = 2$ there is the equation $|(x_1 - a_1)\cos\theta + (x_m + a_m)\sin\theta| = L(X, A')\sin\Omega$, and the last summand in (2.11), as X tends to a point of the ray $\Omega = \pi$, different from A', has opposite signs depending on the sign of $(x_1 - a_1)\cos\theta + (x_m + a_m)\sin\theta$.

We study the behavior of the function $H_1(X, A)$ and its derivatives in a domain which satisfies the inequalities $\Omega \le \omega < \pi$, $L(X, A)[L(X, A')]^{-1} \le M < \infty$, where ω and M are fixed. These inequalities hold, in particular, for $x_m > 0$, since now $a_m \ge 0$ and $\cos\Omega \ge -\sin\theta$, and it follows from this that $\Omega \le \theta + \pi/2$ and $L(X, A) \le L(X, A')$. We shall also assume that $\cos\omega \le 0$. Instead of studying the behavior of the derivatives with respect to the variables L and Ω we shall investigate the behavior of the derivatives with respect to the variables L and $x = L\cos\Omega$. Replacing t by $(t + x)L^{-1}$, we rewrite the integral $J(L)$ in the following form:

$$J(L) \equiv F(L, x) = \int_0^{\infty} \frac{\Psi'\left(\sqrt{L^2 + 2tx + t^2}\right)}{\sqrt{L^2 + 2tx + t^2}}\,dt.$$

Since $\Omega \le \omega < \pi$ we have $x \ge L\cos\omega > -L$, from which it follows that $\sqrt{L^2\,2tx\,t^2} \ge L\sin\omega$, since $L^2 + 2tx + t^2 \ge L^2(\sin^2\omega + \cos^2\omega - 2tL\cos\omega + t^2 = L^2\sin\omega + (-L\cos\omega + t)^2 \ge L^2\sin^2\omega$. By direct calculation we find that

$$\frac{\partial^{p+q}F}{\partial L^p \partial x^q} = \sum_{n=0}^{p+q} \int_0^{\infty} \Psi^{(n+1)}\left(\sqrt{L^2 + 2tx + t^2}\right)$$

$$\times\left(\sqrt{L^2 + 2tx + t^2}\right)^{n-1-2p-2q} A_{pqn}(L, x, t)\,dt,$$

where $A_{000} = 1$, and A_{pqn} is a homogeneous polynomial in the variables L, x, and t of degree $p + q$. Since we have

$$\Psi^{(n+1)}\left(\sqrt{L^2 + 2tx + t^2}\right) = O\left[\left(\sqrt{L^2 + 2tx + t^2}\right)^{2-m-n-1}\right],$$

for each summand of the sum we get

$$\int_0^{\infty} \Psi^{(n+1)}\left(\sqrt{L^2 + 2tx + t^2}\right)\left(\sqrt{L^2 + 2tx + t^2}\right)^{n-1-2p-2q} A_{pqn}(L, x, t)\,dt$$

$$= \int_0^{\infty} O\left[\left(\sqrt{L^2 + 2tx + t^2}\right)^{-m-p-q}\right]dt = O\left(L^{1-m-p-q}\right).$$

If $gL \geq 1$, then one can give another estimate, namely

$$\int_0^\infty \Psi^{(n+1)}(\sqrt{L^2+2tx+t^2})(\sqrt{L^2+2tx+t^2})^{n-1-2p-2q} A_{pqn}(L, x, t)\,dt$$

$$= \int_0^\infty O\left(g^{\frac{m-1}{2}+n}\right)(\sqrt{L^2+2tx+t^2})^{n-\frac{m+1}{2}-p-q} \exp(-g\sqrt{L^2+2tx+t^2})\,dt. \qquad (2.13)$$

For negative x one has

$$\sqrt{L^2+2tx+t^2} > L\sin\omega + L^{-1}(t+x)^2(\sqrt{2}-1)$$

for $0 < t < L - x$, and for $x \geq 0$ in the domain $0 < t < L$ we find $\sqrt{L^2+2tx+t^2} > L + L^{-1}t^2(\sqrt{2}-1)$. The integral on the right side of (2.13), taken over those segments on which these inequalities hold, behaves like

$$O\left(g^{\frac{m-1}{2}+n}\right) L^{n-\frac{m+1}{2}-p-q} \exp(-gL\sin\omega) \int_{-\infty}^{+\infty} \exp\left[L^{-1}gt^2(1-\sqrt{2})\right] dt$$

$$= O\left(g^{\frac{m}{2}+n-1}\right) L^{n-p-q-\frac{m}{2}} \exp(-gL\sin\omega).$$

For the rest of the manifold of integration in (2.13) one has

$$\sqrt{L^2+2tx+t^2} > \begin{cases} L\sin\omega + t + x - L, & x < 0, \\ t, & x \geqslant 0, \end{cases}$$

by virtue of which the integral over this part of the manifold of integration behaves like

$$O(g^{(m-3)/2+n}) L^{n-p-q-(m+1)/2} \exp(-gL\sin\omega).$$

Finally, for $\Omega \leq \omega < \pi$, $\cos\omega \leq 0$ we get

$$\frac{\partial^{p+q}F}{\partial L^p \partial x^q} = \begin{cases} O(L^{1-m-p-q}), & L - \text{arbitrary}, \\ O(g^{p+q-1+m/2}) L^{-m/2} \exp(-gL\sin\omega), & gL \geqslant 1, \end{cases}$$

where the constants which occur in the symbol O depend on w but not on g.

Now we return to the function $H_1(X, A)$. It depends on X, A, and θ. We differentiate it p times with respect to the coordinates of point X and point A, and q times with respect to θ, where $p, q \geq 0$. The assertion follows from the estimates made above.

Lemma 2 1. If $\Omega \leq \omega < \pi$ and ω is independent of g, and $L(X, A)[L(X, A')]^{-1}$ are bounded above by a constant independent of g, then the derivative of H_1 of order p behaves like $O(L^{2-m-p})$ for $m + p > 2$ and like $O(\ln 2/gL)$ if $gL \leq 1$ and $m + p = 2$ (i.e., $m - 2 = p = 0$), and for $gL \geq 1$ behaves like $O(g^{m+p-2}) \exp(-gL\sin\omega)$.

Now we show that

$$\Delta H_1 - g^2 H_1 = 0, \quad \frac{\partial H_1}{\partial x_1} \tan\theta - \frac{\partial H_1}{\partial x_m} = 0. \tag{2.14}$$

According to the definition of the function H_1 we have

$$\frac{\partial H_1}{\partial y_m} = \frac{\partial}{\partial y_m} \Psi(L(Y, B)) + \sin 2\theta \frac{\partial}{\partial y_1} \Psi(L(Y, B')) + \cos 2\theta \frac{\partial}{\partial y_m} \Psi(L(Y, B')).$$

It follows from this relation that $\Delta H_1 - g^2 H_1$ is a function which is independent of y_m and tends to zero as $y_m \to \infty$ for all the other y_a fixed. From this it follows that the combination $\Delta H_1 - g^2 H_1$ can differ from zero only on rays of lines issuing from points A and A', and parallel to the Oy_m axis. Since on a ray issuing from point A the function H_1 is analytic, the first relation is proved.

By direct calculation we find

$$\frac{\partial H_1}{\partial y_m} = \frac{\partial H_1}{\partial x_1} \sin\theta - \frac{\partial H_1}{\partial x_m} \cos\theta = \Psi'(L(X, A))\,[(x_1 - a_1)\sin\theta$$
$$- (x_m - a_m)\cos\theta]\,[L(X, A)]^{-1} - \Psi'(L(X, A'))\,[(x_1 - a_1)\sin\theta$$
$$+ (x_m + a_m)\cos\theta]\,[L(X, A')]^{-1}. \tag{2.15}$$

Obviously $L(X, A') = L(X, A)$ for $x_m = 0$, so for $x_m = 0$ it follows from (2.15) that

$$\frac{\partial H_1}{\partial x_1} \sin\theta - \frac{\partial H_1}{\partial x_m} \cos\theta = 0.$$

Thus the second relation of (2.14) is also proved.

We note a few more properties of the functions $H_1(X, A)$. If we interchange the roles of points X and A, then the function $H_1(A, X)$ will satisfy the relations which are obtained from (2.14) by replacing θ by $-\theta$. This follows directly from the representation (2.11) of function H_1 and $L(X, A') = L(X', A)$, if we denote by X' the point $(x_1, \ldots, x_{m-1}, -x_m)$. Further, if $x_m \geq 0$ and $a_m \geq 0$, then H_1 is positive in a neighborhood of point A, since at this point $\Psi(L(X, A))$ goes to $+\infty$, and the other summands which occur in expression (2.11) for function H_1 are bounded at point A. At infinity, H_1 vanishes. At points of the hyperplane $x_m = 0$, according to (2.14) the derivative of function H_1 in some direction not tangent to this hyperplane is equal to zero. By the Zaremba–Giraud principle, at points of this hyperplane the function H_1 cannot achieve either a negative minimum or a positive maximum, and for $x_m > 0$ it cannot achieve a negative minimum by virtue of the maximum principle. From this it follows that $H_1 > 0$ everywhere for $x_m \geq 0$.

We apply Green's formula to the functions $H_1(X, A)$ and $\exp(-ga_m)$ in the domain defined by the inequalities $a_m > 0$, $L(0, A) < R$ and $L(X, A) > \eta$. Here O is the origin. Afterward, letting R tend to infinity and η to zero, we get

$$\int_\Gamma H_1(X, B)\, dS_B = g^{-1} \exp(-g x_m), \tag{2.16}$$

$$x_m \geqslant 0,$$

where Γ is the plane $b_m = 0$, and $B = (b_1, \ldots, b_{m-1}, 0)$ is a variable point of this plane. Application of Green's formula to the functions $H_1(X, A)$ and 1 gives the relation

$$\int_Q H_1(x, A)\, dV_A = g^{-2}, \quad x_m \geqslant 0, \tag{2.17}$$

where Q is the half-space $a_m > 0$.

We show that a solution of the problem (2.5), (2.6) is given by the formula

$$u(X) = -\int_Q H_1(X, A) f(A)\, dV_A + \int_\Gamma H_1(X, B)\, \varphi(B)\, dS_B. \tag{2.18}$$

By the first relation of (2.14), the second summand on the right side of (2.18) is a solution of the homogeneous equation corresponding to (2.5), if $X \in Q$, and the first summand coincides with the right side of (2.12). Consequently, the first summand in (2.18) is a solution of (2.5), which satisfies the homogeneous boundary condition corresponding to (2.6), by virtue of (2.14) and the fact that $H_1(X, A)$ has the same singularity at point A as the fundamental solution of (2.5). We consider the second summand of (2.18):

$$v(X) = \int_\Gamma H_1(X, B)\, \varphi(B)\, dS_B.$$

by direct calculation we find that

$$\frac{\partial v}{\partial x_1} \tan\theta - \frac{\partial v}{\partial x_m} = -2\int_\Gamma \frac{x_m \Psi'(L(X, B))}{L(X, B)}\, \varphi(B)\, dS_B.$$

The right side of this equation is a solution of the homogeneous equation corresponding to (2.5) which, for $x_m = 0$, assumes the value $\varphi(X)$ [51]. Thus, (2.18) gives a solution of (2.5), which satisfies (2.6).

It remains to show that $u(X)$ from (2.18) vanishes at infinity. We take a number $R_1 > 0$ such that $|f(X)| < \varepsilon g^2$ for $L(0, A) \geq R_1$ and $|\varphi(X)| < \varepsilon g$ for $L(0, B) \leq R_1$. We split the integrals which appear in (2.18) into the parts over the balls $L(0, A) < R_1$ and $L(0, B) < R_1$, respectively, and the parts over the complements of these balls. By (2.16) and (2.17) the integrals over the complements of the balls do not exceed ε in absolute value. Now the integrals taken over the balls tend to zero as $L(0, X)$ tends to infinity due to the behavior of H_1 at infinity, so one can find an $R > 0$ such that the sum of the integrals taken over the balls will be less than ε in absolute value when $L(0, X) > R$. Consequently, for such X we will have $|u(X)| < 3\varepsilon$, and this means that $u(X)$ vanishes at infinity.

If the problem (2.5), (2.6) had two solutions $u_1(X)$ and $u_2(X)$, then the difference of these solutions $w(X) = u_1(X) - u_2(X)$ would satisfy the relations

$$\Delta w - g^2 w = 0, \quad x_m > 0; \quad \frac{\partial w}{\partial x_1} \tan\theta - \frac{\partial w}{\partial x_m} = 0, \quad x_m = 0,$$

and would vanish at infinity. It follows from the relations given, the maximum principle, and the Zaremba–Giraud principle, that $w \equiv 0$. Thus the solution of the problem (2.5), (2.6) is unique and is given by (2.18).

We consider the more general elliptic operator

$$L(u) = \sum_{\alpha,\beta=1}^{m} a_{\alpha\beta} \frac{\partial^2 u}{\partial x_\alpha \partial x_\beta} - g^2 u, \quad g > 0, \quad m \geqslant 2,$$

with constant coefficients $a_{\alpha\beta}$, where the quadratic form $\Sigma_{\alpha,\beta=1}^{m} a_{\alpha\beta}\xi^\alpha\xi^\beta = \chi(\xi)$ is assumed to be positive definite. In the half-space H, bounded by the hyperplane T, which is given by

$$H : \left\{\sum_{\alpha=1}^{m} c_\alpha x_\alpha < k\right\}, \quad T : \left\{\sum_{\alpha=1}^{m} c_\alpha x_\alpha = k\right\}, \quad \sum_{\alpha=1}^{m} c_\alpha^2 = 1,$$

we investigate the problem

$$L(u) = f(X), \quad X \in H; \quad \sum_{\alpha=1}^{m} d_\alpha \frac{\partial u}{\partial x_\alpha} = \varphi(X), \quad X \in T,$$

where f and φ are given functions and the constant vector $d = (d_1, \ldots, d_m)$ makes an acute nonzero angle with the vector $c = (c_1, \ldots, c_m)$. One can formally reduce this problem to the preceding one (2.5), (2.6). By a linear change of variables one can transform the operator $L(u)$ into the operator $\Delta u - h^2 u$. After this transformation one can rotate the coordinate axes so that the image of the half-space H coincides with the half-space $x_m' > 0$. Further, one can make a rotation about the axis Ox_m' so that the direction of vector d goes into the direction of vector $(\sin\theta', 0, \ldots, 0, -\cos\theta')$, $0 < 2\theta' \leq 2\pi$. As a result we arrive at the problem (2.5), (2.6). Here we consider only the case $d_\alpha = \Sigma_{\beta=1}^{m} a_{\alpha\beta} c_\beta$.

We denote by D the determinant of the matrix $\|(a_{\alpha\beta} + a_{\beta\alpha})/2\|$, and by $A_{\alpha\beta}$ the quantities defined by the system of equations

$$\sum_{\gamma=1}^{m} (a_{\beta\gamma} + a_{\gamma\beta}) A_{\alpha\gamma} = 0, \quad \beta \neq \alpha,$$

$$\sum_{\gamma=1}^{m} (a_{\alpha\gamma} + a_{\gamma\alpha}) A_{\alpha\gamma} = 1, \quad \alpha, \beta = 1, \ldots, m.$$

Let $A_{\alpha\beta}'$ be the constants defined analogously with respect to the new variables y_α, $\alpha = 1, \ldots, m$, μ a positive constant. We choose new independent variables $y_1, \ldots, y_m$ so that

$$\sum_{\alpha=1}^{m} c_\alpha x_\alpha - k = -\mu y_m, \quad \mu > 0,$$

$$A'_{\alpha\beta} = 0, \ \alpha \neq \beta, \quad A'_{\alpha\alpha} = 1, \ \alpha, \beta = 1, \ldots, m; \ \theta(u) \equiv$$

$$= \sum_{\alpha,\beta=1}^{m} a_{\alpha\beta} c_\beta \frac{\partial u}{\partial x_\alpha} = \mu \left(\frac{\partial u}{\partial y_1} \tan \theta - \frac{\partial u}{\partial y_m} \right), \quad \tan \theta > 0.$$

To define μ, y_m, $\tan \theta$, and dy_1 we have the equations

$$\sum_{\alpha,\beta=1}^{m} a_{\alpha\beta} c_\alpha c_\beta = \mu^2, \qquad \sum_{\alpha,\beta,\gamma,\delta=1}^{m} a_{\alpha\gamma} A_{\gamma\delta} a_{\delta\beta} c_\alpha c_\beta = \mu^2 (1 - \tan^2 \theta),$$

$$\sum_{\alpha,\beta,\gamma=1}^{m} c_\beta a_{\beta\gamma} A_{\gamma\alpha} dx_\alpha = -\mu (\tan \theta dy_1 + dy_m),$$

which follow from the invariance of the left sides of the equations under linear changes of the independent variables and the geometric meaning of the right sides. From these equations we find

$$\begin{aligned} \mu &= \left[\sum_{\alpha,\beta=1}^{m} a_{\alpha\beta} c_\alpha c_\beta \right]^{1/2}, \quad y_m = \left[k - \sum_{\alpha=1}^{m} c_\alpha x_\alpha \right] \left[\sum_{\alpha,\beta=1}^{m} a_{\alpha\beta} c_\alpha c_\beta \right]^{-1/2}, \\ \tan \theta &= \left[\sum_{\alpha,\beta=1}^{m} \left(a_{\alpha\beta} - \sum_{\gamma,\delta=1}^{m} a_{\alpha\gamma} A_{\gamma\delta} a_{\delta\gamma} \right) c_\alpha c_\beta \right]^{1/2} \left[\sum_{\alpha,\beta=1}^{m} a_{\alpha\beta} c_\alpha c_\beta \right]^{-1/2}, \\ dy_1 &= \left[\sum_{\alpha=1}^{m} \left(c_\alpha - \sum_{\beta,\gamma=1}^{m} c_\beta a_{\beta\gamma} A_{\gamma\alpha} \right) dx_\alpha \right] \left[\sum_{\alpha,\beta=1}^{m} \left(a_{\alpha\beta} - \sum_{\gamma,\delta=1}^{m} a_{\alpha\gamma} A_{\gamma\delta} a_{\delta\beta} \right) c_\alpha c_\beta \right]^{-1/2} \end{aligned} \tag{2.19}$$

Now we construct the Green's function $H_2(X, \Xi)$ for the problem considered. Let Y and Z be the images of points X and Ξ, respectively, under the change of variables considered. By definition we set $\sqrt{D} H_2(X, \Xi) = H_1(Y, Z)$, where H_1 is the Green's function constructed above for the problem (2.5), (2.6). In order to represent H_2 in more explicit form, we introduce the notation

$$\begin{aligned} \Lambda_1 &= \left[\sum_{\alpha,\beta=1}^{m} A_{\alpha\beta} (x_\alpha - \xi_\alpha)(x_\beta - \xi_\beta) \right]^{1/2}, \\ \Lambda_2 &= \left\{ \sum_{\alpha,\beta=1}^{m} A_{\alpha\beta} (x_\alpha - \xi_\alpha)(x_\beta - \xi_\beta) \right. \\ &\quad \left. + 4 \left[\sum_{\alpha,\beta=1}^{m} a_{\alpha\beta} c_\alpha c_\beta \right]^{-1} \left[\left(k - \sum_{\alpha=1}^{m} c_\alpha x_\alpha \right) \left(k - \sum_{\beta=1}^{m} c_\beta \xi_\beta \right) \right] \right\}^{1/2}, \\ \Lambda_3 &= 2 \left\{ \sum_{\alpha=1}^{m} \left(c_\alpha - \sum_{\beta,\gamma=1}^{m} c_\beta a_{\beta\gamma} A_{\gamma\alpha} \right) (x_\alpha - \xi_\alpha) \cos^2 \theta \right. \\ &\quad \left. + \left[2k - \sum_{\alpha=1}^{m} c_\alpha (x_\alpha + \xi_\alpha) \right] \sin^2 \theta \right\} \\ &\quad \times \left[\sum_{\alpha,\beta=1}^{m} \left(2a_{\alpha\beta} - \sum_{\gamma,\delta=1}^{m} a_{\alpha\gamma} A_{\gamma\delta} a_{\delta\beta} \right) c_\alpha c_\beta \right]^{-1/2}, \\ \Lambda_2 \cos \Omega &= \left\{ 2k - \sum_{\alpha=1}^{m} c_\alpha (x_\alpha + \xi_\alpha) - \sum_{\alpha=1}^{m} \left(c_\alpha - \sum_{\beta,\gamma=1}^{m} c_\beta a_{\beta\gamma} A_{\gamma\alpha} \right) (x_\alpha - \xi_\alpha) \right\} \\ &\quad \times \left[\sum_{\alpha,\beta=1}^{m} \left(2a_{\alpha\beta} - \sum_{\gamma,\delta=1}^{m} a_{\alpha\gamma} A_{\gamma\delta} a_{\delta\beta} \right) c_\alpha c_\beta \right]^{-1/2}, \quad 0 \leqslant \Omega \leqslant \pi. \end{aligned} \tag{2.20}$$

Now one can write the Green's function in the form

$$H_2(X, \Xi) = D^{-1/2}\left[\Psi(\Lambda_1) + \Psi(\Lambda_2)\cos 2\theta - \Lambda_3 \int\limits_{\cos\Omega}^{\infty} \frac{\Psi'\left(\Lambda_2\sqrt{\sin^2\Omega + t^2}\right)}{\sqrt{\sin^2\Omega + t^2}}\,dt\right]. \quad (2.21)$$

We show that a solution of the problem considered is given by the formula

$$u(X) = -\int\limits_H H_2(X, A) f(A)\,dV_A + \int\limits_T H_2(X, B)\,\varphi(B)\,dS_B, \quad (2.22)$$

where H is the half-space, T is its boundary, given by the relations

$$H:\left\{\sum_{\alpha=1}^{m} c_\alpha a_\alpha < k\right\}, \quad T:\left\{\sum_{\alpha=1}^{m} c_\alpha b_\alpha = k\right\},$$

and dV_A and dS_B are the volume elements of H and T, respectively. $D^{-1/2}dV$ is invariant under linear changes of variables, so

$$\int\limits_H H_2(X, A) f(A)\,dV_A = \int\limits_{z_m>0} H_1(X, Z) f(Z)\,dV_Z,$$

where Z is the image of point A. Let dS_B and $d\Sigma_z$ be the corresponding volume elements of manifolds T and $z_m = 0$. Comparing the expressions for a certain cylindrical volume in the different coordinates, by (2.19) we get $\mu dS_B = D^{1/2}d\Sigma_z$, from which it follows that

$$\mu \int\limits_T H_2(X, B)\,\varphi(B)\,dS_B = \int\limits_{z_m=0} H_1(X, Z)\,\varphi(Z)\,d\Sigma_Z.$$

By virtue of the properties of the solution of the problem (2.5), (2.6), we get from these equations that $u(X)$ of (2.22) satisfies the boundary condition and vanishes at infinity, i.e., $u(X)$ is a solution of the more general problem considered.

We note that the expressions for $\tan\theta$ in (2.19) and expressions (2.20) are homogeneous with degree zero with respect to c_α, $\alpha = 1, \ldots, m$, and also k, so the condition $\Sigma_{\alpha=1}^{m} c_\alpha^2 = 1$ has no value for (2.21) which expresses H_2, but the given condition is essential for the validity of (2.22), which is false without it. With the help of (2.21) it is easy to verify that upon replacing $a_{\alpha\beta}$ by $a_{\beta\alpha}$ (this changes the operator θ but does not change L) the new Green's function coincides with the function $H_2(\Xi, X)$.

In the Euclidean space R^m, $m \geq 2$ we consider a bounded domain D with boundary S, which can be covered by a finite number of subdomains S_j, so that some jth coordinate of a point of such a subdomain can be expressed as a function of the other $m - 1$ coordinates, satisfying a Hölder condition with exponent $h \leq 1$, where in R^m one takes rectangular coordinates, and to be definite we set $j = m$. The direction cosines of the normal to S at the point $Y \in S$, which is outward with respect to D, will be denoted by $\omega_\alpha(Y)$, $\alpha = 1, \ldots, m$. Let $a_{\alpha\beta}(X)$, $\alpha, \beta = 1, \ldots, m$ be given functions, which in the closed domain $D \cup S$ satisfy a Hölder condition with exponent h. The quadratic form $\Omega = \Sigma_{\alpha,\beta=1}^{m} a_{\alpha\beta}(X)\zeta_\alpha\zeta_\beta$ is positive definite for all X from the closed domain $D \cup S$. In addition, let $c(X), f(X), b_\alpha(X), \alpha = 1, \ldots, m$ be some other functions defined in $D \cup S$, and $k(Y)$ and $h(Y)$ be continuous functions of point Y of surface S. For any twice continuously differentiable function in D which is continuously differentiable in $D \cup S$, u, we set

$$M(u) = \sum_{\alpha,\beta=1}^{m} a_{\alpha\beta}(X) \frac{\partial^2 u}{\partial x_\alpha \partial x_\beta} + \sum_{\alpha=1}^{m} b_\alpha(X) \frac{\partial u}{\partial x_\alpha} + c(X) u, \tag{2.23}$$

$$\theta(u) = \sum_{\alpha,\beta=1}^{m} a_{\alpha\beta}(Y)\, \omega_\beta(Y) \frac{\partial u}{\partial x_\alpha} + p(Y) u, \tag{2.24}$$

where Y is a point of surface S. We shall study the following problem: find a function u, continuously differentiable in $D \cup S$, twice continuously differentiable in D, which satisfies the relations

$$M(u) = f \text{ in } D, \quad \theta(u) = \varphi \text{ on } S. \tag{2.25}$$

To investigate this problem we construct an auxiliary function $H(X, \Xi)$ of two points X and Ξ, which is continuous when these points lie in $D \cup S$ and are different, but when X tends to Ξ, and Ξ is considered fixed in D, function H tends to infinity in the same way as the Green's function of operator M (if $\Xi \in S$, then the singularity can be of a different character). If X and Ξ do not coincide, then H is twice continuously differentiable and one has

$$\begin{aligned} M[H(X, \Xi)] &= O[L^{h-m}(X, \Xi)], \\ \theta[H(Y, \Xi)] &= O[L^{1+h-m}(Y, \Xi)], \end{aligned} \tag{2.26}$$

where h is a number $0 < h \leq 1$. Here and below we shall assume that the operators M and θ are taken with respect to the coordinates of the first point. The function which we construct below as the Green's function will further depend on a parameter g. This dependence gives a number of useful properties of the function H.

Before constructing function H, we make some auxiliary constructions. Let $w(\Xi)$ be a continuous function in $D \cup S$, which is negative in D and equal to zero on S. We shall assume that the first derivatives of this function exist and satisfy a Hölder condition with an exponent $h > 0$, and the derivative of w in the direction of the outer normal to S relative to D exists and is strictly positive. As w one can take, for example, a solution of the following problem: $\Delta w = 1$ in D, $w = 0$ on S.

At each point Ξ of the closed domain $D \cup S$ we consider the expressions

$$c_\alpha(\Xi) = \frac{\partial w}{\partial \xi_\alpha}, \quad \alpha = 1, \ldots, m, \quad k(\Xi) = \sum_{\alpha=1}^{m} \xi_\alpha \frac{\partial w}{\partial \xi_\alpha} - w(\Xi).$$

If at the point Ξ not all of the $c_\alpha(\Xi)$ vanish, then the manifold

$$\sum_{\alpha=1}^{m} x_\alpha c_\alpha (\Xi) = k(\Xi) \tag{2.27}$$

will be called the manifold associated with the point Ξ. According to our assumptions, $\partial w/\partial n|_S > 0$, so for points Ξ sufficiently close to S, at least one $c_\alpha(\Xi)$ is nonzero. If $\Xi \in S$, then the manifold (2.27) is tangent to the surface S at point Ξ, while if Ξ does not lie on S, then $\Sigma_{\alpha=1}^{m} \xi_\alpha c_\alpha(\Xi) - k(\Xi) = w(\Xi) < 0$.

Let σ be the shortest distance from point Ξ to points of surface S. We show that the ratio $-w(\Xi)/\sigma(\Xi)$ is bounded above and below. It suffices to show this for small σ. Let Y be a point of S which is closest to Ξ. Obviously the line segment $Y\Xi$ lies on the normal to S at point Y and all the points of this segment closer to Ξ lie in D. By the mean value theorem

$$w(\Xi) = w(Y) + \sum_{\alpha=1}^{m} (\xi_\alpha - z_\alpha) \frac{\partial w}{\partial \xi_\alpha}(Z),$$

where Z is a point of the segment $Y\Xi$. Since $w|_S = 0$ and the first derivatives of this function satisfy a Hölder condition, from the preceding relation we find

$$-w(\Xi) = \sum_{\alpha=1}^{m} (z_\alpha - \xi_\alpha) \frac{\partial w}{\partial \xi_\alpha}(Y) + O(\sigma^{1+h}).$$

The expression $\Sigma_{\alpha=1}^{m}(z_\alpha - \xi_\alpha)\, \partial w/\partial \xi_\alpha$ is the scalar product of the vector with components $z_1 - \xi_1, \ldots, z_m - \xi_m$ of length σ and the vector grad w, so it has the form $\sigma \times w(\Xi, Y)$, where ω is bounded both above and below. The boundedness of the ratio $-w/\sigma$ follows directly from this.

We take an angle θ_1, $0 < 2\theta_1 < \pi$, and we consider a point X of the manifold (2.27) such that the angle between the vectors ΞX and grad w does not exceed θ_1. We show that if $\sigma(\Xi)$ is sufficiently small, then the shortest distance from X to S behaves like $O[\sigma^{1+h}(\Xi)]$. Assuming that the vector ΞX lies entirely in D, we have

$$w(X) = w(\Xi) + (\Xi X, \operatorname{grad} w) + O[L^{1+h}(X, \Xi)],$$

and by (2.27) we get

$$w(X) = O[L^{1+h}(X, \Xi)].$$

If $\sigma(\Xi)$ is sufficiently small that $\Sigma_{\alpha=1}^{m} c_\alpha^2(\Xi)$ is between fixed limits, then it follows from the relation

$$\sum_{\alpha=1}^{m} \xi_\alpha c_\alpha(\Xi) - k(\Xi) = w(\Xi) = O(\sigma)$$

that the distance between Ξ and the manifold (2.27) has the form $O(\sigma)$. In turn, $L(X, \Xi)$ is the ratio of this distance to $\cos\theta_1$, so

$$w(X) = O[L^{1+h}(X, \Xi)] = O[\sigma^{1+h}(\Xi)].$$

Now let the segment ΞX have points in common with the surface S. We denote by Y the closest such common point to the point Ξ. Since $w(Y) = 0$ and $L(Y, \Xi) \leq L(X, \Xi) = O(\sigma)$, we have

$$(\Xi Y, \operatorname{grad} w(\Xi)) + w(\Xi) = O[L^{1+h}(Y, \Xi)] = O(\sigma^{1+h}).$$

This relation means that the distance from Y to the manifold (2.27) is equal to $O(\sigma^{1+h})$. Since $L(Y, X)$ does not exceed this distance divided by $\cos\theta_1$, our assertion is also proved in this case.

Through the point Ξ we draw a ray whose direction cosines are proportional to

$$\sum_{\beta=1}^{m} [a_{\alpha\beta}(\Xi) + a_{\beta\alpha}(\Xi)\, c_\beta(\Xi)],\ \alpha = 1, \ldots, m.$$

By the ellipticity of the operator M the given ray makes an acute angle with grad w. Consequently, the line drawn through the point Ξ intersects the manifold (2.27) in a point X. We denote by Ξ' a point for which the vector $\Xi\Xi'$ is equal to the vector $2\Xi X$. We shall say that this point is associated with the point Ξ. If $\sigma(\Xi)$ is sufficiently small, then Ξ' does not belong to the closed domain $D \cup S$, except for the case when Ξ lies on S.

First we show that for sufficiently small $\sigma(\Xi)$ on the segment $\Xi\Xi'$ there exists a point, not one of the end-points, and lying on S. If there were not such a point, then

$$\begin{aligned} w(\Xi') &= w(\Xi) + (\Xi\Xi', \operatorname{grad} w(\Xi)) + O[L^{1+h}(\Xi\Xi')] \\ &= w(\Xi) + 2(\Xi X, \operatorname{grad} w(\Xi)) + O[L^{1+h}(X, \Xi)] = -w(\Xi) + O[L^{1+h}(X, \Xi)]. \end{aligned}$$

From this, by virtue of the properties of function w, for sufficiently small $\sigma(\Xi)$ we have $w(\Xi') > Q_1\sigma + Q_2\sigma^{1+h}$, where Q_1 and Q_2 are positive constants. It follows from the given inequality that $w(\Xi') > 0$, and this leads to a contradiction. Consequently, on the segment $\Xi\Xi'$ there is a point $Y \in S$. The angle between the normal to S at point Y and vector grad $w(\Xi)$ is equal to $O(\sigma_h)$, since the square of its sine does not exceed the product of some constant by

$$\sum_{\alpha,\beta=1}^{m} |c_\alpha(\Xi)\, c_\beta(Y) - c_\beta(\Xi)\, c_\alpha(Y)| = O(\sigma^{2h}).$$

Let this angle not exceed θ_2 for sufficiently small σ. By virtue of the conditions which S satisfies, the half-line l issuing from point Y and passing through point Ξ' may again intersect S only at a point separated from Y by a distance $a > 0$, where a depends only on S and θ_2. If σ is sufficiently small, then Ξ' now lies between

Y and this new point of intersection, if it exists in general. Consequently, Ξ' does not belong to $D \cup S$.

Obviously if a is sufficiently small, then the distance between Y or Ξ and any point of $D \cup S$, lying on a ray l issuing from the point Ξ' with direction cosines proportional to

$$\sum_{\beta=1}^{m} a_{\alpha\beta}(\Xi)\, c_{\beta}(\Xi),$$

is always greater than a under the condition that such common points for this ray and $D \cup S$ exist in general.

We construct a fundamental solution for the problem (2.25). For this we take the function H_2, represented by (2.21), and we replace the quantities k, $a_{\alpha\beta}$, c_α, and D, which participate in the construction of this function, in (2.19) and (2.20) by the functions $a_{\alpha\beta}(\Xi)$, $c_\alpha(\Xi)$, and $k(\Xi)$ introduced in (2.27), and we define $A_{\alpha\beta}(\Xi)$ and $D(\Xi)$ with the help of these functions by analogy with $A_{\alpha\beta}$ and D. Now the D, $A_{\alpha\beta}$, θ are functions of Ξ, and Λ_1, Λ_2, Λ_3, and Ω are functions of X and Ξ. We denote by $H_3(X, \Xi)$ the function which one gets from H_2 as a result of such substitutions in (2.21). Let $\mu > 0$ be a sufficiently small constant such that if $w(\Xi) \geq -2\mu$ then the $\sigma(\Xi)$ which occurs in the definition of the associated manifold and point is sufficiently small. We set

$$H_4(X, \Xi) = \begin{cases} H_3(X, \Xi), & w(\Xi) \geqslant -\mu, \\ \dfrac{w(\Xi) + 2\mu}{\mu} H_3(X, \Xi) - \dfrac{w(\Xi) + \mu}{\mu} \dfrac{\Psi(\Lambda_1)}{\sqrt{D(\Xi)}}, & -\mu \geqslant w(\Xi) \geqslant -2\mu, \\ \Psi(\Lambda_1)/\sqrt{D(\Xi)}, & w(\Xi) \leqslant -2\mu. \end{cases}$$

We choose $a > 0$ as follows: if $w(\Xi) \geq -2\mu$ and Ξ' is the point associated with Ξ, then the distance between Ξ and the closest point of intersection of a ray l issuing from Ξ' with S is bigger than a if there is such an intersection in general. We set

$$H(X, \Xi) = \begin{cases} [1 - a^{-2}(L(X, \Xi))^2]^3 H_4(X, \Xi), & L(X, \Xi) \leqslant a, \\ 0, & L(X, \Xi) \geqslant a. \end{cases} \tag{2.28}$$

This function has all the properties characteristic of the function H_2.

We investigate the properties of function H. First we note that if $\Xi \in D \cup S$, and $X \neq \Xi$ and $X \in D \cup S$ and the inequalities $w(\Xi) \leq -2\mu$, $L(X, \Xi) \leq a$ hold, then the function H_3 is twice continuously differentiable with respect to X, since this property can only fail when X is on a ray l issuing from the associated point Ξ' or $X = \Xi$, but both these possibilities are excluded by our hypotheses. Now if in the domain $0 > w(\Xi) \geq -2\mu$ the point X tends to Ξ, then H_3 tends to infinity like the Green's function. It follows from the expression for $H_4(X, \Xi)$ that this function also has the properties indicated. Consequently, if X and Ξ lie in $D \cup S$ and $X \neq \Xi$, then $H(X, \Xi)$ is twice continuously differentiable with respect to X, and when X

and Ξ tend to the same point of domain D, function H tends to infinity like a Green's function.

Now we occupy ourselves with the first relation of (2.26). It follows from the definition of function H_4 that

$$\sum_{\alpha,\beta=1}^{m} a_{\alpha\beta}(\Xi)\cdot\frac{\partial^2 H_4(X,\Xi)}{\partial x_\alpha \partial x_\beta} - g^2 H_4(X,\Xi) = 0,$$

if X and Ξ lie in D and $0 < L(X, \Xi) \leq a$. Obviously this implies the relations

$$\sum_{\alpha,\beta=1}^{m} a_{\alpha\beta}(\Xi)\,\frac{\partial^2 H(X,\Xi)}{\partial x_\alpha \partial x_\beta} - g^2 H(X,\Xi) = \begin{cases} O\left[L^{1-m}(X,\Xi)\right], \\ O(g^{m-1})\exp\left[-\nu g L(X,\Xi)\right], \quad gL \geqslant 1, \end{cases}$$

for any X and Ξ from domain D. Here the constant ν and the constants which figure in the symbols O are independent of X and of Ξ and of g. The relation given is obvious for the subdomain $w(\Xi) \leq -2\mu$; now if $w(\Xi) \geq -2\mu$, then it follows from the fact that the ratio of $L(X, \Xi)$ to $L(X, \Xi')$ is bounded and continuous, and $\sup \Omega < \pi$ for the same reasons. From this expression we easily get

$$M\left[H(X,\Xi)\right] - g^2 H(X,\Xi) = \begin{cases} O\left[L^{h-m}(X,\Xi)\right], \\ O(g^m)\exp\left[-\nu g L(X,\Xi)\right], \quad gL \geqslant 1, \end{cases} \tag{2.29}$$

by virtue of the Hölder continuity with exponent h of the coefficients of operator M.

Now we proceed to the second relation of (2.25). Let Y be a point of surface S, and $\Xi \neq Y$ be a point of domain $D \cup S$. We have already established that for $\sigma(\Xi) \leq L(Y, \Xi)$ we have

$$k(\Xi) - \sum_{\alpha=1}^{m} \xi_\alpha c_\alpha(\Xi) = -w(\Xi) = O(\sigma) = O\left[L(Y,\Xi)\right].$$

One also has

$$k(Y) - \sum_{\alpha=1}^{m} \xi_\alpha c(Y) = \sum_{\alpha=1}^{m} (y_\alpha - \xi_\alpha)\frac{\partial w}{\partial y_\alpha}(Y) - w(Y) = O\left[L(Y,\Xi)\right],$$

since $w(Y) = 0$. Now we estimate

$$k(\Xi) - \sum_{\alpha=1}^{m} y_\alpha c_\alpha(\Xi) = \sum_{\alpha=1}^{m} (\xi_\alpha - y_\alpha)\, c_\alpha(\Xi) - w(\Xi).$$

If the segment ΞY lies entirely in D, then by the finite increments theorem this quantity is equal to $O[L^{1+h}(Y, \Xi)]$ considering that $w(Y) = 0$. If $Y\Xi$ does not lie entirely in D, then we denote by Z the closest common point of the given segment with the surface S to Ξ. We have

$$k(\Xi) - \sum_{\alpha=1}^{m} z_\alpha c_\alpha(\Xi) = O\left[L^{1+h}(Z, \Xi)\right] = O\left[L^{1+h}(Y, \Xi)\right].$$

The quantity $\Sigma_{\alpha=1}^{m}(z_\alpha - y_\alpha)c_\alpha(\Xi)$ is the scalar product of vectors $O[L(Y, \Xi)]$ and $O(1)$, the angle between which is equal to $\pi/2 - O[L^h(Y, \Xi)]$, so this product is equal to $O[L^{1+h}(Y, \Xi)]$, whence

$$k(\Xi) - \sum_{\alpha=1}^{m} y_\alpha c_\alpha(\Xi) = O\left[L^{1+h}(Y, \Xi)\right].$$

Considering that $w(Y) = 0$, the last equation can be rewritten as follows:

$$k(\Xi) - k(Y) - \sum_{\alpha=1}^{m} y_\alpha\left[c_\alpha(\Xi) - c_\alpha(Y)\right] = O\left[L^{1+h}(Y, \Xi)\right].$$

Further, we find

$$k(\Xi) - k(Y) - \sum_{\alpha=1}^{m} \xi_\alpha\left[c_\alpha(\Xi) - c_\alpha(Y)\right] = k(\Xi) - k(Y)$$

$$- \sum_{\alpha=1}^{m} y_\alpha\left[c_\alpha(\Xi) - c_\alpha(Y)\right] - \sum_{\alpha=1}^{m} (\xi_\alpha - y_\alpha)\left[c_\alpha(\Xi) - c_\alpha(Y)\right] = O\left[L^{1+h}(Y, \Xi)\right].$$

We consider the expression Λ_2 in (2.20). In it we take the values of c_α, k at point Y once and then at point Ξ we set $X = Y$, and we determine the difference of these two expressions where $a_{\alpha\beta}$ and $A_{\alpha\beta}$ are taken at point Ξ both times. With the help of the finite increment theorem we find that the difference is $O[L^{1+h}(Y, \Xi)]$. If we consider the difference between the two values of Λ_3 and $\Lambda_2 \cos\Omega$ analogously, then we find that these differences are $O[L^{1+h}(Y, \Xi)]$. One verifies directly that the difference between the values of the derivatives with respect to y_α of the functions $\Lambda_2(Y, \Xi)$, $\Lambda_3(Y, \Xi)$, $\Lambda_2(Y, \Xi)\cos\Omega(Y, \Xi)$, which figure in the construction of the func-tion H_3, and those expressions into which these derivatives are transformed upon replacing $k(\Xi)$ and $c_\alpha(\Xi)$ by $k(Y)$ and $c_\alpha(Y)$ is $O[L^h(Y, \Xi)]$.

In the expression for function $H_3(X, \Xi)$ we replace all the functions $c_\alpha(\Xi)$ and $k(\Xi)$ by their values at point $Y \in S$, and we denote the function so obtained by $H_3{}^*(X, \Xi)$. At the given point Y we have

$$\sum_{\alpha,\beta=1}^{m} a_{\alpha\beta}(Y)\,\omega_\beta(Y)\,\frac{\partial H_3^*(Y, \Xi)}{\partial x_\alpha} = 0,$$

and from this for the same point Y and $L(Y, \Xi) \le a$ one gets

$$\theta H_3(Y, \Xi) = \begin{cases} O\left[L^{1+h-m}(Y, \Xi)\right], & L \leqslant a, \\ O\left(g^{m-1}\right) L^h(Y, \Xi) \exp\left[-\nu g L(Y, \Xi)\right], & gL \geqslant 1. \end{cases}$$

If the conditions $w(\Xi) \geq -2\mu$, $L(Y, \Xi) \leq a$ hold, the derivative of the difference $H_3(X, \Xi) - H_3{}^*(X, \Xi)$ with respect to x_λ, $\lambda = 1, \dots, m$ for X tending to Y behaves like $O[L^{1+h-m}(Y, \Xi)]$ for $L \leq a$, $O(g^m)L^{1+h}(Y, \Xi) \exp[-\nu g L(Y, \Xi)]$ for $gL \geq 1$. Consequently, for $w(\Xi) \geq -2\mu$, $L(Y, \Xi) \leq a$ the same type of estimates hold for $\theta[H_3(Y, \Xi)]$, and for $L(Y, \Xi) \leq a$ these estimates are also valid for $\theta[H_4(Y, \Xi)]$. It follows from these estimates that

$$\theta[H(Y, \Xi)] = \begin{cases} O[L^{1+h-m}(Y, \Xi)], \\ O(g^m) L^{1+h}(Y, \Xi) \exp[-\nu g L(Y, \Xi)], \quad gL \geqslant 1. \end{cases} \tag{2.30}$$

This is the second relation of (2.26).

In the formulation of the problem (2.25) we change the notation slightly. Instead of the operator $M(u)$ we consider the operator $M(u) - g^2 u$, which only leads to the replacement of the function $c(X)$ by the function $c(X) + g^2$. Now (2.25) assumes the form

$$M(u) - g^2 u = f \text{ in } D, \quad \theta(u) = \varphi \text{ on } S. \tag{2.31}$$

We seek its solution in the form

$$u(X) = -\int_D H(X, A)\rho(A)\,dV_A + \int_S H(X, B)\sigma(B)\,dS_B, \tag{2.32}$$

where H is the function constructed above, and ρ and σ are unknown functions, which must be chosen so that (2.32) is a solution of the problem (2.31). Substituting (2.32) into (2.31), after the usual transformations of the potential-type integrals which arise [26], for the determination of ρ and σ we get a system of integral equations

$$\begin{aligned} \rho(X) - \int_D K_{11}(X, A)\rho(A)\,dV_A - \int_S K_{12}(X, B)\sigma(B)\,dS_B &= f(X), \\ \sigma(Y) - \int_D K_{21}(Y, A)\rho(A)\,dV_A - \int_S K_{22}(Y, B)\sigma(B)\,dS_B &= \varphi(Y), \end{aligned} \tag{2.33}$$

where we have introduced the notation $K_{11}(X, A) = M[H(X, A)] - g^2 H(X, A)$, $K_{12}(X, B) = K_{11}(X, B)$, $K_{21}(Y, A) = \theta[H(Y, B)]$, $K_{22}(Y, B) = -K_{21}(Y, B)$. By (2.29) and (2.30), for the kernels of system (2.33) we have

$$K_{1\alpha} = O(L^{h-m}), \quad K_{2\alpha} = O(L^{1+h-m}), \quad \alpha = 1, 2.$$

Since $h > 0$ the system of integral equations is Fredholm. By iterations one can reduce it to a system of integral equations with continuous kernels [55].

Generally the question of the equivalence of problem (2.31) and system (2.33) remains open, i.e., it is not clear whether any solution of this problem can be represented by (2.32), and it is unclear whether different solutions of (2.33) correspond to different solutions of (2.31). However, in a number of important cases we are able to prove this equivalence. If $c - g^2 < 0$ and $p > 0$, then with the help of the maximum principle and the Zaremba–Giraud principle it is easy to prove the uniqueness of a solution of (2.31). We show that for sufficiently large g^2 the system (2.33) can be solved for any f and φ. From this, by virtue of the Fredholm alternative the equivalence of system (2.33) and problem (2.31) will follow.

We prove that for sufficiently large g^2 the integrals

$$\int_D |K_{11}(A, \Xi)|\, dV_A, \quad \int_S |K_{21}(B, \Xi)|\, dS_B \tag{2.34}$$

are arbitrarily small. For the first integral this follows from (2.29). We split it into two integrals over the part D_1 of domain D in which $gL(A, \Xi) < 1$ and, over the rest, D_2 of domain D. On D_1 the function K_{11} has a bound which is independent of g, so the given integral is proportional to the volume of D_1, which, in its own right, tends to zero with g^{-1}. Now the integral over D_2 tends to zero with g^{-1} like $O(g^m) \times \exp(-\nu g)$ by virtue of the second estimate of (2.29). One proves that the second integral of (2.34) is small analogously with the help of (2.30). It follows from the fact that these integrals are small for sufficiently large g that one can solve system (2.33) for sufficiently large g by the method of successive approximations [30], while the solution exists and is unique for any f and φ.

One can write the solution of a system of Fredholm integral equations, if it exists and is unique, in the form of certain integrals by means of resolving kernels [30]. For system (2.33) for sufficiently large g^2 we have

$$\rho(X) = f(X) - \int_D R_{11}(X, A)\, f(A)\, dV_A - \int_S R_{12}(X, B)\, \varphi(B)\, dS_B,$$

$$\sigma(Y) = \varphi(Y) - \int_D R_{21}(Y, A)\, f(A)\, dV_A - \int_S R_{22}(Y, B)\, \varphi(B)\, dS_B,$$

where the resolving kernels have the same singularities as the kernels of system (2.33) [30], i.e.,

$$R_{1\alpha} = O(L^{h-m}), \quad R_{2\alpha} = O(L^{1+h-m}), \quad \alpha = 1, 2.$$

Substituting these expressions for ρ and σ in (2.32), we find

$$u(X) = -\int_D G(X, A)\, f(A)\, dV_A + \int_S G(X, B)\, \varphi(B)\, dS_B, \tag{2.35}$$

$$G(X, A) = H(X, A) - \int_D H(X, C)\, R_{11}(C, A)\, dV_C + \int_S H(X, B)\, R_{21}(B, A)\, dS_B. \tag{2.36}$$

The function G, defined by (2.36), satisfies the relations $M[G(X, \Xi)] - g^2 G(X, \Xi) = 0$, $X \in D$, $X \neq \Xi$, $\theta[G(Y, \Xi)] = 0$, $Y \in S$, $Y \neq \Xi$. Now if X and Ξ tend to the same point of domain D, then G becomes infinite exactly like a fundamental solution, i.e., abstracting from the dependence on g, we have $G(X, \Xi) = O[L^{2-m}(X, \Xi)]$, $m > 2$. One can always choose the functions $c_1(X)$ and $q(Y)$ so that the problem $M(u) - c_1 u = f$ in D, $\theta(u) - qu = \varphi$ on S is solvable uniquely for any f and φ. We rewrite the problem (2.25) as follows: $M(u) - c_1 u = f - c_1 u$ in D, $\theta(u) - qu = \varphi - qu$ on S. Using (2.35), we find

$$u(X) - \int_D G(X, A)[c_1(A)u(A) - f(A)]\,dV_A + \int_S G(X, B)[\varphi(B) - q(B)u(A)]\,dS_B. \tag{2.37}$$

For any $Y \in S$ we set $v(Y) = u(Y)$, and we introduce the notation

$$F(X) = -\int_D G(X, A) f(A)\,dV_A + \int_S G(X, B)\varphi(B)\,dS_B.$$

Now one can write (2.37) in the form of the system

$$u(X) - \int_D G(X, A) c_1(A) u(A)\,dV_A + \int_S G(X, B) q(B) v(B)\,dS_B = F(X),$$
$$v(Y) - \int_D G(Y, A) c_1(A) u(A)\,dV_A + \int_S G(Y, B) q(B) v(B)\,dS_B = F(Y),$$

for which the Fredholm alternative is valid, since its kernels have only integrable singularities. Thus we have

Theorem 2.1. The problem (2.25) is always Fredholm.

Remark. In order that the function $G(X, \Xi)$ be differentiable with respect to the coordinates of point Ξ, it suffices, in the construction of function H_3, instead of $a_{\alpha\beta}(\Xi)$ to take

$$\overset{*}{a}_{\alpha\beta}(X, \Xi) = \lambda^q \Gamma(m/2)\left[2\pi^{m/2}\Gamma(q) L^{\lambda}(X, \Xi)\right]^{-1} \times \int_{L(\Xi, A) \leqslant L(X, \Xi)} a_{\alpha\beta}(A) L^{\lambda - m}(\Xi, A)\left[\ln \frac{L(X, \Xi)}{L(\Xi, A)}\right]^{q-1} dV_A,$$

where λ and q are positive constants. It is also assumed that for any X and Ξ on $D \cup S$ the coefficients $a_{\alpha\beta}$ are defined in domain $L(\Xi, A) \leq L(X, A)$. The functions $a_{\alpha\beta}{}^*$ are differentiable with respect to the coordinates of both points to order r, if $\lambda > p$, $q \geq p$, since $a_{\alpha\beta}$ are continuous. If $q > 0$ is fixed and λ tends to zero, then $a_{\alpha\beta}{}^*(X, \Xi)$ tends to $a_{\alpha\beta}(\Xi)$.

2. REDUCTION OF THE OBLIQUE DERIVATIVE PROBLEM FOR HARMONIC FUNCTIONS TO FREDHOLM EQUATIONS

We consider the application and some modifications of the methods of the preceding section to the investigation of Laplace's equation in R^3. Suppose we have three real constants $\beta_1, \beta_2, \beta_3$ such that

$$\beta_1^2 + \beta_2^2 + \beta_3^2 = 1, \quad 0 < \delta \leqslant \beta_3 \leqslant 1. \tag{2.38}$$

We shall seek a harmonic function $u(x_1, x_2, x_3)$, which is regular in the half-space $x_3 > 0$ and vanishes at infinity, satisfying the condition

$$\sum_{j=1}^{3} \beta_j \frac{\partial u}{\partial x_j} = F(x_1, x_2), \quad |F| < C(1 + x_1^2 + x_2^2)^{-\alpha}, \quad \alpha > 1/2. \tag{2.39}$$

We shall determine a solution in the form

$$u(X) = \frac{1}{2\pi} \iint_{-\infty}^{+\infty} f(y_1, y_2) \frac{\partial}{\partial y_3} \ln \left[R + \sum_{j=1}^{3} \beta_j (x_j - y_j) \right] dy_2 dy_1,$$

$$R^2 = \sum_{j=1}^{3} (x_j - y_j)^2, \quad X = (x_1, x_2, x_3).$$

One can rewrite this formula as

$$u(X) = -\frac{1}{2\pi} \iint_{-\infty}^{+\infty} \frac{(\beta_3 R + x_3) f(y_1, y_2)\, dy_2 dy_1}{R \left[R + \sum_{j=1}^{3} \beta_j (x_j - y_j) \right]}. \tag{2.40}$$

If $|f| < C(1 + y_1^2 + y_2^2)^{-\alpha}$, $\alpha > 1/2$, then (2.40) is a harmonic function which is regular in the half-space $x_3 > 0$ and vanishes at infinity. By direct calculation one verifies that

$$\lim_{x_3 \to 0} \sum_{j=1}^{3} \beta_j \frac{\partial u}{\partial x_j} = -\frac{1}{2\pi} \lim_{x_3 \to 0} \iint_{-\infty}^{+\infty} f(y_1, y_2) \frac{\partial}{\partial x_3} (R^{-1})\, dy_2 dy_1 = f(x_1, x_2).$$

Consequently, setting $f = F$, we get a solution of the oblique derivative problem considered for harmonic functions.

Now let β_j in the boundary condition (2.39) be a function which satisfies (2.38) and a Hölder condition:

$$|\beta_j(y_1', y_2') - \beta_j(y_1, y_2)| < C[(y_1' - y_1)^2 + (y_2' - y_2)^2]^{h/2},$$

where $C > 0$ and h, $0 < h \leq 1$, are fixed constants. Again we consider the function

$$u(X) = \frac{1}{2\pi} \iint_{-\infty}^{+\infty} f(y_1, y_2) \frac{\partial}{\partial y_3} \ln \left[R + \sum_{j=1}^{3} \beta_j (y_1, y_2)(x_j - y_j) \right] dy_2 dy_1.$$

Here f is a continuous function which decreases sufficiently rapidly at infinity. By direct calculation one verifies that

$$\sum_{j=1}^{3} \beta_j(x_1, x_2) \frac{\partial}{\partial x_j} \frac{\partial}{\partial x_3} \ln\left[R + \sum_{j=1}^{3} \beta_j (x_j - y_j)\right] = \frac{\partial}{\partial x_3} R^{-1}$$

$$+ \sum_{j=1}^{3} \{[\beta_j(x_1, x_2) - \beta_j(y_1, y_2)] N_j/D\}$$

$$+ [\beta_3(x_1, x_2) - \beta_3(y_1, y_2)] \left\{R\left[R + \sum_{j=1}^{3} \beta_j(y_1, y_2)(x_j - y_j)\right]\right\}^{-1},$$

$$D = R^3\left[R + \sum_{j=1}^{3} \beta_j(y_1, y_2)(x_j - y_j)\right]^2, N_j = -x_3(x_j - y_j)\left[R\right.$$

$$\left. + \sum_{j=1}^{3} \beta_j(y_1, y_2)(x_j - y_j)\right] - R[x_3 + \beta_3(y_1, y_2)R][x_j - y_j + \beta_j(y_1, y_2)R].$$

Using these relations, we get

$$\sum_{j=1}^{3} \beta_j(x_1, x_2) \frac{\partial u}{\partial x_j} = \frac{1}{2\pi} \iint_{-\infty}^{+\infty} \frac{x_3 f(y_1, y_2)\, dy_2\, dy_1}{[(x_1 - y_1)^2 + (x_2 - y_2)^2 + x_3^2]^{3/2}}$$

$$+ \frac{1}{2\pi} \iint_{-\infty}^{+\infty} N(x_1, x_2, x_3, y_1, y_2) f(y_1, y_2)\, dy_2 dy_1, \tag{2.41}$$

where we have introduced the notation

$$N(X, y_1, y_2) = [\beta_3(x_1, x_2) - \beta_3(y_1, y_2)] R^{-1}$$

$$\times \left[R + \sum_{j=1}^{3} \beta_j(x_j - y_j)\right]^{-1} + \sum_{j=1}^{3} [\beta_j(x_1, x_2) - \beta_j(y_1, y_2)] N_j/D.$$

In (2.41) we pass to the limit as $x_3 \to 0$. We have

$$F(x_1, x_2) = f(x_1, x_2) + \frac{1}{2\pi} \iint_{-\infty}^{+\infty} K(x_1, x_2, y_1, y_2) f(y_1, y_2)\, dy_2 dy_1, \tag{2.42}$$

where $K(x_1, x_2, y_1, y_2) = N(x_1, x_2, y_1, y_2)$. We write this kernel explicitly:

$$K = [\beta_3(x_1, x_2) - \beta_3(y_1, y_2)] [(x_1 - y_1)^2 + (x_2 - y_2)^2]^{-1/2}$$

$$\times [\sqrt{(x_1 - y_1)^2 + (x_2 - y_2)^2} + \beta_1(x_1 - y_1) + \beta_2(x_2 - y_2)]^{-1}$$

$$- \sum_{j=1}^{3} \beta_3(y_1, y_2) [\beta_j(x_1, x_2) - \beta_j(y_1, y_2)] [x_j - y_j + \beta_j(y_1, y_2)$$

$$\times \sqrt{(x_1 - y_1)^2 + (x_2 - y_2)^2}]$$

$$\times [(x_1 - y_1)^2 + (x_2 - y_2)^2]^{-1/2} [\sqrt{(x_1 - y_1)^2 + (x_2 - y_2)^2}$$

$$+ \beta_1(y_1, y_2)(x_1 - y_1) + \beta_2(y_1, y_2)(x_2 - y_2)]^{-2}.$$

If the functions β_j satisfy a Hölder condition with exponent $h > 0$, then it is clear that the kernel K is Fredholm. Consequently, (2.42) is a Fredholm equation for determining the function f.

Problem (2.39) has a unique solution equal to zero at infinity. Obviously for constant β_j all solutions of (2.39) can be represented in the form (2.40). For variable β_j the equivalence of Eq. (2.42) to problem (2.39) is proved above, but it can also be proved directly. Analogously one also reduces the oblique derivative problem with the boundary condition

$$\sum_{j=1}^{3} \beta_j (X) \frac{\partial u}{\partial x_i} = F(X), X = (x_1, x_2, x_3) \in \Gamma,$$

for harmonic functions which are regular in a bounded convex domain D, whose boundary Γ has continuous principal curvatures, to a Fredholm equation, where it is assumed that the coefficients in the boundary condition are Hölder continuous and satisfy the condition

$$\sum_{j=1}^{3} [\beta_j (X)]^2 = 1, \quad \sum_{j=1}^{3} \beta_j (X) \nu_j (X) \neq 0, X \in \Gamma, \tag{2.43}$$

where ν_j are the components of the normal vector to Γ. We shall seek a solution of this problem in the form

$$u(X) = \frac{1}{2\pi} \int_{\Gamma} \frac{f(Y) \sum_{j=1}^{3} \nu_j (Y) [x_j - y_j + \beta_j (Y) R]}{R \left[R + \sum_{j=1}^{3} \beta_j (Y) (x_j - y_j) \right]} d\Sigma_Y. \tag{2.44}$$

Here $d\Sigma_Y$ is the area element of the surface Γ, and

$$R = [(x_1 - y_1)^2 + (x_2 - y_2)^2 + (x_3 - y_3)^2]^{1/2}.$$

Substituting $u(X)$ from (2.44) into the boundary condition, and letting point X tend to a point of the surface Γ, we get a Fredholm equation for function f.

In order to reduce the oblique derivative problem to a Fredholm equation in general, we proceed as follows. From point $P \in \Gamma$ we draw a half-line which lies outside D in a neighborhood of P, with direction cosines $-\beta_1, -\beta_2, -\beta_3$. This half-line can intersect the surface Γ at other points. Let P_1 be the closest such point to P. The distance between P and P_1 is a function of point P. By virtue of the conditions imposed on Γ in the preceding section, this distance has a positive greatest lower bound $2l$. We consider the function

$$\ln \left[R + \sum_{j=1}^{3} \beta_1 (Y) (x_j - y_j) \right] - \ln \left[R_1 + \sum_{j=1}^{3} \beta_j (Y) (X_j - y_j + l\beta_j (Y)) \right],$$

$$Y = (y_1, y_2, y_3), R_1^2 = \sum_{j=1}^{3} (x_j - y_i + l\beta_j)^2.$$

This function is regular everywhere except the points of the segment $x_j = y_j - s\beta_j(Y)$, $0 \le s \le l$, $j = 1, 2, 3$. If the vector $(\beta_1, \beta_2, \beta_3)$ makes an acute angle with the direction of the normal to Γ which is inner with respect to D, then it is regular everywhere in the closed domain $D \cup \Gamma$, except for point Y, where it has the singularity defined by the function

$$\ln\left[R + \sum_{j=1}^{3} \beta_j(Y)(x_j - y_j)\right].$$

Now we seek a solution of the oblique derivative problem in the form

$$u(X) = \frac{1}{2\pi}\int_{\Gamma} \frac{f(Y)\sum_{j=1}^{3} \nu_j(Y)\left[x_j - y_j + \beta_j(Y) R\right]}{R\left[R + \sum_{j=1}^{3} \beta_j(Y)(x_j - y_j)\right]} d\Sigma_Y$$

$$- \frac{1}{2\pi}\int_{\Gamma} \frac{f(Y)\sum_{j=1}^{3} \nu_j(Y)\left[x_j - y_j + \beta_j(Y) R_1 + l\beta_j(Y)\right]}{R_1\left[R_1 + \sum_{j=1}^{3} \beta_j(Y)(x_j - y_j + l\beta_j(Y))\right]} d\Sigma_Y. \tag{2.45}$$

Substituting this expression for u into the boundary condition of the oblique derivative problem, and letting the point $X = (x_1, x_2, x_3)$ tend to a point of the surface Γ, we get a Fredholm equation for determining function f.

The oblique derivative problem for harmonic functions can also be reduced in a different way to a Fredholm integral equation. We extend the vector field $Q(\beta_1, \beta_2, \beta_3)$ continuously to a neighborhood of the surface Γ. Since $Q \neq 0$ everywhere on Γ, after the extension one can find a neighborhood Ω_1 of the surface Γ in which the field Q vanishes nowhere, so the trajectories of field Q intersect nowhere in Ω_1. We set $\Omega_0 = \Omega_1 \cap D$, and denote by Ω the intersection of Ω_1 with the complement of domain D. If at a point $Y \in \Gamma$, the vector of field Q does not lie in the tangent plane to Γ, then from this point there issues a piece of a trajectory $l(Y)$ of field Q, which lies entirely in Ω; now if the vector of the field Q at the point Y lies in the tangent plane to Γ, then the trajectory $l(Y)$ can lie entirely in Ω or entirely in Ω_0, or pass at this point from Ω_0 to Ω_1. From the point $Y \in \Gamma$ we draw a trajectory $l(Y)$ of field Q lying in Ω and we consider the integral

$$H(X, Y) = \int_{l(Y)} \left[(x_1 - z_1)^2 + (x_2 - z_2)^2 + (x_3 - z_3)^2\right]^{-1/2} ds_Z.$$

This integral defines a harmonic function which is regular in the whole space except for points of an arc of the trajectory $l(Y)$ of field Q. If the vector field Q is such that one can choose a neighborhood Ω of the surface Γ sufficiently small that the trajectory of the field Q, drawn from point Y of the surfaces Γ in Ω, has no other points in common with Γ within the limits of Ω, then the function $H(X, Y)$ has a unique singular point Y in $D \cup \Gamma$. This is obviously the situation when the vector field Q does not go into the tangent plane to Γ at any point of Γ and Q and Γ themselves are sufficiently smooth. We shall seek a solution of the oblique derivative problem in the form of the integral

$$u(X) = \frac{1}{2\pi} \int_{\Gamma} f(Y) \frac{\partial}{\partial_{Y} n} H(X, Y)\, d\Sigma_Y. \tag{2.46}$$

Substituting this expression for u in the boundary condition of the oblique derivative problem, and letting the point X tend to a point of the surface Γ from within D, we get the Fredholm equation

$$f(X) + \int_{\Gamma} K(X, Y) f(Y)\, d\Sigma_Y = F(X) \tag{2.47}$$

for determining the function f.

When the direction in which one takes the derivative does not lie in the tangent plane to the boundary at any point of the boundary of the domain, the oblique derivative problem was reduced in the preceding section and in this one to Fredholm equations. The kernels of these equations have the same singularity for $X = Y$. These equations are always equivalent to the problem. Equation (2.47) also remains valid in some cases when the vector field in whose direction one takes the derivative goes into the tangent plane to the boundary, but now (2.47) is not equivalent to the oblique derivative problem, since it is not clear whether any harmonic function can be represented by (2.46). This question is of particular concern for nontrivial solutions of the homogeneous problem.

For harmonic functions $u(x, y, z)$, which are regular in the ball Σ: $\{x^2 + y^2 + z^2 < R^2\}$, we consider the oblique derivative problem with the boundary condition

$$u_z|_S = F(X), \quad S: \{x^2 + y^2 + z^2 = R^2\}. \tag{2.48}$$

The Green's function for the Dirichlet problem for the ball Σ has the form [19]

$$G(X, Y) = [(x - x_0)^2 + (y - y_0)^2 + (z - z_0)^2]^{-1/2} - R\rho^{-1} [(x - R^2\rho^{-2}x_0)^2 + (y - R^2\rho^{-2}y_0)^2 + (z - R^2\rho^{-2}z_0)^2]^{-1/2}, \rho^2 = x_0^2 + y_0^2 + z_0^2, Y = (x_0, y_0, z_0).$$

We consider the function H of the pair of points $X = (x, y, z)$, $Y = (x_0, y_0, z_0)$:

$$H(X, Y) = \ln \{z_0 - z + [(x - x_0)^2 + (y - y_0)^2 + (z - z_0)^2]^{1/2}\}$$
$$- \frac{R}{\rho} \ln \left\{\frac{R^2}{\rho^2} z_0 - z + \left[\left(x - \frac{R^2}{\rho^2} x_0\right)^2 + \left(y - \frac{R^2}{\rho^2} y_0\right)^2 + \left(z - \frac{R^2}{\rho^2} z_0\right)^2\right]^{1/2}\right\}, \quad z_0 \geqslant 0,$$

$$H(X, Y) = \ln \{z - z_0 + [(x - x_0)^2 + (y - y_0)^2 + (z - z_0)^2]^{1/2}\}$$
$$- \frac{R}{\rho} \ln \left\{z - \frac{R^2}{\rho^2} z_0 + \left[\left(x - \frac{R^2}{\rho^2} x_0\right)^2 + \left(y - \frac{R^2}{\rho^2} y_0\right)^2 + \left(z - \frac{R^2}{\rho^2} z_0\right)^2\right]^{1/2}\right\}, \quad z_0 < 0.$$

We calculate $\partial H/\partial\rho$ and let ρ tend to R. As a result we get a function $E(X, Y)$, which has the form

$$E(X, Y) = \frac{1}{R} \left\{ 2 \frac{z_0 + \dfrac{x_0(x_0 - x) + y_0(y_0 - y) + z_0(z_0 - z)}{\sqrt{(x - x_0)^2 + (y - y_0)^2 + (z - z_0)^2}}}{z_0 - z + \sqrt{(x - x_0)^2 + (y - y_0)^2 + (z - z_0)^2}} + \right.$$

$$+ \ln\left[z_0 - z + \sqrt{(x-x_0)^2 + (y-y_0)^2 + (z-z_0)^2}\right]\Bigg\}, \quad z_0 \geqslant 0;$$

$$E(X, Y) = \left\{ 2\,\frac{-z_0 + \dfrac{x_0(x_0 - x) + y_0(y_0 - y) + z_0(z_0 - z)}{\sqrt{(x-x_0)^2 + (y-y_0)^2 + (z-z_0)^2}}}{z - z_0 + \sqrt{(x-x_0)^2 + (y_0-y_0)^2 + (z-z_0)^2}} \right.$$

$$+ \ln\left[z - z_0 + \sqrt{(x-x_0)^2 + (y-y_0)^2 + (z-z_0)^2}\right]\Bigg\}, \quad z_0 < 0.$$

It follows from the construction of function E that

$$\frac{\partial}{\partial z} E(X, Y) = \frac{\partial}{\partial \rho} G(X, Y)\,|_{\rho=R},$$

and from the given equation we have

$$u(X) = \frac{1}{4\pi} \int_S E(X, Y)\, F(Y)\, dS_Y. \tag{2.49}$$

Formula (2.49) gives a solution of problem (2.48). However, the function $v(X) = u(X) + \gamma(x, y)$, where u is given by (2.49) and γ is an arbitrary harmonic function of two independent variables which is regular in the disk $x^2 + y^2 < R^2$, will also be a solution of it. This is related to the fact that the harmonic function $\gamma(x, y)$ cannot be represented by (2.49). The example considered shows that, when the direction in which one takes the oblique derivatives goes into the tangent plane to the boundary of the domain, there can occur a loss of equivalence of the problem to the Fredholm equation, although one can apply such a reduction to construct some particular solutions. The structure of the set of all solutions of the homogeneous problem needs special study.

3. SIMPLEST PROPERTIES OF THE NON-FREDHOLM OBLIQUE DERIVATIVE PROBLEM

In what follows we shall consider the oblique derivative problem for harmonic functions of three independent variables in the following formulation almost exclusively: in domain D with sufficiently smooth boundary Γ find a regular harmonic function $u(X)$, which on Γ satisfies the condition

$$a(X)u_x + b(X)u_y + c(X)u_z = f(X), \quad X \in \Gamma, \tag{2.50}$$

where a, b, c, and f are continuously differentiable functions defined on Γ such that

$$[a(X)]^2 + [b(X)]^2 + [c(X)]^2 > 0, \quad X \in \Gamma.$$

We seek a solution $u(X)$ in the class of continuously differentiable functions on $D \cup \Gamma$ and we assume that the principal curvatures of surface Γ are continuous. The ball

Σ: $\{x^2 + y^2 + z^2 < R^2\}$ will figure most frequently as the domain D. In the simplest examples we study those effects which are caused by the vector field Γ going into the tangent plane to $P(X) = \{a(X), b(X), c(X)\}$ at some points of the surface Γ.

We find out for which a, b, and c the function $v = au_x + bu_y + cu_z$ will be harmonic along with u. In order that v be harmonic for arbitrary harmonic u, the functions a, b, and c must be harmonic and satisfy

$$a_x = b_y = c_z, \quad a_y + b_x = 0, \quad a_z + c_x = 0, \quad b_z + c_y = 0, \tag{2.51}$$

which one verifies by direct substitution of v in Laplace's equation. It follows from (2.51) that a, b, and c are linear functions of the form $a = \alpha x + \beta y + \gamma z + A$, $b = -\beta x + \alpha y + \delta z + B$, $c = -\gamma x - \delta y + \alpha z + C$, where α, β, γ, δ, A, B, C are arbitrary constants. We consider the oblique derivative problem (2.50) with the boundary condition

$$(\alpha x + \beta y + \gamma z + A)u_x + (-\beta x + \alpha y + \delta z + B)u_y + (-\gamma x - \delta y + \alpha z + C)u_z = f. \tag{2.52}$$

This problem is equivalent to finding all harmonic solutions of the first-order equation

$$(\alpha x + \beta y + \gamma z + A)u_x + (-\beta x + \alpha y + \delta z + B)u_y + (-\gamma x - \delta y + \alpha z + C)u_z = F(x, y, z), \tag{2.53}$$

where F is a harmonic function which is regular in the domain D and satisfies the condition $F = f$ on Γ.

When the domain D is the ball Σ: $\{x^2 + y^2 + z^2 < R^2\}$, the vector field P which appears in the boundary condition for problem (2.52) goes into the tangent plane to the sphere S: $\{x^2 + y^2 + z^2 = R^2\}$ on the manifold K which is defined by the equations $\alpha R^2 + Ax + By + Cz = 0$, $x^2 + y^2 + z^2 = R^2$. The manifold K is the intersection of the sphere S with a plane, so K can be a circle, point, or the empty set. If $\alpha^2 R^2 > A^2 + B^2 + C^2$, then the distance from the origin to the plane T: $\{\alpha R^2 + Ax + By + Cz = 0\}$ is greater than R. In this case K is empty and problem (2.52) is Fredholm. Now if

$$\alpha^2 R^2 \leqslant A^2 + B^2 + C^2, \tag{2.54}$$

then set K is nonempty.

In the investigation of (2.53) an important role is played by those points at which all three coefficients of the equation vanish simultaneously, i.e., solutions of the system [7]

$$\alpha x + \beta y + \gamma z + A = 0, \quad -\beta x + \alpha y + \delta z + B = 0, \quad -\gamma x - \delta y + \alpha z + C = 0. \tag{2.55}$$

If $\alpha \neq 0$, then (2.55) can always be solved and has a unique solution, i.e., for $\alpha \neq 0$ all the coefficients of (2.53) vanish simultaneously at a unique point which can lie either in domain D or in the closure of its complement. Now if $\alpha = 0$ and $\delta A - \gamma B + \beta C \neq 0$, then the coefficients of (2.53) do not vanish simultaneously at any point, and for $\alpha = 0$, $\delta A - \gamma B + \beta C = 0$ they vanish on a line.

We consider a number of special cases of problem (2.52) and Eq. (2.53). Suppose for the unit ball Σ: $\{x^2 + y^2 + z^2 < 1\}$ the boundary condition (2.52) has the form

$$u_z = f \quad \text{on} \quad S: \{x^2 + y^2 + z^2 = 1\}. \tag{2.56}$$

Now it is necessary to find all harmonic solutions of the equation

$$u_z = F(x, y, z), \tag{2.57}$$

which are regular in the ball Σ, where F is a regular harmonic function in Σ. The set K is the circle $z = 0$, $x^2 + y^2 = 1$. From (2.57) we find

$$u(x, y, z) = \int_0^z F(x, y, t)\, dt + \varphi(x, y).$$

By direct calculation one verifies that in order that u be harmonic, φ must satisfy the equation $\Delta\varphi = -F_z|_{z=0}$, from which we find [34]

$$\varphi(x, y) = -\int_M G(x - \xi, y - \eta)\, F_z(\xi, \eta, 0)\, d\xi d\eta + \psi(x, y),$$

where M is the disk $x^2 + y^2 < 1$, G is the Green's function of the Dirichlet problem for disk M, and ψ is an arbitrary regular harmonic function of two variables in disk M. The particular solution

$$u_0(x, y, z) = \int_0^z F(x, y, t)\, dt - \int_M G(x - \xi, y - \eta)\, F_z(\xi, \eta, 0)\, d\xi d\eta$$

of (2.56) vanishes on circle K, along which the vector field $P = (0, 0, 1)$ goes into the tangent plane to the sphere S. All solutions of (2.56) are given by the formula $u(x, y, z) = u_0(x, y, z) + \psi(x, y)$, where ψ is an arbitrary regular harmonic function of two independent variables in the disk M: $\{x^2 + y^2 < 1\}$. It follows from this that the following modified problem is proper: Find a regular harmonic function u in the ball Σ which satisfies (2.56) on sphere S and assumes preassigned continuously differentiable values on the circle K: $\{z = 0, x^2 + y^2 = 1\}$. Obviously this modified problem can always be solved and has a unique solution.

We investigate (2.56) in the exterior Ω: $\{x^2 + y^2 + z^2 > 1\}$ of ball Σ. Now (2.57) is considered in the domain Ω. With solutions of the homogeneous problem corresponding to (2.56) we associate harmonic solutions of the equation $u_z = 0$, which are regular on the entire plane of the variables x and y, and are equal to zero at infinity. Such harmonic functions are identically zero by Liouville's theorem [19], so if (2.56) is solvable in Ω, then its solution is unique. Since in (2.57) the harmonic function F is regular in the domain Ω, it can be represented by a series

$$F(x, y, z) = \sum_{n=0}^{\infty} Q_n(x, y, z)\, r^{-2n-1}, \tag{2.58}$$

where $r^2 = x^2 + y^2 + z^2$, and Q_n is a homogeneous polynomial of degree n. One has the equation [14]

$$Q_n(x, y, z)\, r^{-2n-1} = (-1)^n Q_n\left(\frac{\partial}{\partial x}, \frac{\partial}{\partial y}, \frac{\partial}{\partial z}\right)\frac{1}{r}. \tag{2.59}$$

Now any harmonic function Q_n can be represented uniquely in the form

$$Q_n(x, y, z) = P_n(x, y, z^2) + zR_{n-1}(x, y, z^2).$$

Here P_n is a homogeneous polynomial of degree n with respect to x, y, and z, and R_{n-1} is a homogeneous polynomial of degree $n-1$. Further, we have [46]

$$p_n(x, y, z^2) = \sum_{k=0}^{[n/2]} \frac{(-1)^k}{(2k)!} z^{2k}\Delta^k q_{0n}(x, y),$$

where $q_{0n}(x, y)$ is a homogeneous polynomial of degree n (not harmonic). Obviously q_{0n} can be represented as follows:

$$q_{0n}(x, y) = \sum_{l=0}^{[n/2]} (x^2 + y^2)^l\, p_{n-2l}(x, y),$$

where p_{n-2l} are completely definite homogeneous harmonic polynomials of degree $n-2l$. We set

$$q_{0n}(x, y) = (x^2 + y^2) q_{0n-2}(x, y) + p_n(x, y),$$

where p_n is a homogeneous harmonic polynomial of degree n. Consequently,

$$P_n(x, y, z^2) = p_n(x, y) + (x^2 + y^2)\, q_{0n-2}(x, y) + \sum_{k=1}^{s} z^{2k} q_k(x, y), \quad s = [n/2].$$

One has the relations

$$zR_{n-1}(x, y, z^2)\, r^{-2n-1} = \frac{\partial}{\partial z}\left\{R_{n-1}\left(\frac{\partial}{\partial x}, \frac{\partial}{\partial y}, \frac{\partial^2}{\partial z^2}\right)\frac{1}{r}\right\},$$

$$P_n(x, y, z^2)\, r^{-2n-1} = p_n(x, y)\, r^{-2n-1}$$

$$+ \frac{\partial}{\partial z}\left\{-\frac{\partial}{\partial z} q_{0n-2}\left(\frac{\partial}{\partial x}, \frac{\partial}{\partial y}\right)\frac{1}{r} + \frac{\partial}{\partial z}\sum_{k=1}^{s} \frac{\partial^{2k-2}}{\partial z^{2k-2}} q_k\left(\frac{\partial}{\partial x}, \frac{\partial}{\partial y}\right)\frac{1}{r}\right\},$$

since due to the harmonicity of the function r^{-1}

$$\left(\frac{\partial^2}{\partial x^2}+\frac{\partial^2}{\partial y^2}\right)r^{-1}=-\frac{\partial^2}{\partial z^2}r^{-1}.$$

It follows from these relations that if $p_n(x, y) \equiv 0$, then there exists a homogeneous harmonic polynomial $w_{n-1}(x, y, z)$

$$\frac{\partial}{\partial z}\left[r^{-2n}w_{n-1}(x, y, z)\right]=Q_n(x, y, z)\,r^{-2n-1}.$$

Consequently, (2.57) has a regular solution in Ω for those $F(x, y, z)$ whose trace f on sphere S is orthogonal to all homogeneous harmonic polynomials $p_n(x, y)$ of two independent variables x and y, i.e., is orthogonal with respect to S to all harmonic functions of two variables, which are regular in the unit disk $x^2 + y^2 < 1$ [5].

Now suppose we have

$$F(x, y, z)=p_n(x, y)\,r^{-2n-1}=p_n\left(\frac{\partial}{\partial x}, \frac{\partial}{\partial y}\right)\frac{1}{r}.$$

From (2.57) we find

$$u(x, y, z)=p_n\left(\frac{\partial}{\partial x}, \frac{\partial}{\partial y}\right)\ln\left(z+\sqrt{x^2+y^2+z^2}\right).$$

This function has a singularity on the ray $x = y = 0$, $z < 0$, so it cannot be regular in domain Ω. Consequently, the orthogonality of the trace f of function F on sphere S to the traces on S of all regular harmonic functions of two variables x and y in ball Σ is not only sufficient but also necessary for the solvability of (2.57) in the class of regular harmonic functions in Ω. Thus, problem (2.56) for regular harmonic functions in domain Ω has an infinite set of necessary and sufficient conditions for solvability of the type of orthogonality conditions.

For regular harmonic functions in the unit ball Σ we consider the oblique derivative problem with the boundary condition [5]

$$xu_x+yu_y+(z-z_0)u_z=f \text{ on } S. \tag{2.60}$$

This problem is equivalent to finding harmonic solutions of the equation

$$xu_x+yu_y+(z-z_0)u_z=F(x, y, z), \tag{2.61}$$

where F is a regular harmonic function in ball Σ, which coincides with f on sphere S.

As was established in Chapter 1, regular harmonic functions F and u in axially symmetric domains can be represented by series

$$\begin{aligned} F(x, y, z) &= \sum_{l=0}^{\infty}\left[p_l(x^2+y^2, z)\rho^l\cos l\varphi\right.\\ &\quad \left.+\,q_l(x^2+y^2, z)\rho^l\sin l\varphi\right],\quad \rho^2=x^2+y^2,\\ u(x, y, z) &= \sum_{l=0}^{\infty}\left[g_l(x^2+y^2, z)\rho^l\cos l\varphi+h_l(x^2+y^2, z)\rho^l\sin l\varphi\right], \end{aligned} \tag{2.62}$$

where the functions $p_l(\sigma, z)$, $q_l(\sigma, z)$, $g_l(\sigma, z)$, $h_l(\sigma, z)$ are uniquely determined by the functions $P_l(z) = p_l(0, z)$, $Q_l(z) = q_l(0, z)$, $G_l(z) = g_l(0, z)$, $H_l(z) = h_l(0, z)$ of a complex variable z, analytic in a domain which coincides with a meridional section of the domain of regularity of functions F and u. Substituting (2.62) into (2.61), equating the coefficients of $\rho^l \cos l\varphi$ and $\rho^l \sin l\varphi$ on both sides of the equation, and setting $x = y = 0$ in the equations obtained, we find

$$(z - z_0)\frac{\partial G_l}{\partial z} + lG_l = P_l, \quad l = 0, 1, \ldots,$$
$$(z - z_0)\frac{\partial H_l}{\partial z} + gH_l = Q_l, \quad l = 1, 2, \ldots. \tag{2.63}$$

The investigation of (2.61) in the ball Σ reduces to finding holomorphic solutions in the disk $|z| < 1$ of (2.63), and the investigation of (2.61) in the exterior of ball Σ to finding holomorphic solutions in domain $|z| > 1$ of (2.63), decreasing at infinity in a specific way.

We investigate the first equation of (2.63), since the second essentially coincides with the first. If $|z_0| < 1$, then the coefficient of the first derivative of the function G_l vanishes at the point z_0 of the disk $|z| < 1$. For $l = 0$ the equation assumes the form $(z - z_0)G_0' = P_0$, and it is only solvable for $P_0(z_0) = 0$; the corresponding homogeneous equation has the holomorphic solution $G_0 = \text{const}$ in the disk $|z| < 1$. For all $l \geq 1$ the corresponding homogeneous equation has no solutions which are not identically zero and are holomorphic in the disk $|z| < 1$; the inhomogeneous equation is always solvable and a holomorphic solution of it is given by the formula

$$G_l(z) = (z - z_0)^{-l} \int_{z_0}^{z} (t - z_0)^{l-1} P_l(t)\, dt. \tag{2.64}$$

Obviously the condition $P_0(z_0) = 0$ is equivalent to the condition

$$F(0, 0, z_0) = \int_S \Omega(0, 0, z_0; \xi, \eta, \zeta) f(\xi, \eta, \zeta)\, dS = 0, \tag{2.65}$$

where $\Omega(X, Y)$ is the kernel of the Poisson formula for the ball Σ. Consequently, this condition is equivalent to one orthogonality condition on the function f.

If $|z_0| < 1$, then for (2.60) to be solvable in the ball Σ it is necessary and sufficient to impose an orthogonality condition (2.65) on the function f, where the corresponding homogeneous problem has one linearly independent solution. In this case problem (2.60) is Fredholm. At no point of sphere S does the vector field $P = \{x, y, z - z_0\}$ go into the tangent plane to S, since the set of such points satisfies the relations $1 - zz_0 = 0$, $x^2 + y^2 + z^2 = 1$. The set of points where the vector field P goes into the tangent plane to S for $|z_0| > 1$ is a circle.

If $|z_0| > 1$, the homogeneous equation corresponding to the first equation of (2.63) has a solution $(z - z_0)^{-l}$, which is holomorphic in the unit disk, which defines a solution of the homogeneous problem corresponding to (2.60):

$$u_l(x, y, z) = [(z - z_0)^2 + x^2 + y^2]^{-l/2} \rho^l \cos l\varphi$$
$$\times F\left(\frac{l}{2}, \frac{l+1}{2}, l + 1; \frac{x^2 + y^2}{x^2 + y^2 + (z - z_0)^2}\right),$$

which is regular in ball Σ. Now the inhomogeneous equation is always solvable, and a particular solution of it is given by (2.64). Analogous facts hold for the second equation of (2.63) also. Now the problem (2.60) is unconditionally solvable and the homogeneous problem corresponding to it has an infinite set of linearly independent solutions.

Now we consider problem (2.60) for regular harmonic functions in the exterior Ω of ball Σ. Harmonic functions of the form $u_l = g_l(x^2 + y^2, z)\rho^l \cos l\varphi$ which are bounded at infinity correspond only to functions $g_l(0, z)$ which decrease faster than z^{-2l} and are holomorphic at infinity. Consequently, in (2.63) the P_l and Q_l decrease at infinity faster than z^{-2l}, and we seek solutions of these equations in the same class of functions. Again we consider the first equation of (2.63). We introduce a new independent variable $\zeta = z^{-1}$ and we set $P_l(z) = z^{-2l-1}R_l(z^{-1})$. We have

$$\zeta(1-\zeta z_0)\frac{\partial G_l}{\partial \zeta} - lG_l = -\zeta^{2l+1}R_l(\zeta). \tag{2.66}$$

The function $\omega_l(\zeta) = \zeta^l(1-\zeta z_0)^{-l}$ is a solution of the homogeneous equation corresponding to (2.66), and a particular solution of the inhomogeneous equation (2.66) is given by the formula

$$G_l(\zeta) = -\left(\frac{\zeta}{1-\zeta z_0}\right)^l \int_0^{\zeta} (1-tz)\, t^l R_l(t)\, dt.$$

If $|z_0| < 1$, then the function G_l is holomorphic in the domain $|z| > 1$ and decreases at infinity faster than z^{-2l} for any holomorphic function $R(z^{-1})$ in the domain $|z| > 1$; now if $|z_0| > 1$, then the function G_l has the analogous properties only if

$$\int_0^{z_0^{-1}} (1-tz_0)^{l-1} R_l(t)\, dt = 0, \quad l = 0, 1, \ldots \tag{2.67}$$

It follows from these facts that the problem (2.60) for regular harmonic functions in the domain Ω always has a unique solution, for $|z_0| < 1$ it is solvable if one of the orthogonality conditions imposed on function f holds, for $|z_0| > 1$ there is an infinite set of solvability conditions corresponding to the conditions (2.67) and the analogous conditions which arise from the second equation of (2.63).

One can investigate the oblique derivative problem analogously for regular harmonic functions in a bounded axially symmetric domain, with boundary condition

$$u_z + \lambda(xu_y - yu_x) = f. \tag{2.68}$$

Instead of (2.63) here one gets the following system:

$$\frac{\partial G_l}{\partial z} + \lambda l H_l = P_l, \quad \frac{\partial H_l}{\partial z} - \lambda l G_l = Q_l, \quad l = 0, 1, \ldots.$$

It has a holomorphic solution for any holomorphic P_l, Q_l. A solution of the corresponding homogeneous system has the form $G_l = C_1 \cos \lambda lz + C_2 \sin lz$, $H_l =$

$C_1 \sin \lambda lz - C_2 \cos \lambda lz$, where C_1 and C_2 are arbitrary constants. To the given functions there corresponds an infinite set of linearly independent solutions of the homogeneous problem corresponding to the problem (2.68):

$$u_l = \rho^l E_l(\rho^2)[C_1 \cos l(\lambda z - \varphi) + C_2 \sin l(\lambda z - \varphi)],$$

where $E_l(\sigma)$ is the entire function

$$E_l(\sigma) = \sum_{k=0}^{\infty} \frac{l!}{k!\,(l+k)!} \left(\frac{\lambda l}{4}\right)^k \sigma^k.$$

The examples considered here show that when the field of directions in which the oblique derivative is taken goes into the tangent plane to the boundary at certain points of this boundary, the oblique derivative problem can have an infinite set of solvability conditions of the type of orthogonality conditions, or the corresponding homogeneous problem can have an infinite set of linearly independent solutions. Such a situation is typical for the oblique derivative problem when the field of directions of differentiation goes into the tangent plane to the boundary of the domain. This situation is well illustrated by the oblique derivative problem for regular harmonic functions in the ball Σ: $\{x^2 + y^2 + z^2 < 1\}$ with the boundary condition

$$\alpha(z)(xu_x + yu_y) + \beta(z)u_z + \gamma(z)(xu_y - yu_x) = f, \tag{2.69}$$

defined on the unit sphere S: $\{x^2 + y^2 + z^2 = 1\}$. In this boundary condition α, β, γ, and f are sufficiently smooth given functions.

We shall seek a solution of problem (2.69) in the form of the potential of a simple layer

$$u(x, y, z) = \frac{1}{2\pi} \int_S \frac{w(x_0, y_0, z_0)\, dS}{\sqrt{(x - x_0)^2 + (y - y_0)^2 + (z - z_0)^2}}. \tag{2.70}$$

Substituting (2.70) into (2.69), and letting the point (x, y, z) tend from within Σ to a point on the sphere S, considering the properties of the derivatives of the potential of a simple layer, we get a singular integral equation for the function

$$f(z, \varphi) = [\alpha(z)(1 - z^2) + z\beta(z)]\, w(z, \varphi)$$

$$+ \frac{1}{2\pi} \int_{-1}^{1}\int_0^{2\pi} \frac{[\beta(z) - z\alpha(z)](z - z_0) + (1 - zz_0)\gamma(z)\sin(\psi - \varphi)}{2\sqrt{2}(1 - zz_0)^{3/2}[1 - \lambda\cos(\psi - \varphi)]^{3/2}}\, w(z_0, \psi)\, d\psi\, dz_0, \tag{2.71}$$

where $x_0 = \sqrt{1 - z_0^2}\cos\psi$, $y_0 = \sqrt{1 - z_0^2}\sin\psi$, $\varphi = \arg(x + iy)$, $\lambda = (1 - zz_0)^{-1}\sqrt{(1 - z^2)(1 - z_0^2)}$, $0 \le \lambda \le 1$. The operator on the right side of (2.71) will be denoted by $B(w)$. We introduce the adjoint operator

$$B^*(v) = [\alpha(z)(1 - z^2) + z\beta(z)]\, v(z, \varphi) +$$

$$+\frac{1}{2\pi}\int_{-1}^{1}\int_{0}^{2\pi}\frac{[\beta(z_0)-z_0\alpha(z_0)](z_0-z)+(1-zz_0)\gamma(z_0)\sin(\varphi-\psi)}{2\sqrt{2}(1-z_0z)^{3/2}[1-\lambda\cos(\psi-\varphi)]^{3/2}}v(z_0,\psi)\,d\psi dz_0$$

for the operator $B(w)$. These operators are related by

$$\int_{-1}^{+1}\int_{0}^{2\pi} vB(w)\,d\psi dz=\int_{-1}^{+1}\int_{0}^{2\pi} wB^*(v)\,d\psi dz,$$

from which it follows that for the equation $B(w)=f$ to be solvable it is necessary that the function f be orthogonal to all solutions of the homogeneous adjoint equation $B^*(v)=0$.

We shall assume $\gamma(z)\equiv 0$ and we introduce the notation $P(z)=\alpha(z)(1-z^2)+z\beta(z)$, $2Q(z)=i[z\alpha(z)-\beta(z)]\sqrt{1-z^2}$, $G(z)=[P(z)-Q(z)][P(z)+Q(z)]^{-1}$. In (2.71) we separate the variables. For this we expand the potential

$$I(w)=\int_{-1}^{1}\int_{0}^{2\pi}\frac{w(z_0,\psi)\,d\psi dz_0}{2\sqrt{1-zz_0}\,[1-\lambda\cos(\psi-\varphi)]^{1/2}}$$

in a Fourier series in the variable φ. We have

$$I(w)=\sum_{l=0}^{\infty}\int_{-1}^{1}\frac{2^l l!}{\Gamma(l+1/2)}F\left(\frac{l}{2}+\frac{1}{4},\frac{l}{2}+\frac{3}{4};l+1;\lambda^2\right)\times\lambda^l[u_l(z_0)\cos l\varphi+v_l(z_0)\sin l\varphi]\,dz_0, \tag{2.72}$$

where F is the Gauss hypergeometric function, and

$$u_l(z_0)=\int_0^{2\pi}w(z_0,\psi)\cos l\psi d\psi,\quad v_l(z_0)=\int_0^{2\pi}w(z_0,\psi)\sin l\psi d\psi.$$

We shall assume the functions $w(z,\varphi)$ and $f(z,\varphi)$ sufficiently smooth for the convergence of the series (2.72) and the series obtained from it by term-by-term differentiation, and for the absolute convergence of the series

$$w(z,\varphi)=\sum_{l=0}^{\infty}[u_l(z)\cos l\varphi+v_l(z)\sin l\varphi],$$
$$f(z,\varphi)=\sum_{l=0}^{\infty}[g_l(z)\cos l\varphi+h_l(z)\sin l\varphi]. \tag{2.73}$$

For this it suffices that the functions $w(z, \varphi)$ and $f(z, \varphi)$ have the first three derivatives with respect to φ [3]. Substituting (2.72) and (2.73), we get

$$\sum_{l=0}^{\infty} [g_l(z) \cos l\varphi + h_l'(z) \sin l\varphi] = P(z) \sum_{l=0}^{\infty} [u_l(z) \cos l\varphi + v_l(z) \sin l\varphi]$$
$$+ \sum_{l=0}^{\infty} \frac{\Gamma(l+3/2)}{2^l \pi l!} \int_{-1}^{1} \frac{\lambda^2}{z-z_0} F\left(\frac{l}{2}+\frac{1}{4}, \frac{l}{2}-\frac{1}{4}; l+1; \lambda^2\right)$$
$$\times \sqrt{1-zz_0}\, [\beta(z) - z\alpha(z)]\, [u_l(z_0) \cos l\varphi + v_l(z_0) \sin l\varphi]\, dz_0. \tag{2.74}$$

Comparing the coefficients of $\cos l\varphi$ and $\sin l\varphi$ on both sides of (2.74), we get two families of equations

$$g_l(z) = P(z)\, u_l(z) + \frac{\Gamma(l+3/2)}{2^l \pi l!} [z\alpha(z) - \beta(z)]$$
$$\times \int_{-1}^{1} \lambda^l \sqrt{1-zz_0}\, F\left(\frac{l}{2}+\frac{1}{4}, \frac{l}{2}-\frac{1}{4}; l+1; \lambda^2\right) \frac{u_l(z_0)}{z_0-z}\, dz_0, \tag{2.75}$$

$$h_l(z) = P(z)\, v_l(z) + \frac{\Gamma(l+3/2)}{2^l \pi l!} [z\alpha(z) - \beta(z)]$$
$$\times \int_{-1}^{1} \lambda^l \sqrt{1-zz_0}\, F\left(\frac{l}{2}+\frac{1}{4}, \frac{l}{2}-\frac{1}{4}; l+1; \lambda^2\right) \frac{v_l(z_0)}{z_0-z}\, dz_0 \tag{2.76}$$

with respect to the functions $u_l(z)$ and $v_l(z)$.

If analogously we separate variables in the equation adjoint to (2.71), then instead of (2.75) and (2.76) we get

$$g_l(z) = P(z)\, \omega(z) - \frac{\Gamma(l+3/2)}{2^l \pi l!} \int_{-1}^{1} \frac{z_0\alpha(z_0) - \beta(z_0)}{z_0 - z}$$
$$\times \lambda^l \sqrt{1-zz_0}\, F\left(\frac{l}{2}+\frac{1}{4}, \frac{l}{2}-\frac{1}{4}; l+1; \lambda^2\right) \omega(z_0)\, dz_0. \tag{2.77}$$

The characteristic parts of these three equations are independent of l. The characteristic part of (2.75) has the form

$$\frac{Q(z)}{\pi i} \int_{-1}^{1} \frac{u(z_0)}{z_0 - z}\, dz_0 + P(z)\, u(z).$$

The normality of (2.75) and (2.77) is guaranteed by the condition

$$D(z) \equiv [P(z)]^2 - [Q(z)]^2 \neq 0 \tag{2.78}$$

holding [29]. We shall consider (2.71) only under the condition that (2.78) holds. The following facts are known [29]. The homogeneous equation corresponding to (2.75) has a finite number of linearly independent solutions and for (2.75) to be solvable it is necessary and sufficient that the function $g_l(z)$ be orthogonal to all the solutions of the homogeneous adjoint equation; the difference between the number of linearly independent solutions of the homogeneous equation corresponding to (2.75) and the number of linearly independent solutions of the adjoint homogeneous equation is equal to n, where $2\pi n$ is the increment of the argument of the function $G(z) = [P(z) - Q(z)]/[P(z) + Q(z)]$ on the segment $-1 \le z \le 1$.

Let $p_{1l}(z), \ldots, p_{ml}(z)$, $m \ge n$, be a complete linearly independent system of solutions of the homogeneous equation corresponding to (2.75), and $q_{1l}(z), \ldots, q_{rl}(z)$, $r = m - n$, be a complete linearly independent system of solutions of the homogeneous equation corresponding to (2.77). The functions

$$p_{1l}(z) \cos l\varphi, \; p_{1l}(r) \sin l\varphi, \; \ldots, \; p_{ml}(z) \cos l\varphi, \\ p_{ml}(z) \sin l\varphi, \quad l = 0, 1, \ldots, \tag{2.79}$$

are solutions of the homogeneous equation corresponding to (2.71), and the functions

$$q_{1l}(z) \cos l\varphi, \; q_{1l}(z) \sin l\varphi, \; \ldots, \; q_{rl}(z) \cos l\varphi, \\ q_{rl}(z) \sin l\varphi, \quad l = 0, 1, \ldots, \tag{2.80}$$

of the adjoint homogeneous equation $B^*(v) = 0$. We denote by $H_1(S)$ the subspace of the space $H(S)$ of Hölder continuous functions on sphere S, generated by the functions of (2.79). Obviously $H_1(S)$ is the space of solutions of the homogeneous equation corresponding to (2.71). We denote by $H_2(S)$ the subspace of space $H(S)$ generated by the functions of (2.80). For (2.71) to be solvable, it is necessary that the function $f(z, \varphi)$ be orthogonal to the space $H_2(S)$, since this space consists of solutions of the adjoint homogeneous equation $B^*(v) = 0$. Formula (2.70) gives a one-to-one correspondence between the space $H_1(S)$ and the space $H_0(\Sigma)$ of solutions of the homogeneous oblique derivative problem corresponding to problem (2.69) for $\gamma(z) \equiv 0$.

If $n > 0$, then the space $H_1(S)$, and with it the space $H_0(\Sigma)$ as well, are infinite-dimensional, i.e., the homogeneous problem corresponding to (2.69) has an infinite set of linearly independent solutions. Now if $n < 0$, then the space $H_2(S)$ is infinite-dimensional, i.e., for (2.69) to be solvable it is necessary to impose a countable set of orthogonality conditions on the function $f(z, \varphi)$. Thus, one has

Theorem 2.2. If $n > 0$, then the homogeneous problem corresponding to (2.69) for $\gamma(z) \equiv 0$ has an infinite set of linearly independent solutions, and for $n < 0$ in order that problem (2.69) be solvable, it is necessary to impose a countable set of orthogonality conditions on the function $f(z, \varphi)$.

Problem (2.69) can only be Fredholm for $n = 0$; in the other cases it is not Fredholm and not Noetherian. In particular, if $P(z)$ does not vanish at any point of the segment $-1 \le z \le 1$, then $n = 0$. The vector field $\{x\alpha(z), y\alpha(z), \beta(z)\}$ goes into the tangent plane to the sphere S: $\{x^2 + y^2 + z^2 = 1\}$ at the set of points which satisfies the conditions

$$P(z) \equiv \alpha(z)(1 - z^2) + z\beta(z) = 0, \quad x^2 + y^2 + z^2 = 1,$$

so $n = 0$ if $P(z) \neq 0$ on the segment $-1 \leq z \leq 1$, and the direction in which one takes the derivative in (2.69) is not in the tangent plane to S. In this case problem (2.69) is Fredholm.

We consider the following special case of (2.69):

$$\alpha(z)(xu_x + yu_y + zu_z) + \gamma(z)(xu_y - yu_x) = f. \tag{2.81}$$

Now (2.71) assumes the form

$$f(z, \varphi) = \alpha(z) w(z, \varphi)$$
$$+ \frac{1}{2\pi} \int_{-1}^{1} \int_{0}^{2\pi} \frac{(1 - zz_0)\gamma(z) \sin(\psi - \varphi) w(z_0, \psi)\, d\psi dz_0}{2\sqrt{2}(1 - zz_0)^{3/2} [1 - \lambda \cos(\psi - \varphi)]^{3/2}}. \tag{2.82}$$

After separation of the variables from (2.82) we get a system of Fredholm integral equations of the third kind

$$g_l(z) = \alpha(z) u_l(z) + \frac{2^{l-2} l!\, l}{\pi \Gamma(l + 1/2)} \int_{-1}^{1} F\left(\frac{l}{2} + \frac{1}{4}, \frac{l}{2} + \frac{3}{4}; l+1; \lambda^2\right) v_l(z_0)\, dz_0, \tag{2.83}$$

$$h_l(z) = \alpha(z) v_l(z) - \frac{2^{l-2} l!\, l}{\pi \Gamma(l + 1/2)} \int_{-1}^{1} F\left(\frac{l}{2} + \frac{1}{4}, \frac{l}{2}\right.$$
$$\left. + \frac{3}{4}; l + 1; \lambda^2\right) u_l(z_0)\, dz_0, \quad l = 1, 2, \ldots,$$
$$g_0(z) = \alpha(z) u_0(z). \tag{2.84}$$

If $\alpha(z) \neq 0$ on the segment $-1 \leq z \leq 1$, this system of equations is always Fredholm, while if $\alpha(z)$ vanishes at some points of this segment, then the system of equations is not of normal type.

Of the function f we required the existence of third derivatives. This requirement is unnecessary for a continuously differentiable solution of the oblique derivative problem in a closed domain, however, and Hölder continuity of the function f alone is insufficient [16, 23, 49]. The following example clearly illustrates the effects which arise here. We consider the oblique derivative problem for regular harmonic functions in the unit ball Σ with the boundary condition $u_x = \mathrm{Re}\,(1 - y - iz)^\varepsilon$ on the unit sphere S, where $\varepsilon \leq 1$ is a positive number, where one takes the branch of the function $(1 - y - iz)^\varepsilon$, which is equal to one for $y = z = 0$. Obviously, everywhere in Σ we have $u_x = \mathrm{Re}\,(1 - y - iz)^\varepsilon$, where the trace of u_x on sphere S satisfies a Hölder condition with exponent ε. The function u has the form $u = x\,\mathrm{Re}\,(1 - y - iz)^\varepsilon + \gamma(y, z)$, where γ is an arbitrary regular harmonic function of two variables in the disk $y^2 + z^2 < 1$. Further, we find

$$u_y = x \frac{\partial}{\partial y} \mathrm{Re}\,(1 - y - iz)^\varepsilon + \gamma_y,$$
$$u_z = x \frac{\partial}{\partial z} \mathrm{Re}\,(1 - y - iz)^\varepsilon + \gamma_z.$$

The functions $u_y - \gamma_y$, $u_z - \gamma_z$ behave in a neighborhood of the point (0, 1, 0) like $x(1 - y - iz)^{\varepsilon-1}$, so as the point (x, y, z) approaches the point (0, 1, 0) along any path not tangent to sphere S, they tend to zero. For the traces of these functions on sphere S we have the expressions

$$u_y - \gamma_y = \sqrt{1 - y^2 - z^2}\,\frac{\partial}{\partial y}\,\mathrm{Re}\,(1 - y - iz)^{\varepsilon},$$
$$u_z - \gamma_z = \sqrt{1 - y^2 - z^2}\,\frac{\partial}{\partial z}\,\mathrm{Re}\,(1 - y - iz)^{\varepsilon}.$$

It follows from this that these functions are unbounded at the point (0, 1, 0) for $\varepsilon < 1/2$ and for $\varepsilon > 1/2$ they are Hölder continuous with exponent $h = \varepsilon - 1/2$. Obviously it is impossible to improve the behavior of u_y, u_z by the choice of the function γ of two variables.

We consider some simple examples of the Poincaré problem for regular harmonic functions in the ball Σ with boundary condition

$$l(u) \equiv l_1 u_x + l_2 u_y + l_3 u_z + hu = f \tag{2.85}$$

on S. Let $l_1 = -2xz$, $l_2 = -2yz$, $l_3 = x^2 + y^2 - z^2$, $h = -z$. We introduce new independent variables

$$\xi = xR^{-2},\ \eta = yR^{-2},\ \zeta = zR^{-2},\ R^2 = x^2 + y^2 + z^2.$$

In the new variables the expression $l(u)$ assumes the form

$$l\,(u) = u_\zeta - \zeta\rho^{-2}u = \rho\,\frac{\partial}{\partial \zeta}(\rho^{-1}u),$$

where $\rho^2 = \xi^2 + \eta^2 + \zeta^2$. The function

$$u(\xi,\ \eta,\ \zeta) = \rho^{-1}u(\xi\rho^{-2},\ \eta\rho^{-2},\ \zeta\rho^{-2}) = \rho^{-1}u(x,\ y,\ z)$$

is regular and harmonic in the domain $\rho > 1$. For a function U which is regular in the exterior of the unit ball, on the unit ball we have the condition $N_\zeta = f$. As already established, this problem has an infinite set of solvability conditions of the type of orthogonality conditions.

It is easy to verify that if the function u is harmonic, then so is the function $F = -2z(xu_x + yu_y) + (x^2 + y^2 - z^2)u_z - zu$, so that writing the representations (2.62) for the functions F and u, in this case we reduce problem (2.85) to the system of equations

$$\begin{aligned} &-z^2G_l' - (2l + 1)\,zG_l = P_l, \quad l = 0, 1, \ldots, \\ &-z^2H_l' - (2l + 1)\,zH_l = Q_l, \quad l = 1, 2, \ldots. \end{aligned} \tag{2.86}$$

Obviously the homogeneous equations corresponding to (2.86) have only zero as holomorphic solution, and for the inhomogeneous equations (2.86) to be solvable it is necessary and sufficient that $P_l(0) = Q_l(0) = 0$, which means that in the first expansion of (2.62) we have

$$\sum_{l=0}^{\infty} \left[p_l(0, 0)\rho^l \cos l\varphi + q_l(0, 0)\rho^l \sin l\varphi\right] = 0,$$

and from this it follows that the trace f of the function F on the unit sphere must be orthogonal to all functions $\rho^l \cos l\varphi$, $\rho^l \sin l\varphi$, $l = 0, 1, \ldots$. These conditions are necessary and sufficient for the solvability of the inhomogeneous problem

$$-2z(xu_x + yu_y) + (x^2 + y^2 - z^2)u_z - zu = f. \tag{2.87}$$

If we consider problem (2.87) for regular harmonic functions in the exterior Ω of the unit ball Σ, then in (2.86) it is necessary to set

$$P_l(z) = z^{-2l-1}R_l(z^{-1}),\quad Q_l(z) = z^{-2l-1}T_l(z^{-1}),$$

where R_l and T_l are holomorphic functions in the domain $|z| > 1$. We seek functions G_l and H_l in the analogous class, i.e.,

$$G_l(z) = z^{-2l-1}E_l(z^{-1}),\quad H_l(z) = z^{-2l-1}K_l(z^{-1}).$$

The equations of (2.86) for the functions $E_l(\zeta)$ and $K_l(\zeta)$ give

$$\frac{\partial}{\partial\zeta} E_l(\zeta) = \zeta R_l(\zeta),\quad \frac{\partial}{\partial\zeta} K_l(\zeta) = \xi T_l(\zeta),$$

where the functions R_l and T_l are holomorphic in the disk $|\zeta| < 1$, and we seek solutions of these equations in the same class of functions. To solutions of the homogeneous problem there correspond functions of the form Az^{-2l-1}, where A is an arbitrary constant. Consequently, in this case problem (2.87) behaves exactly like the problem $u_\zeta = f$ for regular harmonic functions in the unit ball, to which it reduces by a change of independent variables.

In investigation problem (2.87) one uses the fact that if u is harmonic, so is the combination which occurs on the left side of (2.87). It is easy to verify that for any harmonic function u, the function

$$v = l_1u_x + l_2u_y + l_3u_z + hu \tag{2.88}$$

is harmonic, if the coefficients l_1, l_2, l_3, and h have the form

$$\begin{aligned} l_1 &= -\alpha x + \beta y + \gamma z + A(-x^2 + y^2 + z^2) - 2x(By + Cz),\\ l_2 &= -\beta x - \alpha y + \delta z + B(x^2 - y^2 + z^2) - 2y(Ax + Cz),\\ l_3 &= -\gamma x - \delta y - \alpha z + C(x^2 + y^2 - z^2) - 2z(Ax + By),\\ h &= -\alpha - (Ax + By + Cz), \end{aligned} \tag{2.89}$$

where α, β, γ, δ, A, B, C are arbitrary constants. If in (2.88) one makes the change of variables

$$\xi = xR^{-2},\ \eta = yR^{-2},\ \zeta = zR^{-2},\ R^2 = x^2 + y^2 + z^2,$$

then it assumes the form

$$\rho^{-1}v(\xi\rho^{-2}, \eta\rho^{-2}, \zeta\rho^{-2}) = (\alpha\xi + \beta\eta + \gamma\zeta + A)w_\xi$$
$$+ (-\beta\xi + \alpha\eta + \delta\zeta + B)w_\eta + (-\gamma\xi - \delta\eta + \alpha\zeta + C)w_\zeta,$$

where $w(\xi, \eta, \zeta) = \rho^{-1}u(\xi\rho^{-2}, \eta\rho^{-2}, \zeta\rho^{-2})$, $\rho^2 = \xi^2 + \eta^2 + \zeta^2$. The combination of derivatives appearing on the right side of this equation is harmonic for any harmonic function w, and w is harmonic simultaneously with the function u. It is easily verified that the functions l_1, l_2, and l_3 satisfy (2.51). A number of concrete examples of overdeterminedness as well as nonoverdeterminedness of the Poincaré problem are given by the problem with boundary condition (2.85), where l_1, l_2, and l_3, and h have the form (2.89), and are defined on the boundary of the domain D.

4. GLOBAL METHODS OF INVESTIGATION OF THE NON-FREDHOLM OBLIQUE DERIVATION PROBLEM

It was noted at the end of Section 2 of this chapter that one can try to use the method that Giraud used to reduce the oblique derivative problem to a Fredholm integral equation under broader assumptions, permitting the direction of differentiation to go into the tangent plane to the boundary. In this section we show that one can use the method of Bouligand–Giraud in a more general situation, although the equations one gets are no longer equivalent to the original problem. We give an outline of the method on the example of a convex domain, altering it according to the singularities of the problem which arise due to the influence of direction of differentiation in the boundary condition of the oblique derivative problem going into the tangent plane to the boundary.

We consider a strictly convex domain D with smooth boundary Γ. Suppose given at each point of Γ a unit vector l with components $\alpha(X)$, $\beta(X)$, $\gamma(X)$, depending on the point $X \in \Gamma$, which makes a nonobtuse angle with the outer normal $n(X)$ of the surface Γ at the point $X = (x, y, z)$. We take the function

$$H(X, A) = \ln\{r - [(x - a)\alpha(A) + (y - b)\beta(A) + (z - c)\gamma(A)]\},$$

where $A = (a, b, c)$ is a point of the surface Γ, $X = (x, y, z)$ is a point of domain D, and $r^2 = (x - a)^2 + (y - b)^2 + (z - c)^2$, i.e., r is the distance between points A and X. Obviously $H(X, A)$ is a regular harmonic function of the variables x, y, z in the whole space, except for the points of the ray $\sigma(A)$: $\{x = a + \alpha(A)t, y = b + \beta(A)t, z = c + \gamma(A)t, 0 \le t < \infty\}$, and at points of this ray it admits logarithmic singularities. By virtue of the strict convexity of domain D, this ray has only one common point A with the surface Γ, and in domain D there are no points of this ray, so the function $H(X, A)$ is regular in domain D. The harmonic function

$$\omega(X, A) = [L(X, A) - (x - a)\alpha(A) - (y - b)\beta(A)$$
$$- (z - c)\gamma(A)]^{-1}\{\nu_1(A)\alpha(A) + \nu_2(A)\beta(A) +$$

$$+\nu_3(A)\gamma(A) - [(x-a)\nu_1(A) + (y-b)\nu_2(A) + (z-c)\nu_3(A)][L(X, A)]^{-1}\}, \quad L(X, A) = r,$$

where ν_1, ν_2, ν_3 are the direction cosines of the outer normal to Γ, and $L(X, A) = r$, is also regular in domain D with respect to the coordinates of point X for any point $A \in \Gamma$. With the help of this function we form the regular harmonic function in domain D

$$u(X) = \int_\Gamma \omega(X, A)\mu(A)\, dS. \tag{2.90}$$

We try to find a solution of the oblique derivative problem for regular harmonic functions in domain D with boundary condition

$$\alpha(X)u_x + \beta(X)u_y + \gamma(X)u_z = f(X), \tag{2.91}$$

defined on boundary Γ of domain D, in the form (2.90). Substituting (2.90) into (2.91), and letting point X tend to boundary Γ of domain D, for the function $\mu(X)$ we get the integral equation

$$\mu(X) + \int_\Gamma K(X, A)\, \mu(A)\, dS = \frac{1}{2\pi} f(X), \tag{2.92}$$

where $K(X, A) = -\frac{\partial}{\partial_A n}(r^{-1}) + E(X, A)$, and n is the outer normal to Γ at the point A, $r = L(X, A)$ is the distance between X and A, and, finally,

$$\begin{aligned} E(X, A) &= [r - (x-a)\alpha(A) - (y-b)\beta(A) - (z-c)\gamma(A)]^{-2}\{r^{-1}[(x\\ &-a)\nu_1(A) + (y-b)\nu_2(A) + (z-c)\nu_3(A) - \nu_1(A)\alpha(A) - \nu_2(A)\beta(A) - \nu_3\\ &\times(A)\gamma(A)][(\alpha(X)-\alpha(A))^2 + (\beta(X)-\beta(A))^2 + (\gamma(X)-\gamma(A))^2]\cdot 2^{-1}\\ &+ r^{-1}[(x-a)(\alpha(X)-\alpha(A)) + (y-b)(\beta(X)-\beta(A)) + (z-c)(\gamma(X)\\ &-\gamma(A))]\} + \{(x-a)[\alpha(X)-\alpha(A)] + (y-b)[\beta(X)-\beta(A)] + (z-c)[\gamma(X)\\ &-\gamma(A)]\}[r-(x-a)\alpha(A)-(y-b)\beta(A)-(z-c)\gamma(A)]\frac{\partial}{\partial_A n}(r^{-1}) - \{\nu_1\\ &\times(A)[\alpha(X)-\alpha(A)] + \nu_2(A)[\beta(X)-\beta(A)] + \nu_3(A)[\gamma(X)-\gamma(A)]\}r^{-1}\\ &\times[r-(x-A)\alpha(A)-(y-b)\beta(A)-(z-c)\gamma(A)]^{-1}. \end{aligned}$$

If the direction l in which one takes the derivative in (2.91) makes an acute angle with the outer normal to Γ everywhere on Γ, i.e.,

$$\nu_1(X)\alpha(X) + \nu_2(X)\beta(X) + \nu_3(X)\gamma(X) > 0, \tag{2.93}$$

then the kernel $K(X, A)$ of (2.92) is Fredholm, when the functions α, β, and γ are Hölder continuous. $\rho = r - (x-a)\alpha(A) - (y-b)\beta(A) - (z-c)\gamma(A)$ is the difference of the distance between points X and A and the projection of this distance to the plane T: $\{(x-a)\alpha(A) - (y-b)\beta(A) + (z-c)\gamma(A) = 0\}$, passing through point

A and perpendicular to the direction l. When the scalar product (n, l) is strictly bounded from zero, i.e., $(n, l) \geq \delta > 0$, we have $r \geq \rho \geq \delta r$, so as $X \to A$, ρ has the same order of smallness as r. From this one gets easily that the kernel $E(X, A)$ is Fredholm.

In Section 1 of this chapter a modified method of reducing the oblique derivative problem to a Fredholm integral equation suitable for any domain D with sufficiently smooth boundary Γ was given. One can require that there exist a fixed positive number d such that a segment of length d drawn from point A in the direction of the vector $l(A)$ not intersect the surface Γ in more than one point. Let this segment $\lambda(A)$ be defined by the equations $x = a + \alpha(A)t$, $y = b + \beta(A)t$, $z = c + \gamma(A)t$, $0 \leq t \leq \Delta$, and let $a_1 = a + \alpha(A)\Delta$, $b_1 = b + \beta(A)\Delta$, $c_1 = c + \gamma(A)\Delta$. Instead of the function $H(X, A)$ we consider the following function:

$$H_1(X, A) = \ln[r - (x-a)\alpha(A) - (y-b)\beta(A) - (z-c)\gamma(A)]$$
$$- \ln[r_1 - (x-a_1)\alpha(A) - (y-b_1)\beta(A) - (z-c_1)\gamma(A)], \qquad r_1 = L(X, A_1),$$
$$r_1^2 = (x-a_1)^2 + (y-b_1)^2 + (z-c_1)^2, \qquad A_1 = (a_1, b_1, c_1).$$

With respect to X the function H_1 is harmonic and regular in the whole space, except for points of the segment $\lambda(A)$. Further, instead of the function ω we construct the function

$$\omega_1(X, A) = \omega(X, A) - [r_1 - (X - a_1)\alpha(A) - (y - b_1)\beta(A) - (z - c_1)\gamma$$
$$\times (A)]^{-1}\{\nu_1(A)\alpha(A) + \nu_2(A)\beta(A) + \nu_3(A)\gamma(A) - r_1^{-1}[(x - a_1)\nu_1(A)$$
$$+ (y - b_1)\nu_2(A) + (z - c_1)\nu_3(A)]\},$$

which is also regular everywhere except the segment $\lambda(A)$. Now we seek a solution of problem (2.91) in the form

$$u(X) = \int_\Gamma \omega_1(X, A)\mu(A)\,dS. \tag{2.94}$$

With the help of the representation (2.94) the problem (2.91) reduces to the Fredholm integral equation

$$\mu(X) + \int_\Gamma K_1(X, A)\mu(A)\,dS = \frac{1}{2\pi} f(X), \tag{2.95}$$

whose kernel has the form $K_1(X, A) = K(X, A) + M(X, A)$, and

$$2\pi M(X, A) = \alpha(X)\frac{\partial}{\partial x}[\omega_1(X, A) - \omega(X, A)] + \beta(X)\frac{\partial}{\partial y}[\omega_1(X, A)$$
$$- \omega(X, A)] + \gamma(X)\frac{\partial}{\partial z}[\omega_1(X, A) - \omega(X, A)].$$

Obviously the kernel $M(X, A)$ is continuous on Γ, so the kernel K_1 has the same singularities as K.

If the principal curvatures of the surface Γ are continuous, then one can take the length d of the segment $\lambda(A)$, used in constructing the function H_1, sufficiently small that the segments $\lambda(A_1)$ and $\lambda(A_2)$, corresponding to different points, do not intersect. One can achieve this by using properties of the normals to the surface Γ and (2.93). The existence of points of intersection of such segments arbitrarily close to the surface Γ can lead to the appearance in the kernel K_1 of new singularities, which can turn out to be nonintegrable as usually occurs in the investigation of the oblique derivative problem by reducing it to integral equations by representing solutions by potentials of a simple layer. We give an example in which one is not always able to construct segments $\lambda(A)$ of fixed length. As the domain D we take the exterior $\Omega(R)$: $\{x^2 + y^2 + z^2 > R\}$of a ball, and as the direction l of differentiation in the boundary condition of the oblique derivative problem, the direction of the Oz axis. Obviously a segment of arbitrarily small length, issuing from any point of the circle L: $\{x^2 + y^2 = R^2, z = 0\}$, parallel to the Oz axis, lies in domain D, so the length of the segment $\lambda(A)$, issuing from the point A parallel to the Oz axis and lying in the ball Σ: $[x^2 + y^2 + z^2 < R^2\}$, tends to zero when A tends to the circle L. If as domain D one takes the ball Σ and takes the condition of the oblique derivative problem in the form $2z(xu_x + yu_y) - (x^2 + y^2 - z^2)u_z = f$, then here one can construct segments $\lambda(A)$ of fixed length d, lying outside the ball, but the points of the circle L: $\{x^2 + y^2 + z^2 = R^2, z = 0\}$ are limit points for points of intersection of different segments $\lambda(A)$.

If (2.93) does not hold, then we shall assume that the function $T(X) = \nu_1(X)\alpha(X) + \nu_2(X)\beta(X) + \nu_3(X)\gamma(X)$ vanishes to first order on a finite number of closed disjoint lines $Q_1, \ldots, Q_m$. These lines divided the surface Γ into a finite number of pieces, on each of which $T(X)$ is either positive or negative. Where $T(X)$ is positive, with each vector $l(A)$, if this is possible, we associate a segment $\lambda(A)$ just as is done in satisfying (2.93), and where $T(X) < 0$, we construct the analogous segment with respect to the vector $-l(A)$. The segments thus constructed lie outside domain D. The existence of a number $d > 0$ and the possibility of such a construction now depends not only on surface Γ, but also on the behavior of the field of directions l in a neighborhood of a point where this field goes into the tangent plane to Γ. Obviously, for such a construction to be possible, it is necessary that the line drawn through point A in the direction of the vector $l(A)$ have no other points in common with surface Γ in a neighborhood of point A. We define the analog ω_2 of the function ω_1 as follows: On the part of Γ where $T(X) \geq 0$ we set $\omega_2(X, A) = \omega_1(X, A)$, and where $T(X) < 0$ we set $\omega_2(X, A) = -\omega_1(X, A)$, where ω_1 after the construction of the segments $\lambda(A)$ is constructed in exactly the same way as in the case when (2.93) holds. We seek a solution of the oblique derivative problem in the form

$$u(X) = \int_{\Gamma} \omega_2(X, A)\, \mu(A)\, dS, \tag{2.96}$$

and in place of (2.95) we get the equation

$$\mu(X) + \int_{\Gamma} K_2(x, A)\,\mu(A)\,dS = \frac{1}{2\pi} f(X), \tag{2.97}$$

whose kernel is Fredholm everywhere away from the manifolds $Q_1, \ldots, Q_m$.

To study the behavior of the kernel K_2 in a neighborhood of the manifolds $Q_1, \ldots, Q_m$ is difficult, even if one is able to construct this kernel. The complications here are connected with the fact that $\rho(X, A)$ behaves in a neighborhood of these manifolds like the square of the distance between points A and X, if these points lie on the line of intersection of the plane passing through the segment $\lambda(b)$, $B \in Q_j$, and the normal to Γ at point B, with surface Γ. Now if X tends to a point $B \in Q_j$ and A to a point $C \in Q_j$ and $B \neq C$, then ρ behaves like the first power of the distance between B and C. For this reason a rough estimate of the kernel K_2 does not permit us to show that it is Fredholm. However, in a special case one can establish that the kernel K_2 is Fredholm very simply. Suppose that in some neighborhood $V_j(Q_j)$ which is open in Γ of each manifold Q_j, $j = 1, \ldots, m$, the functions $\alpha(X)$, $\beta(X)$, and $\gamma(X)$ are constant. Now on each $V_j(Q_j)$ we have

$$K_2(X, A) = -\frac{\partial}{\partial_A n}\left(\frac{1}{r}\right) + M(X, A),$$

since in these neighborhoods $E(X, A) = 0$, and on the rest of Γ the kernel K_2 is Fredholm since (2.93) holds on it.

Remark. The kernel K_2 is Fredholm even under more general assumptions. It suffices to require that in each $V_j(Q_j)$ the functions $\alpha(X)$, $\beta(X)$, and $\gamma(X)$ are constant only along trajectories of the component of the field of directions $P = l$, tangent to Γ. Under such assumptions one can even get rid of the requirement that at no point is the direction l tangent to the line Q_j when it goes into the tangent plane to Γ.

When the field of directions l with respect to which one takes the derivative in the boundary condition (2.91) does go into the tangent plane to Γ, along with the complications which arise in establishing that the kernel of (2.97) is Fredholm, the investigation of the equivalence of this equation with the original problem (2.91) is a hard problem. In general, it does not follow that any regular harmonic function in domain D, which is sufficiently smooth in the closed domain, can be represented by (2.96); this means that there does not always correspond a solution of (2.97) to any solution of (2.91). However, to any solution of (2.97) there corresponds a solution of (2.91). Thus, if (2.97) is Fredholm, then (2.91) can be solved, provided no more than a finite number of orthogonality conditions imposed on the function f hold.

If (2.91) reduces to the Fredholm equation (2.97) and the homogeneous problem corresponding to it has an infinite set of linearly independent solutions, then it is not the case that any solution of the homogeneous problem corresponding to (2.91) can be represented in the form (2.96), since to each such solution of the homogeneous problem there must be associated a solution of the homogeneous equation corresponding to (2.97). This is the situation, for example, when in the boundary condition (2.91) the coefficients α, β, and γ are constant. Hence the investigation of the homogeneous problem corresponding to (2.91) must be made by a different method. In some cases one can investigate the homogeneous problem completely. Suppose the field of directions l in which one takes the derivative in the boundary condition (2.91), in a neighborhood $V_j(Q_j)$ of each connected component Q_j of the set of points of the surface Γ in which the field l goes into the tangent plane to Γ, coincide with the same constant field $l_0 = (A, B, C)$ of directions. All solutions of the homogeneous oblique derivative problem with boundary condition $Au_x + Bu_y + Cu_z = 0$ have the form $\varphi(\xi, \eta)$, where φ is a regular harmonic function of two independent variables in domain D:

$$\xi = \frac{Bx - Ay}{\sqrt{A^2 + B^2}}, \quad \eta = -\frac{C(Ax + By)}{\sqrt{A^2 + B^2}} + z\sqrt{A^2 + B^2},$$

where here, without loss of generality, it is assumed that $A^2 + B^2 + C^2 = 1$, $A^2 + B^2 \neq 0$. We shall seek a solution of the homogeneous problem corresponding to (2.91) in the form

$$u(X) = \psi(X) + \int_\Gamma \omega(X, A)\sigma(A)\,dS, \tag{2.98}$$

where $\psi(X) = \varphi(\xi, \eta)$. Substituting (2.98) into the homogeneous boundary condition $\alpha(X)u_x + \beta(X)u_y + \gamma(X)u_z = 0$, in the usual way we get a Fredholm integral equation for the function σ

$$\sigma(X) + \int_\Gamma K_2(X, A)\sigma(A)\,dS = \frac{1}{2\pi} g(X).$$

Here $g(X)$ is the value on the surface Γ of the function $\alpha(X)\psi_x + \beta(X)\psi_y + \gamma(X)\psi_z$.

For an example of an oblique derivative problem in which one can apply the method of investigation recounted here, one can consider the problem for regular harmonic functions in the ball Σ: $\{x^2 + y^2 + z^2 < R^2\}$ with boundary condition

$$h(z)(xu_x + yu_y + zu_z) + cu_z = f,$$

where c is a constant and $h(z)$ is a Hölder continuous function which is equal to zero on the interval $-\varepsilon < z < \varepsilon$ ($\varepsilon > 0$ being an arbitrarily small number) and satisfies the inequality $R^2z^{-1}h(z) + c > 0$.

Assuming the boundary Γ of domain D sufficiently smooth, we consider the question of when it is possible to construct a function $\omega_1(X, A)$, i.e., under what restrictions on the behavior of the field of directions l, with respect to which one takes the oblique derivative, do the segment $\lambda(A)$ constructed for points lying in the domain $T(X) > 0$ in a neighborhood of the manifold $T(X) = 0$, not intersect the analogous segments constructed for points lying in the domain $T(X) < 0$. Obviously it is sufficient that the projection of the vector of the direction field l, issuing from point A, $T(A) > 0$, lying in a neighborhood of the set N: $\{T(X) = 0\}$ on Γ, lie entirely in the domain $T(X) > 0$, and for points A, $T(A) < 0$, lying sufficiently close to N, that the analogous projection of the vector $-l$ lie entirely in the domain $T(X) < 0$. This condition is equivalent to the following. Let the point $B \in N$; we draw through the given point the trajectory σ of the projection of the vector field $P(X)$:

$\{\alpha(X), \beta(X), \gamma(X)\}$ to the surface Γ (of the direction field l) and we consider on this trajectory a point A, $T(A) > 0$ and a point C, $T(C) < 0$. The projection of the vector field l at the point A to Γ must be directed to the side $T(X) > 0$ of B, and the projection of $-l$ at the point C to Γ, to the side $T(X) < 0$ of B. Obviously this condition always holds if the rotation of the vector field $P(X)$ along the line σ is negative in a neighborhood of point B. In particular, for problem (2.69) this condition assumes the form of the inequality

$$[\beta(z) - z\alpha(z)]\alpha'(z) > 0, \tag{2.99}$$

which must hold for all z satisfying the equation $\alpha(z)(1 - z^2) + z\beta(z) = 0$, and for such points inequality (2.99) is equivalent to the inequality $\beta(z)\alpha'(z) > 0$.

In Section 2 of this chapter we introduced the function $E(X, Y)$, which has the form

$$\begin{aligned} E(X, Y) &= 2z_0[(x - x_0)^2 + (y - y_0)^2 + (z - z_0)^2]^{-1/2} \\ &\quad + 2[x_0(x_0 - x) + y_0(y_0 - y)][z_0 - z \\ &\quad + \sqrt{(x - x_0)^2 + (y - y_0)^2 + (z - z_0)^2}]^{-1}[(x - x_0)^2 + (y - y_0)^2 \\ &\quad + (z - z_0)^2]^{-1/2} + \ln [z_0 - z + \sqrt{(x - x_0)^2 + (y - y_0)^2 + (z - z_0)^2}], \quad z_0 \geqslant 0, \\ E(X, Y) &= -2z_0[(x - x_0)^2 + (y - y_0)^2 + (z - z_0)^2]^{-1/2} \\ &\quad + 2[x_0(x_0 - x) + y_0(y_0 - y)][z_0 - z \\ &\quad + \sqrt{(x - x_0)^2 + (y - y_0)^2 + (z - z_0)^2}]^{-1}[(x - x_0)^2 + (y - y_0)^2 \\ &\quad + (z - z_0)^2]^{-1/2} + \ln [z - z_0 + \sqrt{(x - x_0)^2 + (y - y_0)^2 + (z - z_0)^2}], \quad z_0 < 0. \end{aligned}$$

The harmonic function

$$u(X) = \frac{1}{4\pi}\int_{\Sigma} E(X, Y)\, F(Y)\, d_Y S + \psi(x, y), \tag{2.100}$$

where Σ is the ball $x^2 + y^2 + z^2 < R^2$, and ψ is an arbitrary regular harmonic function of two independent variables in the ball Σ, is the general solution of the oblique derivative problem $u_z = F(X)$ for regular harmonic functions in the ball Σ. We consider the gradient of the function $u(X)$. We have

$$\begin{aligned} u_x(X) &= \frac{1}{4\pi}\int_S \frac{\partial E(X, Y)}{\partial x} F(Y)\, d_Y S + \psi_x, \\ u_y(X) &= \frac{1}{4\pi}\int_S \frac{\partial E(X, Y)}{\partial y} F(Y)\, d_Y S + \psi_y, \\ u_z(X) &= \frac{1}{4\pi}\int_S \frac{(R^2 - x^2 - y^2 - z^2)\, F(Y)\, d_Y S}{[(x - x_0)^2 + (y - y_0)^2 + (z - z_0)^2]^{3/2}}. \end{aligned} \tag{2.101}$$

If both points X and Y in the first two formulas of (2.101) lie on the sphere S: $\{x^2 + y^2 + z^2 = R^2\}$, then the integrals which figure in these formulas are singular [7]. With the help of these integrals we form the operators

$$J_1(u_z|_S) = [xu_y - yu_x]|_S, \quad J_2(u_z|_S) = [xu_x + yu_y + zu_z]|_S,$$

whose kernels $xE_x - yE_y$ and $xE_x + yE_y + zE_z$ are rather complicated and unperspicuous. With the help of these operators one can reduce the oblique derivative problem for regular harmonic functions in the ball Σ with boundary condition

$$\alpha(X)u_z + \beta(X)(xu_y - yu_x) + \gamma(X)(xu_x + yu_y + zu_z) = f(X), \qquad (2.102)$$

given on the sphere S, to the singular integral equation

$$\begin{aligned} &\alpha(X)\varphi(X) + \beta(X)J_1(\varphi) + \gamma(X)J_2(\varphi) \\ &= f(X) - \beta(X)(x\psi_y - y\psi_x) - \gamma(X)(x\psi_x + y\psi_y) \end{aligned} \qquad (2.103)$$

with kernel $\beta(X)(xE_y - yE_x) + \gamma(X)(xE_x + yE_y + zE_z)$, which has singularities of complicated structure.

With the help of series of spherical functions one can give a different representation for the operators J_1 and J_2. We set $R = 1$ and let

$$u_z|_S = v(\varphi, \theta) = \sum_{l=0}^{\infty} \sum_{m=0}^{l} a_{ml} P_l^m(\cos\theta) \exp(im\varphi), \qquad (2.104)$$

where P_l^m is the adjoint Legendre function. By direct calculation it is easy to verify the equations

$$\begin{aligned} J_1(v) &= \sum_{l=0}^{\infty} \sum_{m=0}^{l} \frac{im}{l+m+1} a_{lm} P_l^m(\cos\theta) \exp(im\varphi), \\ J_2(v) &= \sum_{l=0}^{\infty} \sum_{m=0}^{l} \frac{l+1}{l+m+1} a_{lm} P_{l+1}^m(\cos\theta) \exp(im\varphi). \end{aligned} \qquad (2.105)$$

As is known [27], if the function $v \in W_2^1(S)$, then the series

$$\sum_{l=0}^{\infty} \sum_{m=0}^{l} |la_{lm}|^2 \frac{(l+m)!}{(l-m)!\,(2l+1)!},$$

converges and, conversely, if this series converges, then $v \in W_2^1(S)$. Here $W_2^1(S)$ is the space of functions having first generalized derivatives. We define the norm of the function v as follows:

$$\|v\|^2 = \sum_{l=0}^{\infty} \sum_{m=0}^{l} \frac{(l+m)!}{(l-m)!\,(2l+1)!} |a_{lm}|^2.$$

By virtue of this we have

$$\| J_1 (v) \|^2 = \sum_{l=0}^{\infty} \sum_{m=0}^{l} \frac{m^2}{(l+m+1)^2} \frac{(l+m+1)!}{(l-m+1)!\,(2l+1)!} | a_{lm} |^2$$

$$= \sum_{l=0}^{\infty} \sum_{m=0}^{l} \frac{m^2}{(l+m+1)(l-m+1)} \frac{(l+m)!}{(l-m)!\,(2l+1)} | a_{lm} |^2 \frac{2l+1}{2l+2},$$

$$\| J_2 (v) \|^2 = \sum_{l=0}^{\infty} \sum_{m=0}^{l} \frac{(l+1)^2}{(l+m+1)(l-m+1)} \frac{2l+1}{2l+2} \frac{(l+m)!}{(l-m)!\,(2l+1)} | a_{lm} |^2.$$

Since the coefficients

$$m^2[(l+m+1)(l-m+1)]^{-1}, \quad (l+1)^2[(l+m+1)(l-m+1)]^{-1}$$

are unbounded above (for $l = m$ they grow like l), the operators J_1 and J_2 are unbounded on the spaces $W_2^q(S)$. They carry the space $W_2^q(S)$ into the space $W_2^{q-1/2}(S)$. Due to this fact the theory of singular integral equations of the form (2.103) is much more complicated than the theory of those singular integral equations which Giraud considered [52]. In particular, it is impossible to apply the method of iteration to Eq. (2.103) in the space $W_2^q(S)$.

We consider the set of functions

$$v(\varphi, \theta) = \sum_{l=0}^{\infty} \sum_{m=0}^{l} a_{lm} P_l^m (\cos\theta) \exp(im\varphi) \tag{2.106}$$

such that as $l \to \infty$ we have $a_{lm}^{-1} = O[\Gamma(al+1)]$, $a \geq 3$, and we set

$$\| v(\varphi, \theta) \| = \max_{l,\, m} | a_{lm} |. \tag{2.107}$$

Obviously the operators J_1 and J_2 transform a convergent series of this type into a series which belongs to this same set, although more slowly convergent. If the operators J_1 and J_2 again transform the series (2.106) into convergent ones, in the norm (2.107) we have $\|J_1\| \leq 1/\sqrt{2}$, $\|J_2\| \leq 1$. The loss of smoothness here is due to the fact that the operators J_1 and J_2 shift the index l by one, and the functions $P_{l+1}^m(\cos\theta)$ themselves grow faster in l, although the coefficients of the series (2.106) grow no faster in l than in the original series. One can apply the method of iteration to the class of such functions and show directly that a solution exists. By virtue of the representations (2.105) for the operators J_1 and J_2 one can write a solution of (2.103) for constant α, β, and γ in the form of a series of spherical functions. This lets one investigate some classes of equations of the form (2.103).

It was noted in Section 2 of this chapter that for the reduction of the oblique derivative problem to Fredholm equations one can use potentials of the form

$$E(X, Y) = \int_0^l \frac{[x-\xi(t)]\,\nu_1(X_0) + [y-\eta(t)]\,\nu_2(X_0) + [z-\zeta(t)]\,\nu_3(X_0)}{\{[x-\xi(t)]^2 + [y-\eta(t)]^2 + [z-\zeta(t)]^2\}^{3/2}}\, dt,$$

where $X_0 = (x_0, y_0, z_0)$ is a point of the boundary Γ of domain D, ν_1, ν_2, ν_3 are the direction cosines of the normal to Γ which is outer with respect to D at the point X_0, and ξ, η, ζ are solutions of the system of equations

$$\frac{d\xi}{dt}=a(\xi,\eta,\zeta),\quad \frac{d\eta}{dt}=b(\xi,\eta,\zeta),\quad \frac{d\zeta}{dt}=c(\xi,\eta,\zeta),$$

satisfying the initial conditions $\xi(0) = x_0$, $\eta(0) = y_0$, $\zeta(0) = z_0$; a, b, and c are the components of the vector field P in whose direction one takes the derivative in the boundary condition for the oblique derivative problem [where this field is assumed to be extended to a neighborhood $U(\Gamma)$ of the surface Γ], and l is a sufficiently small fixed number. Obviously $E(X, Y) = \frac{\partial}{\partial n} H(X, Y)$, where

$$H(X,Y)=\int_0^l \{[x-\xi(t)]^2+[y-\eta(t)]^2+[z-\zeta(t)]^2\}^{-1/2}\,dt. \tag{2.108}$$

The properties of such potentials are considered in [4].

One can also construct potentials analogous to (2.108) for more general elliptic equations. We take the fundamental solution $\Gamma(X, Y)$ of the equation $\Delta_2 u = 0$, constructed in Section 6 of Chapter 1, and we consider the integral

$$H(X,Y)=\int_{\lambda(Y)} \Gamma(X,\Xi)]\,d_{\Xi}\lambda, \tag{2.109}$$

where $\lambda(Y)$ is the trajectory of the field P issuing from point Y and lying outside domain D. Formula (2.109) gives a generalization of the potential (2.108) for the operator (1.69). The function $H(X, Y)$ of (2.109) is a solution of the equation $\Delta_2 u = 0$, and with its help one can reduce the oblique derivative problem to a Fredholm equation analogously to the way this was done in Section 1 of the present chapter. If in (2.109), instead of $\Gamma(X, \Xi)$ we take $[s(X, \Xi)]^{2-n}$, where S is the geodesic distance between X and Ξ in the Riemannian metric corresponding to the operator (1.69), then we get the function

$$F(X,Y)=\int_{\lambda(Y)} [s(X,\Xi)]^{2-n}\,d_{\Xi}\lambda, \tag{2.110}$$

which has a singularity of the same type as the function (2.109). In investigating the oblique derivative problem one can use the function $F(X, Y)$ as the Levi function.

One can write the derivative of the function (2.108) in the direction of the vector field P with respect to the coordinates of the point $X = (x, y, z)$ as follows:

$$G(X,X_0)=g(X,X_0)+h(X,X_0)-[(x-x_0)^2+(y-y_0)^2+(z-z_0)^2]^{-1/2},$$

where $g(X, X_0) = \{[x-\xi(l)]^2 + [y-\eta(l)]^2 + [z-\zeta(l)]^2\}^{-1/2}$, and

$$h(X,X_0)=-\int_0^l \frac{[x-\xi(t)][a(X)-a(Y)]+[y-\eta(t)][b(X)-b(Y)]+[z-\zeta(t)][c(X)-c(Y)]}{\{[x-\xi(t)]^2+[y-\eta(t)]^2+[z-\zeta(t)]^2\}^{3/2}}\,dt,$$

$$Y=[\xi(t),\ \eta(t),\ \zeta(t)].$$

Obviously the function $g(X, X_0)$ is regular in the closed domain $D \cup \Gamma$ for any $X_0 \in \Gamma$. Now the behavior of $h(X, X_0)$ in a neighborhood of the point X_0 is determined by the behavior of the function

$$\omega(X, X_0) = (x - x_0)[a(X) - a(X_0)] + (y - y_0)[b(X) - b(X_0)] + (z - z_0)[c(X) - c(X_0)].$$

When $\omega(X, X_0)$ has a zero of sufficiently high order for $X = X_0$, $\frac{\partial}{\partial n} h(X_1, Y_0)$ is a Fredholm kernel.

We note that the question of the kernels just considered being Fredholm has been very little studied at this time.

Chapter 3

OBLIQUE DERIVATIVE PROBLEM WITH DIRECTION OF DIFFERENTIATION GOING INTO THE TANGENT PLANE

In the preceding chapters the Neumann problem and the oblique derivative problem were studied for the case when the direction in which one takes the derivatives does not go into the tangent plane to the boundary at any point of the boundary of the domain. If one seeks a solution of the Neumann problem in the form of the potential of a simple layer, then the problem reduces to a Fredholm integral equation with respect to the density of the potential, while the oblique derivative problem also reduces to a singular integral equation if one seeks its solution in the form of the potential of a simple layer. However, the latter problem can equivalently be reduced to a Fredholm equation, but now it is necessary to seek a solution in the form of potentials of different form, as was done in Chapter 2. One is only able to investigate the singular integral equations to which the oblique derivative problem reduces when the direction of differentiation does not go into the tangent plane to the boundary of the domain anywhere [8]. When this direction does go into the tangent plane, difficulties arise both with the reduction of the problem to Fredholm equations and with the investigation of the singular integral equations. Especially serious difficulties arise in the study of the singular integral equations corresponding to the given type of oblique derivative problem. At the present time the theory of such equations is weakly developed, so it is advisable to try to find other methods of study of the oblique derivative problem, bypassing its reduction to multidimensional singular integral equations.

1. SIMPLEST CONSEQUENCES OF THE MAXIMUM PRINCIPLE

For regular harmonic functions in the unit ball Σ: $\{x^2 + y^2 + z^2 < 1\}$ we consider the oblique derivative problem with boundary condition

$$a(\varphi, \theta)u_r + b(\varphi, \theta)u_\theta + c(\varphi, \theta)u_\varphi = \chi(\varphi, \theta), \tag{3.1}$$

where r, φ, θ are spherical coordinates. We investigate the solution of the homogeneous problem

$$a(\varphi, \theta)v_r + b(\varphi, \theta)v_\theta + c(\varphi, \theta)v_\varphi = 0, \tag{3.2}$$

corresponding to (3.1). Obviously a solution of (3.2) can achieve an extremum only at points of the set N: $\{a(\varphi, \theta) = 0\}$, since by virtue of the Zaremba–Giraud principle, at extremal points $v_r \neq 0$ but, on the other hand, $v_\theta = v_\varphi = 0$, since these points are also extremal points for the trace w on the sphere S: $\{x^2 + y^2 + z^2 = 1\}$ of the function v. Consequently, at any extremal point of the function v we have $a(\varphi, \theta)v_r = 0$, but $v_r \neq 0$, so one must have $a(\varphi, \theta) = 0$.

Let the point $p \in N$ be a maximum point for function v, hence also for function w. Let us assume that $c \neq 0$ at point p. Then from (3.2) we find that

$$\begin{aligned} w_{\varphi\theta} &= -(b/c)w_{\theta\theta} - c^{-1}a_\theta v_r, \\ w_{\varphi\varphi} &= (b/c)w_{\theta\theta} + [bc^{-2}a_\theta - c^{-1}a_\varphi]v_r, \end{aligned} \tag{3.3}$$

assuming that function v is twice continuously differentiable in the closed ball $\Sigma \cup S$. Function w automatically cannot achieve a maximum at point p if the inequalities

$$w_{\varphi\theta}^2 - w_{\varphi\varphi}w_{\theta\theta} > 0, \quad w_{\theta\theta} \leqslant 0,$$

hold at this point [44], and by virtue of (3.3) these inequalities assume the form

$$c^{-2}[(ca_\varphi + ba_\theta)\lambda + (a_\theta)^2] > 0, \quad w_{\theta\theta} \leqslant 0, \tag{3.4}$$

where $\lambda = w_{\theta\theta}(v_r)^{-1} \leq 0$. If we have $c = 0$ at point p, then, at this point, necessarily $b \neq 0$, and analogously we get the inequality

$$b^{-2}[(ca_\varphi + ba_\theta)\mu + (a_\varphi)^2] > 0, \quad w_{\varphi\varphi} \leqslant 0, \tag{3.5}$$

where $\mu = w_{\varphi\varphi}(v_r)^{-1} \leq 0$. If the following inequalities hold at point p:

$$ca_\varphi + ba_\theta < 0, \quad (ba_\varphi)^2 + (ca_\theta)^2 > 0, \tag{3.6}$$

function v cannot achieve a maximum at point p. One can show analogously that it cannot achieve a minimum at point p if (3.6) holds. We note that the second inequality of (3.6) always holds when $a^2 + b^2 + c^2 \neq 0$ and the first inequality of (3.6) holds.

One can give a geometric interpretation to the first inequality of (3.6), which lets us get rid of the requirement of differentiability of the coefficients of boundary condition (3.2) and of the existence of second derivatives of function v in the closed ball $\Sigma \cup S$. On the sphere S we consider the field of tangent vectors $(0, b, c)$ and the trajectories of this field, i.e., the integral curves of the equation

$$b(\varphi, \theta)d\varphi - c(\varphi, \theta)d\theta = 0. \tag{3.7}$$

Through each point of sphere S at which $b^2 + c^2 \neq 0$ there passes a unique integral curve of (3.7) [22]. We choose an orientation on this curve so that the positive direction coincides with the direction of the vector $(0, b, c)$, i.e., with the projection to the tangent plane to sphere S, the direction in which one takes the derivative in boundary condition (3.2). Let p be a point of the set N: $\{a(\varphi, \theta) = 0\}$; by virtue of the condition $a^2 + b^2 + c^2 \neq 0$ at this point we have $b^2 + c^2 \neq 0$, so through p there passes a unique integral curve $J(p)$ of (3.7). Equation (3.6) means that the gradient of the function a makes an obtuse angle with the tangent to $J(p)$ at point p, so function a decrease in the positive direction of the curve $J(p)$ if this curve itself is not entirely contained in set N. In this case, when $J(p) \subset N$ we have $ca_\varphi + ba_\theta = 0$, and it follows from boundary condition (3.2) that along $J(p)$ one has

$$dv/dJ = b(\varphi, \theta)v_\theta + c(\varphi, \theta)v_\varphi = 0,$$

which means that v is constant on curve $J(p)$. In this case curve $J(p)$ is a trajectory of the vector field $p = (a, b, c)$, with respect to which one takes the derivative in boundary condition (3.2).

Definition 3.1. Suppose given on the closed surface Γ a vector field $L = (a, b, c)$, i.e., at each point of this surface we are given the vector with components a, b, and c. Let $\Lambda(\alpha, \beta, \gamma)$ be the projection of the field L to the tangent plane to Γ, i.e., the vector field Λ at a point $X \in \Gamma$ is the projection to the tangent plane to Γ at point X of the vector of field L given at point X. Let N be the set of points of surface Γ at which the vectors of field L go into the tangent plane to Γ, i.e., the set of those points of Γ at which the vectors of fields L and Λ coincide. We shall call the vector field L nondegenerate on the surface Γ if: a) $a^2 + b^2 + c^2 \neq 0$ everywhere on Γ; b) through each point p of set N there passes a unique trajectory $J(p)$ of field Λ where at point p there exists a neighborhood $U(p)$ which is open in Γ, such that within this neighborhood the line $J(p)$ has no other points in common with set N.

Obviously nondegeneracy of the field of directions in which one takes the derivative in (3.2) means that in set N there are no pieces of integral curves of (3.7), although these curves can intersect with lines contained in N, arbitrarily touching them. For nondegenerate vector fields set N cannot contain interior points of surface Γ together with neighborhoods of them. Usually we shall assume that set N consists of no more than a finite number of isolated points and isolated lines.

Theorem 3.1. Let the regular harmonic function v in the unit ball Σ have continuous first derivatives in the closed ball $\Sigma \cup S$ and satisfy, on S, boundary condition (3.2), where a, b, and c are Hölder continuous functions which define a nondegenerate vector field L on Γ. If as the variable point ξ moves along an integral curve $J(p)$ of (3.7) at the point $p \in N$ in the positive direction, function a changes sign from plus to minus or does not change sign, then at point p function v cannot have an extremum.

Proof. We prove the assertion of the theorem for a maximum (the proof is analogous for a minimum). Let point $p \in N$ be a maximum point of function v; then by the Zaremba–Giraud principle, at point p we have $v_r > 0$. We denote by w the trace of function v on the integral curve $J(p)$ of (3.7), passing through point p. This point is also a maximum point for trace w of function v on J, so the derivative of the given function, as the variable point ξ passes through point p, changes sign from plus to minus [15] and up to a positive factor coincides with the trace on $J(p)$

of the expression $(b^2 + c^2)^{-1/2}(bv_\theta + cv_\varphi)$. From boundary condition (3.2) we have $av_r = -bv_\theta - cv_\varphi$, from which it follows that for the condition $v_r > 0$ to hold at point p the function a must change sign from minus to plus as the variable point ξ passes through p along curve $J(p)$, since the right side of the equation changes sign from minus to plus. Consequently, if the hypotheses of the theorem hold at point p, function v cannot achieve a maximum. □

Corollary 1. If function a has continuous first derivatives and at all points of set N one has

$$ca_\varphi + ba_\theta < 0, \tag{3.8}$$

the problem (3.2) has no nonconstant solutions.

Corollary 2. If set N consists of a finite number of points and $a^2 + b^2 + c^2 \neq 0$, (3.2) has no nonconstant solutions.

If (3.8) holds, then so do the hypotheses of Theorem 3.1, and this means that any solution of (3.2) does not achieve either a maximum or a minimum anywhere, so it is identically constant. Under the hypotheses of the second corollary function a does not change sign, so the solution of the homogeneous problem (3.2) is again constant.

For harmonic functions which are regular in the unit ball Σ we consider the oblique derivative problem with the boundary condition

$$\alpha(z)(xu_x + yu_y) + \beta(z)u_z = 0, \tag{3.9}$$

where the Hölder continuous functions α and β satisfy

$$(1 - z^2)\alpha^2 + \beta^2 > 0, \quad \beta(1)\beta(-1) > 0. \tag{3.10}$$

The first condition of (3.10) guarantees the nondegeneracy of the vector field $P = \{x\alpha(z), y\alpha(z), \beta(z)\}$, and the second ensures the possibility of extending the function β to the segment $-1 \leq z \leq 1$ of the Oz axis so that β is continuous and does not vanish on this segment. Formula (3.9) means that everywhere on the sphere S the vectors P and $H = \operatorname{grad} u$ are mutually orthogonal. Through the point (ξ, η, ζ) of sphere S we draw the plane T: $\{\eta x - \xi y = 0\}$, which passes through the Oz axis. This plane intersects the ball Σ in a disk $K(T)$ bounded by a circle $S(T) = S \cap K(T)$. At each point of circle $S(T)$ the vectors of field P lie in plane T, so at each point of $S(T)$ vector P is orthogonal to the projection $H(T)$ of vector H to plane T. We extend the vector field P to the segment I: $\{-1 \leq z \leq 1, x = y = 0\}$ of the Oz axis so that β does not vanish. Together with a semicircle of circle $S(T)$ the segment I forms a closed contour. The rotation of vector field P on this contour coincides with its rotation on the semicircle, since on I the rotation of this field is equal to zero. Let the rotation of field P on the semicircle be equal to m; then its rotation on the whole circle $S(T)$ will be $2m$. Since on $S(T)$ the fields P and $H(T)$ are mutually orthogonal, they are homotopic, and the rotation of field $H(T)$ on circle $S(T)$ is equal to $2m$ if the field $H(T)$ does not vanish at any point of circle $S(T)$.

If there exists a nonconstant solution of (3.9), it defines an extension of vector field H from sphere S to the whole ball. It is continuous everywhere except for a real-analytic set of dimension no higher than 1 [47]. The extension of the field H

defines an extension $H_1(T)$ of field $H(T)$ from circle $S(T)$ to disk $K(T)$. It is also continuous everywhere on $K(T)$, except for the critical points of the trace v_T of function u to plane T, i.e., everywhere on $K(T)$ except for a real-analytic set of dimension no higher than 1 [47]. Obviously the vectors of the field $H_1(T)$ are orthogonal to the lines of intersection of the level surfaces of the function u with the plane T, and these lines are the level lines of the trace v_T of the function u to the plane T. We denote by $G(T)$ the field of tangent vectors to the level lines of function v_T. The vectors of fields $H_1(T)$ and $G(T)$ are orthogonal to one another at each point of the disk $K(T)$, so on $S(T)$ these fields are homotopic to one another and their rotations coincide. Consequently, the rotation of field $G(T)$ on the semicircle $\Gamma(T)$, cut out of circle $S(T)$ by the Oz axis, is equal to m. Let $m > 0$; then the function v_T in disk $K(T)$ has at least one extremal point.

As was established in Chapter 1, any harmonic function which is regular in ball Σ, including a solution of (3.9), can be represented in the form of a series

$$u = \sum_{l=0}^{\infty} [v_l (x^2 + y^2, z) \cos l\varphi + u_l (x^2 + y^2, z) \sin l\varphi]. \tag{3.11}$$

Obviously all the functions $v_l \cos l\varphi$ and $u_l \sin l\varphi$ are harmonic and satisfy (3.9), and the functions $u_l(t, z)$ and $v_l(t, z)$ satisfy

$$4t\omega_{tt} + \omega_{zz} + 4\omega_t - lt^{-1}\omega = 0. \tag{3.12}$$

The trace of a solution of (3.9) of the form $v_l \cos l\varphi$ or $u_l \sin l\varphi$ on the plane T coincides, up to a constant factor, with $v_l(t, z)$ or $u_l(t, z)$. Consequently, if the gradient of such a function on the circle $S(T)$ vanishes nowhere except for points of the Oz axis, and $m > 0$, then these functions have at least one extremal point, which by the maximum principle is either a positive minimal point or a negative maximal point.

To be definite we take the function $v_l(t, z)$ and the half-plane T_0: $\{\varphi = 0\}$. Suppose in the half-disk $K(T_0) = \Sigma \cap T_0$ the function $v_l(t, z)$ has a negative maximal point (t_0, z_0). When φ varies from zero to 2π, this point describes the circle C: $\{x^2 + y^2 = t_0, z = z_0\}$. The traces of the function $v_l(x^2 + y^2, z) \cos l\varphi$ on the half-planes $T(\theta)$: $\{\varphi = \theta\}$ also achieve an extremum at the points of intersection of C with the corresponding $T(\theta)$ (a negative maximum or a positive minimum). As φ varies from zero to $\pi/4l$ the trace $v_l(t_0, z_0) \cos l\varphi$ of the function $v_l \cos l\varphi$ on C increases from the value $v_l(t_0, z_0)$ to zero, and as it varies from $\pi/4l$ to $2\pi/4l$, continues to grow from zero to $-v_l(t_0\ z_0)$, afterwards decreasing from this value to zero. Along with the function $v_l \cos l\varphi$ the function $v_l \sin l\varphi$, whose trace on C decreases for $0 \leq \varphi \leq \pi/2l$, achieves a minimum for $\varphi = \pi/2l$, increases for $\pi/2l < \varphi < 3\pi/2l$ from $v_l(t_0, z_0)$ to $-v_l(t_0, z_0)$, and decreases for $3\pi/2l < \varphi < \pi/l$, is a solution of (3.9). Thus, to negative maxima or positive minima of the traces of the functions $v_l \cos l\varphi$, $v_l \sin l\varphi$ on the half-planes $T(\theta)$ correspond negative minima or positive maxima of the traces of these functions on circle C, and the critical points of the functions $v_l \cos l\varphi$ and $v_l \sin l\varphi$ lying on C will be saddle points. However, for both these functions there are also values of φ for which the vector field $H_1(T)$ corresponding to them vanishes identically. The analogous situation also holds for any linear combination $v_l(x^2 + y^2, z) \times [A \cos l\varphi + B \sin l\varphi]$ of the functions $v_l \cos l\varphi$ and $v_l \sin l\varphi$ with constant coefficients A and B.

The picture for $l = 0$ is completely different. Now (3.12) contains only terms with derivatives, so any solution of it in disk $K(T)$ cannot have a minimum or a maximum. Consequently, for $m > 0$ (3.9) has no axially symmetric solutions whose gradients do not vanish at any point of the circle $S(T)$. Axially symmetric harmonic functions have the following property: let the gradient of the harmonic function $u(x^2 + y^2, z)$, which is regular in ball Σ, not vanish at any point of boundary S of ball Σ; then the rotation of vector field $H = \operatorname{grad} u$ on any meridional semicircle is nonpositive. If this rotation were positive, then the rotation of field H on circle $S(T)$, which is obtained from the given meridional semicircle by completing it to a whole circle, would also be positive and the trace of the function u on disk $K(T)$ would have at least one extremal point, and then function u would also achieve an extremum inside Σ, which is only possible if u is constant. From this property it is easy to get that the inhomogeneous oblique derivative problem with boundary condition

$$\alpha(z)(xu_x + yu_y) + \beta(z)u_z = f(z, \varphi) > 0 \tag{3.13}$$

for $m > 0$ has no solutions which are regular in Σ. In this problem the variables are separable and for the axially symmetric component of f we have

$$f_0(z) = \frac{1}{2\pi}\int_0^{2\pi} f(z, \varphi)\, d\varphi > 0.$$

If (3.13) were solvable, then the problem with the boundary condition

$$\alpha(z)(xu_x + yu_y) + \beta(z)u_z = f_0(z)$$

would also be solvable in the class of harmonic functions which are symmetric with respect to the Oz axis and the vectors of field $H = \operatorname{grad} u$ on sphere S would make acute angles with the vectors of the field $P = \{x\alpha, y\alpha, \beta\}$, from which it would follow that these fields are homotopic. This would mean that the rotation of field H on the meridional semicircles is positive, which is impossible.

2. GENERALIZATIONS OF THE ARGUMENT PRINCIPLE

In the theory of harmonic functions of two independent variables and in the theory of general second-order elliptic equations with two independent variables, an important role for analytic and generalized analytic functions is played by the argument principle [10]. It connects the algebraic sum of the critical and singular points of a function, which lie in a domain, with certain characteristics of the vector field of gradients of this function on the boundary of the domain. The investigation of functions of three independent variables is impeded by the fact that the set of zeros of the gradient of a harmonic function of three independent variables has complicated structure [47] and by the fact that the topological analog of the argument principle which Morse theory gives [58] lets one get considerably less information about the character of the critical points and singularities of a function, which lie in a domain, from the behavior of its gradient on the boundary, than in the two-dimensional case. The example of an axially symmetric oblique derivative problem considered in the preceding section shows that one can try to find a substitute for the argument principle in the three-dimensional case along a somewhat different path. This approach to a generalized argument principle is recounted here.

Definition 3.2. If domain D of space R^3 and the real function $f(X)$, which is analytic in a domain $D_1 \supset D$, satisfy the following conditions:

a) throughout the domain D_1 one has $\operatorname{grad} f \neq 0$ everywhere;

b) in domain D there exists a curve L, joining two points X_1 and X_2 of boundary Γ of domain D, such that trace h of function f on L is a monotone function of the arc length s on L, calculated from point X_1, where L is assumed to be a smooth one-dimensional manifold;

c) the level surfaces M_1: $\{f = h(0)\}$ and M_2: $\{f = h(l)\}$ of function f, passing through the ends X_1 and X_2, respectively, of line L, are tangent to boundary Γ of domain D at points X_1 and X_2, moreover, so that if one takes M_i as the coordinate plane $z = 0$, then Γ in a neighborhood of the point X_i, $i = 1, 2$, is given by the equation $z = Z_i(x, y)$, and at point X_i we have

$$\frac{\partial^2 Z_i}{\partial x^2} \cdot \frac{\partial^2 Z_i}{\partial y^2} - \left(\frac{\partial^2 Z}{\partial x \partial y}\right)^2 > 0, \quad i = 1, 2; \tag{3.14}$$

d) through each point of domain D there passes one and only one level surface $M(\alpha)$: $\{f(X) = \alpha\}$, $h(0) < \alpha < h(l)$ of function f, where the two-dimensional domains of intersection of $M(\alpha)$ with D are homeomorphic to a disk for all α, then we shall say that the analytic function f effects an analytic decomposition of domain D into two-dimensional domains.

Definition 3.3. Let domain D and two functions $P(x, y, z)$ and $Q(x, y, z)$ which are analytic in the domain $D_1 \supset D$ satisfy the following conditions:

a) the functions P and Q have no common factors which vanish at least at one point of domain D_1;

b) P and Q vanish simultaneously on line L, which lies in D and joins two parts X_1 and X_2 of boundary Γ of domain D, where L itself is a smooth manifold and on it we have

$$P_x^2 + P_y^2 + P_z^2 = Q_x^2 + Q_y^2 + Q_z^2 = 1,$$
$$P_x Q_x + P_y Q_y + P_z Q_z = 0;$$

c) everywhere in D except at points of curve L, one has

$$(P_x Q - Q_x P)^2 + (P_y Q - Q_y P)^2 + (P_z Q - Q_z P)^2 > 0;$$

d) through each point of domain D not lying on L there passes one and only one surface $M(\lambda)$: $\{P = \lambda Q\}$, where $M(\lambda_1) \cap D$ and $M(\lambda_2) \cap D$ are homeomorphic to one another for any λ_1 and λ_2.

Then we shall say that the function $f = P/Q$ defines an analytic decomposition of domain D with singular line L.

Suppose that in domain D, on which there is given an analytic decomposition or one with singularity, we have a harmonic function u. Points of domain D, which satisfy

$$(u_y f_z - f_y u_z)^2 + (u_x f_z - f_x u_z)^2 + (u_x f_y - u_y f_x)^2 = 0,$$

will be called points of contact of the function u. Points of contact of function u which lie on surface $M(\alpha)$: $\{f = \alpha\}$ are critical points of the trace u_α of function u on surface $M(\alpha)$. As α varies these points describe certain lines or surfaces. We call a line consisting of points of contact a line of contact. If for all α except possibly a finite number of values, the points of a line of contact are extremal points of function u_α, then we shall call such a line an extremal line of contact. In the remaining cases we speak of a saddle line of contact.

By the multiplicity of a saddle line of contact at the point $X \in M(\alpha)$ we shall mean the multiplicity of the saddle of the function u_α at point X. Point X is a saddle for u_α if there exists a neighborhood V_0 of point X such that after removing X from V_0 the level line of function u_α passing through point X in any neighborhood $V \subset V_0$ of point X splits into $\nu > 2$ connected components, and the number $(\nu - 2)/2$ is called the multiplicity of point X.

Theorem 3.2. The difference between the number of extremal curves of contact and the number of saddle curves of contact (counting multiplicity) of the harmonic function u, which intersect the domain $N(\alpha) = M(\alpha) \cap D$, provided

$$(u_y f_z - f_y u_z)^2 + (u_x f_z - f_x u_z)^2 + (u_x f_y - u_y f_x)^2 > 0 \tag{3.15}$$

on boundary Γ of domain D, is independent of α and is equal to the rotation of the field grad u_α on curve $S(\alpha) = M(\alpha) \cap \Gamma$.

The validity of the theorem follows directly from the properties of gradient vector fields of functions and the continuous dependence of these fields on the parameter α. This theorem serves as the starting point for the generalization of the argument principle.

We find out whether the set E of points of contact of a regular harmonic function has isolated points. We take as function f, effecting the decomposition, the function z, and we consider the harmonic function $u = AZ + \alpha zy + x^3 + 3xy^2 - 6xz^2$, $\alpha < 0$, whose trace u_0 on plane $z = 0$ is equal to $x^3 + 3xy^2$ and has critical point $x = y = 0$, and the trace on any plane $z = \varepsilon \neq 0$, where ε is sufficiently small, has no critical points. Since $u_\varepsilon = \alpha\varepsilon y + x^3 + 3xy^2 + A\varepsilon - 6\varepsilon^2 x$, and the gradient of this function vanishes at points satisfying the equations $\alpha\varepsilon + 6xy = 0$, $x^2 + y^2 - 2\varepsilon^2 x = 0$, setting $x = \varepsilon^2 + \varepsilon^2 \cos\varphi$, $y = \varepsilon^2 \sin\varphi$, for φ we get the equation $\alpha\varepsilon + 6\varepsilon^4 \sin\varphi(1 + \cos\varphi) = 0$, which has no solutions if $|\alpha| > 2|\varepsilon^3|$. Consequently, the origin is an isolated point of contact for this function, and for the function $v = u - Az$ the origin is an isolated critical point. A connected piece of the level surface Q: $\{v = 0\}$ lying on a sufficiently small neighborhood of the origin remains connected after removing the origin from Q, while this piece is homeomorphic to a disk with center deleted, which corresponds to the origin. Obviously, an isolated point of contact cannot be an extremal point of the corresponding function u_α, and as a saddle point it has multiplicity zero, i.e., the level line of the function u_α passing through this point is a Jordan curve. We shall call such points of contact inessential.

Theorem 3.3. Isolated points of the set E of points of contact of a harmonic function can only be inessential points of contact.

Proof. Let X be an isolated point of the set E of points of contact of the harmonic function u, lying on the level surface $M(\alpha_0)$ of function f, effecting an analytic decomposition of domain D in which function u is defined. We construct the ball $\Sigma(X, r)$ with center at point X of sufficiently small radius r that the set $R(\alpha_0, X) = M(\alpha_0) \cap \Sigma(x, r)$ is connected and contains no points of the set E other than X. Through each point of $R(\alpha_0, X)$ we draw the trajectory of the field grad$f$ and we consider the segments of these trajectories included between the surfaces $M(\alpha_0 - \delta)$ and $M(\alpha_0 + \delta)$, $\delta > 0$. The set of points of domain D lying on these segments of trajectories we denote by $W(X, \delta)$ and the boundary of the set $R(\alpha, X) = W(X, \delta) \cap M(\alpha)$ – $\sigma(\alpha, \delta)$. One can represent function u in $W(X, \delta)$ as follows: $u = u_0 + [f(x, y, z) - f(x_0, y_0, z_0)]v(x, y, z)$, where u_0 is the trace of u on $M(\alpha_0)$, v is an analytic function, and x_0, y_0, z_0 are the coordinates of point X. It follows from this representation of function u that the rotation of the field grad u_α on $\sigma(\alpha, \delta)$ is independent of α for sufficiently small δ. Hence the point X can be an isolated point of the set E only if the rotation of the field grad u_{α_0} on $\sigma(\alpha_0, \delta)$ is equal to zero. In this case X is only a saddle point of the function u_{α_0} of multiplicity zero, i.e., the level line of this function passing through point X is a Jordan curve. □

If the harmonic function u has an isolated point of contact, then taking this point as the origin and directing the Oz axis along the normal to the level surface M_0 of function f effecting the decomposition, passing through O we get that the point O must be an isolated critical point of function $v = u - Az$, where A is the value of the derivative u_z at point O. Since the level line of the trace u_0 of function u on M which passes through point O is a Jordan curve, one can find a direction l such that the derivative of u_α in this direction changes sign in a neighborhood of point O, and the derivative of u_α in the direction λ orthogonal to l vanishes only at point O and in a neighborhood of it does not change sign, while both these directions lie in the tangent plane to M_0 at point O. We take l as the direction of the Oy axis and λ as the direction of the Ox axis. Now the derivative u_x vanishes on M_0 only at point O, so at point O we have $u_{xx} = u_{xy} = 0$, and the Taylor expansion of u in a neighborhood of O assumes the form $u = C + Az + \alpha yz + \beta xz + w(x, y, z)$, where w is a harmonic function which vanishes at point O along with all its derivatives of the first two orders. Rotating the coordinate axes about the Oz axis, we transform function u to the form $u = C + Az + \gamma zy + w_1(x, y, z)$, where the new variables are again denoted by x and y. Now

$$u_y = \gamma z + \partial w_1/\partial y, \quad u_x = \partial w_1/\partial x,$$

where for $z = 0$ the derivatives of w_1 with respect to x and y vanish simultaneously only at point O, and for $z \neq 0$ do not vanish simultaneously anywhere in a neighborhood of point O. For the function $v = \gamma zy + w_1$ the point O is an isolated critical point, and the level surface $v = 0$ goes through point O tangent to the planes $z = 0$ and $y = 0$ [55]. It follows from more detailed consideration of the level surface $v = 0$ that after removing point O it remains connected. This shows that the example of an inessential point of contact considered above is typical.

Before studying nonisolated points of the set E of points of contact of the harmonic function u, we introduce some definitions. By $D^+(\alpha)$ we denote the part of domain D in which $f > \alpha$, and by $D^-(\alpha)$ the part of D in which $f < \alpha$. We shall call a curve K consisting of points of contact of function u and lying on surface $M(\alpha)$: $\{f = \alpha\}$ a horizontal curve of contact, and any other curves consisting of points of contact, nonhorizontal. We call a connected horizontal curve of contact K isolated on the surface $M(\alpha)$ if there exists an $\varepsilon > 0$ such that in the set $K(\varepsilon)$ of points of surface $M(\alpha)$, no farther than ε from K, there are no other points of contact than points of curve K.

Theorem 3.4. Let function f, which affects a decomposition, be harmonic in D. At a point of contact $X_0 \in M(\alpha_0)$ there occur the same number (counting multiplicities) of nonhorizontal curves of contact from domain $D^-(\alpha_0)$ and domain $D^+(\alpha_0)$, which are not tangent to $M(\alpha_0)$ at X_0. If at X_0 there enters an extremal curve of contact from $D^-(\alpha_0)$ which is not tangent to $M(\alpha_0)$, then from X_0 there leaves an extremal curve of contact into $D^+(\alpha_0)$ but generally with a different tangent line at point X_0.

Proof. We consider a nonhorizontal curve of contact L^-, coming into the points X_0 from $D^-(\alpha_0)$. Let $X_\alpha = L^- \cap M(\alpha)$, $\alpha_0 - \delta < \alpha < \alpha_0$, and $C(\alpha, \rho)$ be the curve of intersection of the surface $M(\alpha)$ with the sphere of radius ρ with center at point X_α. We transfer the origin to the point X_0 and make the xOy plane coincide with the tangent plane to $M(\alpha_0)$ at point X_0, and we direct the Oz axis along the normal to $M(\alpha_0)$. Without loss of generality, assuming that the harmonic function u vanishes at point X_0, it can be represented as follows [19]:

$$u(x, y, z) = az + \sum_{k=2}^{\infty} p_k(x, y, z), \tag{3.16}$$

where p_k is a homogeneous harmonic polynomial of degree k. Let p_ν be the first polynomial in (3.16) which is not identically zero. We choose ρ sufficiently small that the line $C(\alpha, \rho)$ bounds a connected set, and that in this set there are only points of contact lying on curves of contact, going into X_0 with tangent l, which coincides with the tangent to L^- at point X_0.

It follows from (3.16) that for sufficiently small δ in the neighborhood of X_α bounded by $C(\alpha, \rho)$ one can find exactly as many points of contact of the polynomial p_ν as the sum of the multiplicities of curves of contact of function u going into point X_0 with tangent l, if the gradient of p_ν is different from zero on $C(\alpha, \rho)$. It follows from the homogeneity of the polynomial p_ν that only one line of contact λ, which is tangent to l, and whose multiplicity is equal to the sum of the multiplicities of the curves of contact of function u, going into point X_0 with tangent l, goes through a neighborhood of point X_α. By virtue of the homogeneity of p_ν the line λ goes from $D^-(\alpha_0)$ to $D^+(\alpha_0)$, and it follows from this that at least one curve of contact L^+ of function u with tangent l goes from $D^+(\alpha_0)$ to the point X_0, while the sum of the multiplicities of the curves of contact of function u, going to X_0 from $D^-(\alpha_0)$ with tangent l, is equal to the sum of the multiplicities of the curves of contact going to X_0 from $D^+(\alpha_0)$ with the same tangent line l.

If the gradient of p_ν vanishes on $C(\alpha, \rho)$ for any arbitrarily small ρ, then through point X_α passes a one-dimensional line σ, on which the gradient of p_ν vanishes. Now through point X_0 passes a two-dimensional surface of contact Σ of polynomial p_ν. By virtue of the homogeneity of polynomial p_ν this surface extends from $D^-(\alpha_0)$ through X_0 into domain $D^+(\alpha_0)$, while the curve of contact L^- of function u lies on this surface Σ. Obviously the surface Σ consists of extremal lines of the traces of polynomial p_ν on the corresponding surfaces $M(\alpha)$, since the derivatives of polynomial p_ν, which vanish on Σ, change sign on this surface, so L^- is an extremal curve of contact for function u. To complete the proof of the theorem it now remains to prove the second assertion of the theorem.

Let L^- be an extremal curve of contact of function u, going to point x_0 with tangent l, which does not lie in the tangent plane to $M(\alpha_0)$ at point X_0. We denote by $S(\alpha)$ a sufficiently small neighborhood of the point $X_\alpha = L^- \cap M(\alpha)$ such that in it there are no points of contact of function u other than X_α, and this function everywhere in $S(\alpha)$ assumes values either strictly greater than $u(X_\alpha)$ or strictly smaller. When $|\alpha - \alpha_0|$ is sufficiently small, there passes through $S(\alpha)$ a line of contact λ of the polynomial p_ν, figuring in (3.16). This line also goes to the point X_0 with tangent l. If on λ we have $\operatorname{grad} p_\nu \not\equiv 0$, then going to the point X_0 with tangent l is a unique curve of contact L^- of function u from $D^-(\alpha)$, since in this case, in a neighborhood of point X_0 the level surfaces of functions u and p_ν, in a cone with axis l and vertex x_0 of sufficiently small opening, are constructed identically [47]; i.e., upon approaching point X_0 within the limits of this cone, the deviation of the level surface $p_\nu = c$ from the surface $u = c$ tends to zero faster than the first power of the distance from point x_0. In this case the values of the polynomial p_ν at all points of $S(\alpha)$ are also strictly greater or strictly smaller than its value at point $Y_\alpha = \lambda \cap M(\alpha)$ for α sufficiently close to α_0. By the homogeneity of p_ν its curve of contact λ extends through point X_0 to $D^+(\alpha_0)$ as an extremal curve of contact, so curve L^- also extends into $D^+(\alpha_0)$ as an extremal curve of contact L^+. However, it can happen that on L^- there lie maxima and on L^+ minima of the traces u_α of the function u on $M(\alpha)$. Now let $\operatorname{grad} p_\nu \equiv 0$ on λ, so λ coincides with l by the homogeneity of p_ν, and it follows from the maximum principle that on l there lie saddles of the traces p_ν on $M(\alpha)$. In this case the curve L^- cannot be an extremal curve of contact.

Lemma 3.1. If the extremal curve of contact L^- of function u goes to the point $X_0 \in M(\alpha_0)$ from $D^-(\alpha_0)$, tangent to $M(\alpha_0)$, then it either extends into $D^+(\alpha_0)$ as a saddle curve of contact L^+ or it extends into $D^-(\alpha_0)$ as an extremal curve of contact Q with the same tangent line l. as a saddle curve of contact L^+ or it extends into $D^-(\alpha_0)$ as an extremal curve of contact Q with the same tangent line l.

Proof. Let L^- be an extremal curve of contact of u, going to the point $X_0 \in M(\alpha_0)$ with tangent l. Let us assume that L^- extends smoothly through point X_0, and we prove that only the situations stipulated in the lemma are possible. We denote by ξ and η coordinates on the surface $M(\alpha_0)$. The curve L^- is defined by the equations $u_\xi = 0$, $u_\eta = 0$, and the vector product of the vectors $\operatorname{grad} u_\eta$ and $\operatorname{grad} u_\xi$ is directed along the tangent to L^-. On L^- one has $u_{\xi\eta}{}^2 - u_{\xi\xi}u_{\eta\eta} < 0$, since L^- is an extremal curve of contact. This inequality means that the vectors $(1, 0, 0)$ $(0, u_{\xi\xi}, u_{\xi\eta})$, and $(0, u_{\xi\eta}, u_{\eta\eta})$ form a right-handed orientation. The first of these vectors is directed along the normal to $M(\alpha)$, and the other two are the projections of $\operatorname{grad} u_\xi$

and grad u_η to the surface $M(\alpha)$. After extending the curve of contact L^- past the point X_0 these vectors must also form a right-handed orientation triple; if the extension of L^- lies in $D^-(\alpha_0)$, and if the given extension is found in $D^+(\alpha_0)$, then the vectors cited form a left-handed orientation triple. This means that only the situations stipulated in the assertion of the lemma are possible.

It remains to show that the curve of contact L^- extends past the point X_0 in general. For this we consider the expansion (3.16) of function u. The homogeneous polynomial p_ν also has either a curve of contact λ^- going to point X_0 with the same tangent l as L^-, or a two-dimensional surface of contact σ, passing through the point X_0. If p_ν has a curve of contact λ^-, going to the point X_0, then by virtue of the homogeneity of p_ν this line extends past X_0 by a line of contact λ^+. The level surfaces of the functions u_ξ and u_η , passing through point X_0, are tangent to the level surfaces of the corresponding derivatives of polynomial p_ν [47], so the line of intersection of the level surfaces of u_ξ and u_η extend past the point X_0, since the level surfaces of the derivatives of polynomial p_ν have the analogous property. If through point X_0 there passes a two-dimensional surface of contact p_ν of polynomial $p_{\nu\xi}$, then the level surfaces of the derivatives $p_{\nu\eta}$ and X_0 passing through point X_0 coincide, and the corresponding level surfaces of the derivatives u_ξ and u_η are tangent to one another at point X_0. On L^- one must have $u_{\xi\eta}{}^2 - u_{\eta\eta}u_{\xi\xi} <$ 0 everywhere in a neighborhood of point X_0 except for the point itself, because otherwise on L^- one would have

$$H\,(u) = \begin{vmatrix} u_\xi & u_\eta & u_\zeta \\ u_{\xi\xi} & u_{\xi\eta} & u_{\xi\zeta} \\ u_{\xi\eta} & u_{\eta\eta} & u_{\eta\zeta} \end{vmatrix} = 0,$$

where $\partial/\partial\zeta$ is the derivative with respect to the normal to $M(\alpha)$, from which it would follow that the derivative of u along L^- is equal to zero, i.e., u is constant on L^-. In this case it follows from the maximum principle that L^- cannot be an extremal curve of contact of function u. Since $u_{\xi\eta}{}^2 - u_{\xi\xi}u_{\eta\eta} < 0$ on L^-, the surface $u_\xi = 0$ divides both the domain $u_\eta > 0$ and the domain $u_\eta < 0$, from which it follows that curve L^- of intersection of the surfaces $u_\xi = 0$ and $u_\eta = 0$ cannot stop at point X_0. □

Theorem 3.5. If there goes to the point $X_0 \in M(\alpha_0)$ of the isolated curve of contact K of function u an extremal curve of contact L^- from $D^-(\alpha_0)$ and the function f effecting the decomposition is harmonic, then one can find at least one point $X_1 \in K$ such that an extremal curve of contact from $D^+(\alpha_0)$ goes to it.

Proof. On the surface $M(\alpha_0)$ we take the set of points $K(\varepsilon)$ at a distance less than ε from curve K. We choose $\varepsilon > 0$ sufficiently small that on the boundary $\gamma(\varepsilon)$ of set $K(\varepsilon)$ the gradient of trace u_0 of function u on $M(\alpha_0)$ does not vanish. Through each point of set $K(\varepsilon)$ we draw a trajectory of the field grad f and we extend these trajectories on both sides of $M(\alpha_0)$ up to their intersection with the surface $M(\alpha_0 + \delta)$ or $M(\alpha_0 - \delta)$. For sufficiently small $\delta > 0$ the gradient of trace u_α of function u on $M(\alpha)$, $\alpha_0 - \delta \le \alpha \le \alpha_0 + \delta$, does not vanish on all trajectories of the field grad f, passing through points of $\gamma(\varepsilon)$. We denote by $W(\varepsilon,\delta)$ the domain filled by points of the trajectories of the field grad f, passing through points of

$K(\varepsilon)$, and by $\Gamma(\varepsilon, \delta)$ the set of points of trajectories of this field, passing through points of $\gamma(\varepsilon)$. Obviously the rotation of the field grad u_α on the set $\gamma_\alpha(\varepsilon) = \Gamma(\varepsilon, \delta) \cap M(\alpha)$ is independent of α, so the algebraic sum of the lines of contact going to $W(\varepsilon, \delta)$ through $M(\alpha_0 - \delta)$ coincides with the sum of lines of contact going to $W(\varepsilon, \delta)$ through $M(\alpha_0 + \delta)$.

According to Lemma 3.1, the curve of contact L^- either becomes a saddle curve of contact L_0, passing from $D^-(\alpha_0)$ to $D^+(\alpha_0)$ at the point X_0, or from $D^-(\alpha_0)$ at point X_0 there goes another extremal curve of contact with the same tangent as L^-. Consequently, nonhorizontal curves of contact intersecting K can be reconstructed so that upon passing through $M(\alpha_0)$ their algebraic sum decreases by two, and this is only possible when upon passing from $D^+(\alpha_0)$ to $D^-(\alpha_0)$ the analogous reconstruction also occurs. From this we get that on K there exists a point Y_0 from which at least one extremal curve of contact of the harmonic function u goes into $D^+(\alpha_0)$. □

It follows from Theorem 3.5 that extremal curves of contact of a harmonic function in its domain of regularity cannot be interrupted but must necessarily continue and go to the boundary of the domain of regularity.

Lemma 3.2. If the curve of contact L^- from $D^-(\alpha_0)$, on which there lie maxima of the corresponding functions u_α, goes to the point $X_0 \in M(\alpha_0)$, and the curve of contact L^+, on which lie minima of the corresponding functions u_α, goes from $D^+(\alpha_0)$ to X_0, where L^- and L^+ form a smooth curve L, then at point X_0 all second derivatives of function u vanish.

Proof. We shall assume that the function f effecting a decomposition of domain D is harmonic. We place the origin at point X_0 and as variable ζ we take f, and let the variables ξ and η vary on the surface $M(\alpha_0)$. Since f is harmonic, and $u_\xi = 0$ and $u_\eta = 0$ at points of the curve L, we find from Laplace's equation at points of this curve

$$u_{\zeta\zeta} = -\alpha_1 u_{\xi\xi} - 2\beta u_{\xi\eta} - \alpha_2 u_{\eta\eta}, \tag{3.17}$$

where α_1, β, and α_2 are such that the operator $T(u)$, figuring on the right side of this equation, is elliptic, while $\alpha_1 > 0$ and $\alpha_2 > 0$. We denote by λ, μ, ν the direction cosines of tangent l to curve L. One has

$$\begin{gathered}\lambda u_{\xi\xi} + \mu u_{\xi\eta} + \nu u_{\xi\zeta} = 0,\ \lambda u_{\xi\eta} + \mu u_{\eta\eta} + \nu u_{\eta\zeta} = 0,\\ \lambda u_{\xi\zeta} + \mu u_{\eta\zeta} + \nu u_{\zeta\zeta} = \frac{\partial}{\partial l}(u_\zeta)\end{gathered} \tag{3.18}$$

on curve L. From (3.17) and (3.18) we find

$$-\lambda^2 u_{\xi\xi} - 2\mu\lambda u_{\xi\eta} - \mu^2 u_{\eta\eta} - \nu^2 T(u) = \nu \frac{\partial}{\partial l}(u_\zeta). \tag{3.19}$$

Since maxima of the functions u_α lie on L^- and minima on L^+, at point X_0 any positive or negative definite form whose coefficients are second derivatives of the functions u_α changes sign, i.e., the left side of (3.19) changes sign at point X_0, while at this point $u_{\xi\xi} = u_{\xi\eta} = u_{\eta\eta} = 0$. By (3.17), from this we get $u_{\zeta\zeta} = 0$ at point X_0. If at point X_0 we have $\nu \neq 0$, then it follows from (3.18) that $u_{\xi\xi} = u_{\eta\zeta} = 0$.

However, with the help of (3.17) and (3.18) it is easy to show that these relations also hold at point X_0 under the condition $\nu = 0$, but ν does not change sign at this point. □

It follows from Lemma 3.2 that at those points of the smooth extremal curve of contact L of function u at which the maxima of the functions u_α are replaced by the minima of these functions, the Hessian

$$\det H(u), \quad H(u) = \begin{Vmatrix} u_{\xi\xi} & u_{\xi\eta} & u_{\xi\zeta} \\ u_{\xi\eta} & u_{\eta\eta} & u_{\eta\zeta} \\ u_{\xi\zeta} & u_{\eta\zeta} & u_{\zeta\zeta} \end{Vmatrix}$$

of the harmonic function u vanishes. We show that also at those points through which more than one curve of contact with different tangents pass, the Hessian of the harmonic function also vanishes. We consider the first two equations of (3.18)

$$\lambda u_{\xi\xi} + \mu u_{\xi\eta} + \nu u_{\xi\zeta} = 0, \ \lambda u_{\xi\eta} + \mu u_{\eta\eta} + \nu u_{\eta\zeta} = 0. \tag{3.20}$$

Since through the given point X_0 there pass at least two curves of contact of function u with different tangents, the direction cosines of these tangents give two linearly independent solutions of the system of two linear equations in three unknowns λ, μ, ν (3.20), which is only possible when the vectors $(u_{\xi\xi}, u_{\zeta\eta}, u_{\xi\zeta})$ and $(u_{\xi\eta}, u_{\eta\eta}, u_{\eta\zeta})$ are linearly independent [20]. But this means that the Hessian of function u, expressed in the coordinates ξ, η, ζ, vanishes at point X_0.

Theorem 3.6. If the harmonic function u, which is regular in domain D, has no points of contact on boundary Γ of domain D and throughout D one has

$$[\det H(u)]^2 + u_\xi^2 + u_\eta^2 > 0, \tag{3.21}$$

then function u has no extremal curves of contact in domain D.

Proof. Let function u have an extremal curve of contact L. Since there are no points of contact on Γ, curve L is closed and lies strictly inside D. By (3.21) L does not intersect other manifolds of contact and all points of L are either minimal points or maximal points of the corresponding traces u_α of function u on the surfaces $M(\alpha)$. The trace of function u on L achieves both a minimum and a maximum. Obviously one of these extremal points of the trace of u on L will be an extremal point of function u, and this contradicts the maximum principle for harmonic functions. Consequently, function u cannot have extremal curves of contact in D. □

The assertion of Theorem 3.6 remains valid without the supplementary inequality (3.21) also. However, the proof is essentially more complicated due to the fact that lines of contact can now intersect and one cannot use the maximum principles so simply. Leaving the general case aside, we consider an important particular example. Let Ω be an axially symmetric domain. We make the axis of symmetry of domain Ω coincide with the Oz axis and as functions effecting a decomposition of domain Ω with singular line Oz, we take the functions x and y. Let u be a harmonic function which is regular in domain Ω; its points of contact are given by the equations $u_z = xu_x + yu_y = 0$. We prove the assertion of Theorem 3.6 without (3.21) holding.

Theorem 3.7. Let u be a harmonic function which is regular in the axially symmetric domain Ω, and which satisfies

$$(xu_x + yu_y)^2 + u_z^2 > 0 \tag{3.22}$$

on boundary Γ of domain Ω. In domain Ω the function u cannot have either two-dimensional surfaces of contact bounding a subdomain of domain Ω, or extremal curves of contact.

Proof. If Ω contained two-dimensional surfaces of contact of function u, bounding a subdomain Ω_1, then throughout Ω_1 one would have $u_z = xu_x + yu_y + zu_z = 0$, by the maximum principle for harmonic functions. It would follow from the analyticity of u that these equations hold everywhere in Ω, which contradicts (3.22).

In domain Ω we introduce cylindrical coordinates ρ, z, φ. If function u satisfies (3.21) $[\det H(u)]^2 + u_z{}^2 + u_\rho{}^2 > 0$, then u has no extremal curves of contact by Theorem (3.6). Now if this condition does not hold, then different lines of contact may intersect. At such points of intersection all second derivatives of function u vanish, i.e., these points are critical points of all three functions u_z, u_ρ, u_φ. For the function u_z there always exists a linear function $ax + by + cz$, $a^2 + b^2 + c^2 < \varepsilon$, where ε is an arbitrarily small, previously given fixed positive number such that the function $w_0 = u_z + ax + by + cz$ has only nondegenerate critical points. Fixing a sufficiently small $\varepsilon > 0$, instead of function u we consider function $w = u + axz + byz + (c/2)(z^2 - x^2)$. Suppose that at some point (x_0, y_0, z_0) several different curves of contact of function w intersect. At this point all second derivatives of function w vanish. We transfer the origin to this point and make the Oz axis coincide with the ξ axis; instead of ρ we write η, and we direct the ζ axis along the tangent to the circle $z = z_0$, $\rho^2 = x_0{}^2 + y_0{}^2$. In such coordinates we can represent w by a series

$$w = w_0 + \alpha\zeta + \sum_{k=3}^{\infty} p_k(\xi, \eta, \zeta),$$

where p_k is a harmonic polynomial of degree k, where

$$\frac{\partial p_3}{\partial \xi} = [(\alpha_1\xi + \beta_1\eta + \gamma_1\zeta)^2 + (\alpha_2\xi + \beta_2\eta + \gamma_2\zeta)^2 - 2(\alpha_3\xi + \beta_3\eta + \gamma_3\zeta)^2]\, c,$$

$$\alpha_1^2 + \beta_1^2 + \gamma_1^2 = \alpha_2^2 + \beta_2^2 + \gamma_2^2 = \alpha_3^2 + \beta_3^2 + \gamma_3^2 = 1,$$

$$\alpha_i\alpha_j + \beta_i\beta_j + \gamma_i\gamma_j = 0,\ i,\ j = 1,\ 2,\ 3,\ i \neq j,$$

and c is a constant. Due to this structure of the polynomial p_3 the behavior of the curves of contact of function w in a neighborhood of the point (x_0, y_0, z_0) is uniquely determined by the behavior of the lines of contact of polynomial p_3. Studying the behavior of the lines of contact of function p_3, we find that the extremal line of contact of function w on which the maxima of the traces w_φ of function w on the planes $\varphi = \text{const}$ lie, extends to an extremal line of contact of the same type beyond the point (x_0, y_0, z_0) also. Now one proves the absence of extremal curves of contact for function w exactly as in the proof of Theorem 3.6.

Let us assume that function u has an extremal curve of contact L. If on some piece of this curve $u_{\rho z}{}^2 - u_{\rho\rho}u_{zz} < 0$, then for sufficiently small $\varepsilon > 0$ the function w must also have an extremal curve of contact. Since w has no extremal points of contact, on L one must have $u_{\rho\rho} = u_{zz} = u_{\rho z} = 0$. In cylindrical coordinates (3.19) looks like this:

$$-\gamma^2\rho^2 [u_{\rho\rho} + u_{zz}] - \lambda^2 u_{\rho\rho} - 2\lambda\mu u_{\rho z} - \mu^2 u_{zz} = \nu \frac{\partial}{\partial l} (u_{\varphi}).$$

It follows from this that u_φ is constant on L. Let $u_\varphi = \gamma$ on L so the function $u - \gamma\varphi$ is constant on L, and by the maximum principle L is a saddle curve of contact for this function. Since φ is constant on a meridional plane, line L is also a saddle line of contact for u. Thus, the function u has no extremal curves of contact. □

Corollary. Let the harmonic function u, which is regular in the axially symmetric domain Ω, satisfy (3.22), v_φ being the trace of function u on plane $T(\varphi)$: $\{ax + by = 0\}$, $\varphi = \arg(a + ib)$. Then the rotation of the field grad v_φ on the line $\Gamma(\varphi)$ of intersection of the plane $T(\varphi)$ with boundary Γ of domain Ω is nonpositive for all φ.

Theorem 3.7 and its corollary can serve as a vector analog of the argument principle for regular harmonic functions in three-dimensional axially symmetric domains. With the help of these assertions one can get a test for the overdeterminedness of the oblique derivative problem for harmonic functions. We consider the oblique derivative problem for harmonic functions, regular in the ball Σ: $\{x^2 + y^2 + z^2 < R^2\}$ with boundary condition

$$\alpha(z)(xu_x + yu_y) + \beta(z)u_z = f, \tag{3.23}$$

defined on the sphere S: $\{x^2 + y^2 + z^2 = R^2\}$.

Theorem 3.8. Let the vector field $P = (x\alpha, y\alpha, \beta)$ on sphere S satisfy the condition $\alpha^2(x^2 + y^2) + \beta^2 > 0$, and let its rotation on a meridian of sphere S be positive. Then (3.23) has no solutions which are regular in Σ for any positive function f.

Proof. By the hypotheses of the theorem the vector field P is not perpendicular to the meridional plane at any point of sphere S. For $f > 0$, (3.23) says that the vectors of the field grad u are also not perpendicular to the meridional plane at any point of S and the projection of the vector field grad u to the meridional plane is homotopic to P. Consequently, the rotations of these fields on meridians of the sphere S are identical. Obviously the projection of grad u to a meridional plane coincides with grad v, where v is the trace of u on this meridional plane. Now we get that on a meridional section of the ball Σ the function v has at least one extremum, i.e., u has an extremal point of contact. By Theorem 3.7 this is impossible, so (3.23) has no solutions which are regular in ball Σ. □

Theorem 3.8 shows that the character of the solvability of the oblique derivative problem for harmonic functions which are regular in ball Σ with the boundary condition

$$au_x + bu_y + cu_z = f, \tag{3.24}$$

given on unit sphere S, depends on the extendability properties of the vector field $P = (a, b, c)$ from sphere S to the whole ball Σ. With boundary condition (3.24) one can also associate the first-order partial differential equation

$$Av_x + Bv_y + Cv_z = F, \tag{3.25}$$

the character of whose solvability also depends on the extension properties of vector field P [33].

Let the harmonic function u which is regular in unit ball Σ satisfy, on sphere S, the inequality $f = au_x + bu_y + cu_z > 0$. We set $\omega = u_x{}^2 + u_y{}^2 + u_z{}^2$ and we show that the vector field $P = (a, b, c)$ can be extended to ball Σ so that $F = Au_x + By_y + Cu_z > 0$ in Σ, where equality to zero is only achieved at those points at which $\omega = 0$, and the extended field $P_1 = (A, B, C)$ is continuous and nonzero everywhere in Σ where grad $u \neq 0$, and at those points where $\omega = 0$ the extended field P_1 also vanishes. We construct harmonic functions g, α, b, and γ which are regular in ball Σ, and satisfy on sphere S the following conditions:

$$g|_S = f\omega^{-1},\quad \alpha|_S = a\omega^{-1},\quad \beta|_S = b\omega^{-1},\quad \gamma|_S = c\omega^{-1}.$$

Obviously $g > 0$ everywhere in ball Σ. We extend the vector field P to ball Σ as follows:

$$\begin{aligned} A &= u_x(g - \beta u_y - \gamma u_z) + \alpha(u_y^2 + u_z^2),\\ B &= u_y(g - \alpha u_x - \gamma u_z) + \beta(u_x^2 + u_z^2),\\ C &= u_z(g - \alpha u_x - \beta u_y) + \gamma(u_x^2 + u_y^2). \end{aligned} \tag{3.26}$$

For such an extension of the field P we have

$$\begin{aligned} F &= Au_x + Bu_y + Cu_z = g(u_x^2 + u_y^2 + u_z^2),\\ A^2 + B^2 + C^2 &= (u_x^2 + u_y^2 + u_z^2)[g^2 + (\alpha^2 + \beta^2\\ &+ \gamma^2)(u_x^2 + u_y^2 + u_z^2) - (\alpha u_x + \beta u_y + \gamma u_z)^2]. \end{aligned}$$

Obviously $F \geq 0$, where the equality is achieved only at those points of the ball Σ at which $\omega = 0$. The inequality $F > 0$ means that the vector fields grad u and $P_1 = (A, B, C)$ are homotopic at all those points of ball Σ at which grad $u \neq 0$. Thus we have

Lemma 3.3. If the oblique derivative problem for harmonic functions which are regular in the unit ball, with boundary condition (3.24), also has a regular solution for a positive function f, then the vector field $P = (a, b, c)$ can be extended continuously from sphere S to ball Σ, so that the extended field $P_1 = (A, B, C)$ vanishes only at zeros of the gradient of u and away from these zeros is homotopic to the vector field grad u.

By the maximum principle the harmonic function u has *only critical points of saddle type, i.e., the trajectories of the field* grad u, passing through critical points of function u, fill a manifold of dimension no higher than 2. By Lemma 3.3 the extended field P_1 has the analogous property. In particular, if for some $f > 0$ (3.24) has a solution whose gradient does not vanish in ball Σ, then one can extend the vector field P to the ball so that the extended field P_1 vanishes at no point. If (3.24) has a solution u which is regular in Σ for positive f, then since [58] for any arbitrarily small $\varepsilon > 0$ one can give a linear function $h = \alpha x + \beta y + \gamma z$, $\alpha^2 + \beta^2 + \gamma^2 < \varepsilon^2$ such that $u + h$ has only nondegenerate critical points in Σ, taking a sufficiently small $\varepsilon > 0$, we get that for a positive function f which admits only nondegenerate critical points in u_1, (3.24) has a solution Σ. Consequently, in this case the vector field P can be extended to ball Σ so that the extended field will be homotopic to the field of gradients of a harmonic function which admits only nondegenerate critical points.

Theorem 3.9. If for any continuous extension of the vector field $P = (a, b, c)$ from sphere S to ball Σ, the extended field $P_1 = (A, B, C)$ has at least one closed trajectory L, at all of whose points, with the possible exception of one, $A^2 + B^2 + C^2 \neq 0$, or has at least one singular point of the type of a node or focus, then the oblique derivative problem with boundary condition (3.24) for positive f has no solutions which are regular in Σ.

Proof. Let us assume that (3.24) is solvable and we extend the vector field P by (3.26). As a result of this extension we get (3.25), in which $F \geq 0$. We consider the case of a closed trajectory. We denote by w the trace of the solution u of (3.24) on L. Since $P_1 \neq 0$ everywhere on L, except possibly for one point Q, then everywhere except for this point the tangent to L exists. The function w, which is continuous on the closed curve L, achieves both its minimum and its maximum, so its derivative with respect to the tangent to L changes sign at least at one point M, while $P_1 \neq 0$ at point M. This means that at point M the function $F = Au_x + Bu_y + Cu_z$ changes sign, which contradicts Lemma 3.3. For a singular point Q of the vector field P_1 of the type of a node or a focus, one can find two neighborhoods $U_1(Q) \subset U(Q)$ such that any trajectory of the field P_1, which goes into the neighborhood $U(Q)$, upon continuation to one side or the other, necessarily goes into the neighborhood $U_1(Q)$. Function u achieves a maximum at some point M of the boundary of $U(Q)$, and a minimum at some point N. If the trajectories of the field P_1 upon extension on the positive side go from $U(Q)$ to $U_1(Q)$, then function u must increase along the trajectories passing through point M, since the derivative of function u with respect to the tangent to this trajectory is positive. Consequently, the maximum of u in $U_1(Q)$ is larger than the maximum of u on the boundary of $U(Q)$, which is impossible by the maximum principle. Now if all trajectories of field P_1 extend from the boundary of $U(Q)$ into $U_1(Q)$ on the negative side, then we arrive at a contradiction analogously, considering a trajectory of field P_1 which passes through point N. □

Theorem 3.10. If the oblique derivative problem for harmonic functions u which are regular in the unit ball Σ with boundary condition (3.24) has a solution for a positive function $f \geq \delta > 0$, then vector field $P = (a, b, c)$ can be extended to ball Σ so that the extended field $P_1 = (A, B, C)$ has only isolated nondegenerate singular points of saddle type.

Proof. Suppose a given function $f \geq \delta > 0$ corresponds to a solution u of (3.24). As is known [58], for any arbitrarily small fixed $\varepsilon > 0$ there exist constants α, β, γ such that $\alpha^2 + \beta^2 + \gamma^2 < \varepsilon^2$, and the function $v = u + \alpha x + \beta y + \gamma z$ has only nondegenerate critical points in ball Σ. Obviously for sufficiently small $\varepsilon > 0$ we have $av_x + bv_y + cv_z = f + a\alpha + b\beta + c\gamma > 0$ on sphere S. Starting from function v, we effect an extension P_1 of vector field P by (3.26). By Lemma 3.3 the extension of vector field P constructed has the properties needed. □

Obviously Theorems 3.9 and 3.10 are valid not only for the ball, but also for any domain D with smooth boundary Γ. It follows from Theorem 3.10 that if the oblique derivative problem with boundary condition (3.24) is solvable for some positive function f, then vector field P can be extended to the ball so that the first-order equation

$$Au_x + Bu_y + Cu_z = F, \tag{3.27}$$

where $P_1 = (A, B, C)$ is the extension of field P constructed by Theorem 3.10, will have a solution which is holomorphic in Σ for any holomorphic function F, which vanishes at a finite number of points of ball Σ (at the singular points of field P_1) [33]. In this class of functions occur all functions which assume any preassigned values on S, and their derivatives with respect to the normals to S, independently of the values of the function itself on S, also assume arbitrary preassigned values on S. This means that in (3.27) one can always choose a function F which assumes preassigned values f on S, such that the equation has a solution w, whose normal derivative on S coincides with the normal derivative of the harmonic function u in ball Σ, which assumes the same values as w on S. Obviously u will be a solution of the oblique derivative problem with boundary condition (3.24). Thus we have

Theorem 3.11. If the oblique derivative problem for harmonic functions which are regular in the ball Σ with boundary condition (3.24) is solvable for at least one positive function, then this problem is unconditionally solvable.

Corollary 1. If the oblique derivative problem with the boundary condition $au_x + bu_y + cu_z = f$ is unconditionally solvable, then the oblique derivative problem with the boundary condition $a_1u_x + b_1u_y + c_1u_z = f$, where $\max |a - a_1| < \varepsilon$, $\max |b - b_1| < \varepsilon$, $\max |c - c_1| < \varepsilon$, for sufficiently small ε is also unconditionally solvable.

For the proof it suffices to note that the difference $au_x + bu_y + cu_x - (a_1u_x + b_1u_y + c_1u_z)$ for a fixed function u is arbitrarily small along with ε, so it follows from the inequality $au_x + bu_y + cu_z \geq \delta > 0$ for sufficiently small ε that $a_1u_x + b_1u_y + c_1u_z \geq \delta/2 > 0$.

Corollary 2. Let v be a harmonic function, regular in the closed ball $\Sigma \cup S$, such that $\operatorname{grad} v \neq 0$ on sphere S. The oblique derivative problem with boundary condition $v_xu_x + v_yu_y + v_zu_z = f$, given on S, is unconditionally solvable.

Corollary 3. If the oblique derivative problem with axially symmetric boundary condition $a(z)(xu_x + yu_y) + b(z)u_z = f$, given on S, is unconditionally solvable, then the problem with boundary condition $a(z)(xu_x + yu_y) + b(z)u_z + c(x, y, z)(yu_x - xu_y) = f$ is unconditionally solvable for any continuously differentiable function $c(x, y, z)$.

Let problem (3.24) have a solution v with nondegenerate critical points for some function f_1. With the help of this solution v, using (3.26) we construct an extension $P_1 = (A, B, C)$ of the vector field $P = (a, b, c)$ to ball Σ. The homogeneous equation

$$A\omega_x + B\omega_y + C\omega_z = 0, \tag{3.28}$$

corresponding to (3.25) has two independent integrals which are holomorphic in ball Σ, say ξ and η [33]. We introduce new independent variables ξ, η, and $\zeta = v$. In these variables (3.25) assumes the form

$$g_1(v_x^2 + v_y^2 + v_z^2)u_\zeta = F(\xi, \eta, \zeta), \tag{3.29}$$

where g_1 is a harmonic function which is regular in ball Σ and coincides with f_1 on S. Expressing u in terms of F by (3.29) and substituting the result in Laplace's equation, we get an integrodifferential equation for function F, the investigation of the Dirichlet problem for which is equivalent to the investigation of the oblique derivative problem (3.24) [45].

3. MEASURE OF OVERDETERMINEDNESS OF THE OBLIQUE DERIVATIVE PROBLEM

The measure of overdeterminedness of the oblique derivative problem is characterized by how large the set of solutions of the corresponding homogeneous problem is. We consider the homogeneous oblique derivative problem for harmonic functions which are regular in the unit ball Σ

$$au_x + bu_y + cu_z = 0 \tag{3.30}$$

with coefficients a, b, and c, which are analytic on sphere S. Let us assume that the oblique derivative problem with boundary condition (3.30) has two solutions ξ and η, which are analytic in $\Sigma \cup S$, such that their gradients are not proportional to one another everywhere on S, except possibly for the set of points of sphere S, at which the vectors of the field $P = (a, b, c)$ are directed along the normal to S. It follows from the boundary condition (3.30) that there exists a function $\lambda \neq 0$ on S such that $\lambda a = \xi_z\eta_y - \xi_y\eta_z$, $\lambda b = \xi_x\eta_z - \xi_z\eta_x$, $\lambda c = \xi_y\eta_x - \xi_x\eta_y$. Hence, without loss of generality one can set $a = \xi_z\eta_y - \xi_y\eta_z$, $b = \xi_x\eta_z - \xi_z\eta_x$, $c = \xi_y\eta_x - \xi_x\eta_y$ everywhere in the closed ball $\Sigma \cup S$. For any solution u which is analytic in $\Sigma \cup S$ problem (3.30) on S we have

$$J(u, \xi, \eta) \equiv \begin{vmatrix} u_x & u_y & u_z \\ \xi_x & \xi_y & \xi_z \\ \eta_x & \eta_y & \eta_z \end{vmatrix} = 0. \tag{3.31}$$

We introduce new independent variables ξ, η, $\zeta = ax + by + cz$ in $\Sigma \cup S$ and in (3.31) we set $u = v(\xi, \eta) + (1 - x^2 - y^2 - z^2)w_0(\xi, \eta) + (1 - x^2 - y^2 - z^2)^2 w(x, y,$

z). Substituting this expression into (3.31), we get $w_0(\xi, \eta) \equiv 0$. Consequently, for any solution u of (3.31) one has the representation

$$u(x, y, z) = v(\xi, \eta) + (1 - x^2 - y^2 - z^2)^2 w(x, y, z), \tag{3.32}$$

i.e., for any solution u of the oblique derivative problem with boundary condition (3.31), there exists a function $v(\xi, \eta)$ of two variables such that u itself and its normal derivative on S coincide with the analogous quantities for the function $v(\xi, \eta)$.

Since function u in (3.32) is harmonic in the ball Σ, substituting (3.32) into Laplace's equation, for the functions v and w we get the relation

$$\begin{aligned}(1 - x^2 - y^2 - z^2)^2 \Delta w - 8(1 - x^2 - y^2 - z^2)(xw_x + yw_y + zw_z) \\ + 20(x^2 + y^2 + z^2 - 0,6)w + (\xi_x^2 + \xi_y^2 + \xi_z^2)v_{\xi\xi} \\ + 2(\xi_x\eta_x + \xi_y\eta_y + \xi_z\eta_z)v_{\xi\eta} + (\eta_x^2 + \eta_y^2 + \eta_z^2)v_{\eta\eta} = 0.\end{aligned} \tag{3.33}$$

The arbitrariness in the solution (3.32) of the oblique derivative problem with boundary condition (3.30) is characterized by function v, since for known v the function w is uniquely determined by (3.33). For $v = 0$, (3.33) assumes the form $\Delta[(1 - x^2 - y^2 - z^2)w] = 0$. If this equation had a solution which was bounded on S, which was different from zero and regular in ball Σ, then there would exist a harmonic function which is regular in the ball, which on sphere S vanishes along with its normal derivative, which is only possible if this function is identically equal to zero [19]. In (3.33) we introduce new independent variables $s = \xi$, $t = \eta$, and ζ, and then we set $\zeta = 0$. Now from (3.33) for the function $v|_{\zeta=0} = h(s, t)$ we get the equation

$$A_0(s, t)h_{ss} + 2B_0(s, t)h_{st} + C(s, t)h_{tt} = \omega(s, t), \tag{3.34}$$

where we have introduced the notation

$$A_0 = (\xi_x^2 + \xi_y^2 + \xi_z^2)|_{\zeta=0}, \quad B_0 = (\xi_x\eta_x + \xi_y\eta_y + \xi_z\eta_z)|_{\zeta=0},$$
$$C_0 = (\eta_x^2 + \eta_y^2 + \eta_z^2)|_{\zeta=0}, \quad \omega = \{\Delta[(1 - x^2 - y^2 - z^2)^2 w]\}|_{\zeta=0}.$$

The function ω is uniquely determined by the function w and Eq. (3.34) itself can be considered on the surface M: $\{\zeta = 0\}$.

In Section 1 of this chapter it was shown that a solution u of the homogeneous oblique derivative problem with boundary condition (3.30) can achieve an extremum only at points of the set N, on which the vector field $P = (a, b, c)$ goes into the tangent plane to sphere S, i.e., on the intersection of sphere S with surface M. Consequently, any solution of (3.30) is uniquely determined by its values on N. On the other hand, function v in (3.32) coincides with u everywhere on S and is uniquely determined by u, so it is sufficient to give the values of u on the subset of set N, on which giving the values of function v determines it uniquely. If for a given function ω we find a solution of (3.34), then from this solution h the function v arises uniquely, and if for this function v (3.33) happens to be solvable, then in this way we will construct a solution (3.32) of (3.30). Since ω and w can be expressed uniquely in terms of one another, the arbitrariness in the general solution of (3.30) is no larger than the arbitrariness in the general solution of the equation

$$A_0(s, t)h_{ss} + 2B_0(s, t)h_{st} + C_0(s, t)h_{tt} = 0, \qquad (3.35)$$

given on the surface M: $\{\zeta = 0\}$. This equation is elliptic at all points of the surface M, at which one has

$$(\xi_x\eta_y + \xi_y\eta_x)^2 + (\xi_x\eta_z - \xi_z\eta_x)^2 + (\xi_y\eta_z - \xi_z\eta_y)^2 > 0,$$

and is degenerate at the remaining points. Obviously the trajectories of field P extended to the ball have the form l: $\{\xi = \text{const}, \eta = \text{const}\}$, and at singular points of this field (3.35) degenerates while all singular points of field P belong to set M. To a solution u in (3.32) of the oblique derivative problem with boundary condition (3.30), which is regular in ball Σ, there correspond only single-valued functions v, i.e., to such u there correspond only those solutions of (3.35) which can be continued along trajectories of the field P to functions which are single-valued and analytic in Σ.

Generally it is not the case that any solution of (3.35) can be continued along trajectories of field P to a function $v(\xi, \eta)$, which is single-valued and analytic in ball Σ. In order to get an equation any solution of which could be continued to a function which is single-valued and analytic in Σ, we make additional constructions. The vector field P can be extended by the formulas $a = \xi_z\eta_y - \xi_y\eta_z$, $b = \xi_x\eta_z - \xi_z\eta_x$, $c = \xi_y\eta_x - \xi_x\eta_y$ not only to ball Σ, but by virtue of the analyticity of ξ and η in $\Sigma \cup S$, this vector field extends analytically to a larger domain $T \supset \Sigma$. We denote by D the largest domain containing Σ to which ξ and η can be extended analytically. The vector field P can also be extended analytically to this same domain, as well as the surface M_1: $\{(\zeta = 0)\}$, $\zeta = ax + by + cz$. All points of the set M_1 which one can join to one another by polygonal arcs composed of pieces of trajectories of the field P, not leaving D, we identify. As a result of this identification we get a set M_0 and a map τ: $M_1 \to M_0$. The points of M_0 are the collections of all points of M_1 which can be joined by polygonal arcs composed of pieces of trajectories of field P to a given point X of set M_1. The collection of such points containing X will be denoted by $\tilde{X}$. The map τ assigns to a point X the collection $\tilde{X}$. We denote by R_1 the subset of M_1 which is the image of the subset M of set M_1 under map τ, i.e., in R_1 are all those collections of points from M_1 which contain at least one point of set M. Obviously the coefficients of (3.35) assume identical values at all points of set M_1 corresponding to one point of M_0, so (3.35) gives rise to a second-order equation

$$A_1(s, t)g_{ss} + 2B_1(s, t)g_{st} + C_1(s, t)g_{tt} = 0 \qquad (3.36)$$

on set R_1. We denote by Γ_1 the set of those points of M_0 whose preimages contain at least one point of the set $M_1 \cap S$ and do not contain any interior point of ball Σ. Obviously Γ_1 is the boundary of set R_1.

Any solution g of (3.36) which is single-valued on R_1 extends to a single-valued function along the trajectories of field P, but not all single-valued functions $v(\xi, \eta)$ in ball Σ have single-valued traces on R_1. In order to construct a covering set R of set R_1 such that any solution of (3.36) which is single-valued on R extends along trajectories of field P to a function which is single-valued in Σ, and to each function $v(\xi, \eta)$ which is single valued in Σ there corresponds a function which is single-valued on R, we make a number of additional constructions. In set M_1 we identify all points which one can join to one another by polygonal arcs composed of pieces of trajectories of field P, not leaving ball Σ. After this identification we get a set M_2 and a map σ: $M_1 \to M_2$. Obviously the set M_2 covers set R, which is obtained from M_2 by identifying certain points. We denote this identification by ρ: $M_2 \to R$. Thus we have constructed two sets M_2 and R_1 and a map ρ, which identifies collections of points of set M_1 which make up a point of M_2, which one can join by polygonal arcs composed of pieces of trajectories of field P in the larger domain D. The sets M_1 and R_1 constructed have the following properties: any single-valued solution of (3.36) extends from R_1 along trajectories of field P to a function $v(\xi, \eta)$ which is single-valued in Σ, but it is not so that any function $v(\xi, \eta)$ which is single-valued in Σ has a single-valued trace on R_1; the trace on M_2 of any function $v(\xi, \eta)$ which is single-valued in Σ is single-valued on M_2, but it is not so that any function which is single-valued on M_2 extends to a function $v(\xi, \eta)$ which is single-valued in Σ. Consequently, the trace on M_2 of a function $v(\xi, \eta)$ which is single-valued in Σ can assume identical values at certain different points with identical projections in R_1. As a result of the identification of all such points one gets the set Q we need. These identifications are performed exactly as in the construction of the Riemann surface of a multivalued function.

Due to the analyticity of ξ and η each trajectory of the field P intersects M_1 in a finite number of points or is completely contained in M_1 [47]. Let λ be the set of points of M_1 lying on those trajectories of field P which are completely contained in M_1. These points are singular points of maps τ and σ, while they coincide in M_2 and in R_1. If one discards the set λ from M_1, then on the remaining part of M_1 the maps τ and σ are locally homeomorphic. Consequently, map ρ is also locally homeomorphic on the image of this part of M_1 in M_2. Since in the exterior of the set λ the maps τ and σ carry no more than a finite number of points M_1 into one point outside the image of λ, the map ρ carries no more than a finite number of different points into one point. It follows from this that set M_2 can be covered by a finite number of sets T_j, $j = 1, \ldots, N$ such that map ρ is one-to-one on each T_j. We consider two sets T_k and T_l with nonempty intersection G_{kl}. We take a point $A \in G_{kl}$ and points $X \in T_k$, $Y \in T_l$ which are different but with the same projections in R_1. We join point X in T_k with A by path l_1, and point Y in T_l with A by path l_2. Map ρ carries sets T_k and T_l into a subset Ω_{kl} of set R_1, where the images of the paths l_1 and l_2 form a closed path ω_{kl} in Ω_{kl}. We identify points X and Y if the path ω_{kl} can be contracted to a point in Ω_{kl}. We perform this identification procedure for all different points X and Y of sets T_k and T_l with identical projections in R_1, letting point A run through all of G_{kl}. We perform this operation with all pairs of sets T_k and T_l, which have nonempty intersections. If it happens that the point $X \in T_k$ is identified with point $Y \in T_l$, and Y with $Z \in T_p$, then points X and Z are also identified. As a result we get a set R_2 covering R_1 and covered by set M_2. For the pair of sets R_2 and R_1 we perform the same operation which was just performed for M_2 and R_1. We construct a set R_3, covering R_1 and covered by R_2. After n steps of this construction we find a set R_n covering R_1 and covered by all the sets $M_1, M_2, R_2, \ldots, R_{n-1}$. Obviously one can find an n such that $R_{n+1} = R_n$. The set R_n will be a set on which the trace of any function $v(\xi, \eta)$ which is single-valued on Σ is also single-valued and from which any single-valued function extends to a single-valued function $v(\xi, \eta)$. We denote R_n by Q.

The set Q is obtained by identification of points of set M_1. We denote by Γ the set of those points whose preimages in M_1 consist of points of sphere S only. Γ is the boundary of Q. Formula (3.36) now generates on Q the equation

$$A(s, t)g_{ss} + 2B(s, t)g_{st} + C(s, t)g_{tt} = 0. \tag{3.37}$$

Since (3.37) is an elliptic equation with possible degeneracies, its solution g is uniquely determined by giving the values of this solution on Γ [8]. Thus, the arbitrariness in the general solution of the oblique derivative problem with boundary condition (3.30) is no larger than the arbitrariness in a function which is continuously differentiable on Γ. Due to degeneracy in (3.37) it can happen that it is necessary to free part of Γ from the definition of the boundary conditions, which can restrict the arbitrariness even more. The values of a solution u of (3.30), expressed by (3.32) in terms of the function $v(\xi, \eta)$, defined by a solution of (3.37) assuming given values on Γ, must be given on a subset N' of set N of points of sphere S at which the vector field P goes into the tangent plane to S. This set N' must be mapped in a one-to-one fashion to Γ in the course of constructing set Q. In particular, set N' will not necessarily be connected.

One can perform all the preceding constructions not only for ball Σ, but also for any domain D with analytic boundary B: $\{F(x, y, z) = 0\}$, where the condition grad $F \neq 0$ holds on B. Instead of (3.32) we now have the representation $u(x, y, z) = v(\xi, \eta) + [F(x, y, z)]^2 w(x, y, z)$ of a solution of the homogeneous oblique derivative problem for harmonic functions which are regular in D, with the boundary condition $au_x + bu_y + cu_z = 0$ given on B, in terms of two special particular solutions of this problem ξ and η. One makes the following change of variables: one introduces independent variables ξ, η, and $\zeta = (\xi_z\eta_y - \xi_y\eta_z)F_x + (\xi_x\eta_z - \xi_z\eta_x)F_y + (\xi_y\eta_x - \xi_x\eta_y)F_z$. The rest is an almost word-for-word repetition of the arguments. The phenomena which arise here illustrate the oblique derivative problem with boundary condition $u_z = 0$ well. Now $\xi = x$, $\eta = y$, $u = v(x, y)$, and $w \equiv 0$. The analog of set R_1 will be the projection of domain D to the plane $z = 0$, the analog of set Q is a Riemann surface Ω over this projection, and v is now a harmonic function of two variables which is regular on Ω. In this case the arbitrariness in the general solution of the homogeneous problem coincides precisely with the arbitrariness in a function which is continuously differentiable on the boundary B of the surface Ω [8]. It is also easy to construct domains D for which the subset N' of set N on which the vector field $P = (0, 0, 1)$ goes into the tangent plane to B, which is in one-to-one correspondence with set Γ, is disconnected. The set N here is defined by the system of equations $F = 0$, $F_z = 0$.

4. OBLIQUE DERIVATIVE PROBLEM WITH POLYNOMIAL COEFFICIENTS

In the preceding sections the study of the oblique derivative problem for harmonic functions which are regular in the ball Σ with boundary condition (3.24) was reduced to the investigation of the first-order equation (3.25) in ball Σ. A difficulty of this method is that in advance nothing is known about the function F on the right side of (3.25) and that this equation should have a solution which is harmonic in Σ. However, in some special cases one is able to find other restrictions on the function F, which guarantee the existence of a harmonic solution of (3.25). For harmonic functions which are regular in ball Σ we consider the oblique derivative problem with boundary condition

$$p_1 u_x + p_2 u_y + p_3 u_z = f, \tag{3.38}$$

where p_1, p_2, p_3 are the traces on sphere S of some polynomials in variables x, y, and z. Let the coefficients of this boundary condition be extended to Σ as polynomials of degree no higher than l; in particular they can be extended by harmonic polynomials. After the extension of the coefficients of boundary condition (3.38) we associate with this condition the first-order equation

$$p_1 u_x + p_2 u_y + p_3 u_z = F, \tag{3.39}$$

in which the extensions of the polynomials p_1, p_2, p_3 to the ball are denoted by the same letters again. Obviously

$$\Delta^{l+1} F = 0, \quad F = g_0(x, y, z) + \sum_{k=1}^{l} (1 - x^2 - y^2 - z^2)\, g_k(x, y, z). \tag{3.40}$$

Here g_k, $k = 0, \ldots, l$ are harmonic functions which are regular in ball Σ, while g_0 coincides on S with f, and g_k, $k = 1, \ldots, l$ are harmonic functions and must be chosen so that (3.39) has at least one harmonic solution. We have

$$F_1(x, y, y) = \sum_{k=1}^{l} (1 - x^2 - y^2 - z^2)^k g_k(x, y, z) = \sum_{i=0}^{l} (x^2 + y^2 + z^2)^i h_i(x, y, z),$$

$$h_i(x, y, z) = \frac{1}{i!} \sum_{j=1}^{l} \frac{(-1)^i j!}{(j-i)!} g_j(x, y, z).$$

All h_i are harmonic functions which are regular in ball Σ. By direct calculation we find

$$\Delta^k F_1 = \sum_{j=k-1}^{l} \lambda_{jk}(g_j), \quad k = 0, 1, \ldots, l, \tag{3.41}$$

where the λ_{jk} are completely determined operators which are polynomials of degree $k - 1$ in the operator $x\partial/\partial x + y\partial/\partial y + z\partial/\partial z$ with coefficients depending only on $x^2 + y^2 + z^2$. It follows from this that from (3.41) one can express the harmonic functions h_i or g_j uniquely in terms of F_1, ΔF_1, ..., $\Delta^{l-1} F_1$. Substituting the expressions found for the functions g_j in (3.40), we get the equation $L(F_1) = g_0(x, y, z)$, where L is a completely determined integrodifferential operator. This equation can be transformed to an equivalent differential equation of the form

$$\Lambda_l(F) = \omega(g_0). \tag{3.42}$$

Here Λ_l is a differential operator of order no higher than $2l$, and ω is also a differential operator of no lower order. Obviously the function

$$F_1(x, y, z) = \sum_{k=1}^{l} (1 - x^2 - y^2 - z^2)^k g_k(x, y, z),$$

satisfies the homogeneous equation $\Lambda_l(F) = 0$ corresponding to (3.42), where the g_k are arbitrary harmonic functions which are regular in ball Σ. Now the general solution of (3.42) has the form $F(x, y, z) = F_0(x, y, z) + F_1(x, y, z)$, where F_0 is a function which is determined uniquely by the function g_0, while the first-order equation

$$p_1 u_x + p_2 u_y + p_3 u_z = F_0(x, y, z) \tag{3.43}$$

has a harmonic solution whose singularities can lie only at points satisfying the conditions $p_1, p_2, p_3 = 0$. Now the investigation of the inhomogeneous problem (3.38) reduces to the investigation of (3.43), and the homogeneous oblique derivative problem corresponding to (3.38) reduces to the following problem: find all harmonic functions $g_1, \ldots, g_l$ which are regular in ball Σ, such that the first-order equation

$$p_1 u_x + p_2 u_y + p_3 u_z = F_1(x, y, z) \tag{3.44}$$

has at least one harmonic solution which is regular in ball Σ.

Substituting for F_1 in (3.41) its expression $F_1 = p_1 u_x + p_2 u_y + p_3 u_z - g_0$, we get a system of equations connecting the $l + 2$ harmonic functions $u, g_0, g_1, \ldots, g_l$. This system consists of $l + 1$ equations and contains $l + 1$ unknown functions $u, g_1, \ldots, g_l$. Consequently, the oblique derivative problem is reduced to a problem of the analytic theory of partial differential equations which is finding solutions which are regular in ball Σ of a system of partial differential equations. One can study the present system by eliminating some of the unknown functions. For example, eliminating the functions $g_1, \ldots, g_l$, we get an equation relating u and g_0, which one gets as a result of substituting in (3.42) the expression for F from (3.39). Since we are interested only in harmonic solutions of the equation found, the oblique derivative problem now reduces to the study of the system of two equations

$$\Lambda_l(p_1 u_x + p_2 u_y + p_3 u_z) = \omega(g_0), \quad \Delta u = 0. \tag{3.45}$$

In general the investigation of (3.45) is rather complicated due to the fact that the first equation of the system degenerates.

If the boundary condition of the oblique derivative problem (3.38) has the form $p(z)(xu_x + yu_y + zu_z) + q(z)u_z = f$, i.e., is symmetric with respect to the Oz axis, the expressions (3.41) assume the form

$$\Delta^k [p(z)(xu_x + yu_y + zu_z) + q(z)u_z]$$
$$= \sum_{j=0}^{k} C_{kj} \left[\frac{d^{2k-j} q(z)}{dz^{2k-j}} \frac{\partial^{j+1} u}{\partial z^{j+1}} + \frac{d^{2k-j} p(z)}{\partial z^{2k-j}} \cdot \frac{\partial^j}{\partial z^j} (xu_x + yu_y + zu_z) \right],$$
$$C_{kj} = \frac{2^k}{k!} \prod_{l=0}^{j} (k-l), \quad k = 0, 1 \ldots, \quad 0 \leqslant j \leqslant k.$$

Let $u = h(x^2 + y^2, z)p_l(x, y)$, where p_l is a homogeneous harmonic polynomial of degree l in two variables, and the function $h(\sigma, z)$ satisfies the equation

$$h_{zz} + 4\sigma h_{\sigma\sigma} + 4(l+1)h_z = 0.$$

In this case we have

$$xu_x + yu_y + zu_z = (2\sigma h_\sigma + zh_z + lh)p_l(x, y),$$
$$\Delta^k [p(z)(xu_x + yu_y + zu_z) + q(z)u_z] = p_l(x, y)$$
$$\times \sum_{j=0}^{k} C_{kj} \left[\frac{d^{2k-j} p(z)}{dz^{2k-j}} \cdot \frac{\partial^j}{\partial z^j} (2\sigma h_\sigma + zh_z + lh) + \frac{d^{2k-j} q(z)}{dz^{2k-j}} \cdot \frac{\partial^{j+1} h}{\partial z^{j+1}} \right].$$

One verifies by direct calculation that in (3.14)

$$\lambda_{jk}(g_j) = \sum_{i=k-1}^{j} \frac{2^{k-1}(-1)^i j!}{(j-i)!\,(i-k+1)!} (x^2 + y^2 + z^2)^{i-k+1} \mu_{ik}(g_j),$$

where $\mu_{ik}(g_j) = l_{i-k+2}(\ldots l_i(g_j)\ldots)$ is the composition of the operators

$$l_\alpha = r\partial/\partial r + (\alpha + 1/2), \quad \alpha = i - k + 2, \ldots, i.$$

On functions of the form $g_j = b_j(x^2 + y^2, z)p_l(x, y)$ we get

$$\lambda_{jk}(g_j) = p_l(x, y)\, \nu_{jk}(b_j) = p_l(x, y)$$
$$\times \sum_{i=k-1}^{j} \frac{(-1)^i 2^{k-i} j!}{(j-i)!\,(i-k+1)!} (x^2 + y^2 + z^2)^{i-k+1} \delta_{ik}(b_j),$$

δ_{ik} being the composition of the operators

$$m_\alpha = 2\sigma\partial/\partial\sigma + z\partial/\partial z + (\alpha + l + 1/2), \quad \alpha = i - k + 2, \ldots, i.$$

It follows from these formulas that if the boundary condition of the oblique derivative problem (3.38) is symmetric with respect to the Oz axis, then in (3.41) one can separate variables in cylindrical coordinates. After separating the variable φ the investigation of system (3.41) reduces to the study of the system of ordinary differential equations which one gets for $x^2 + y^2 = \sigma = 0$. Let $h(0, z) = \varphi(z)$, $b_j(0, z) = \psi_j(z)$, $j = 0, \ldots, l$; from (3.41) we find

$$\sum_{j=0}^{k} C_{kj}\left[\frac{d^{2k-j}q(z)}{dz^{2k-j}}\cdot\frac{d^{j+1}\varphi(z)}{dz^{j+1}}+\frac{d^{2k-j}p(z)}{dz^{2k-j}}\cdot\frac{\partial^j}{\partial z^j}[z\varphi(z)+l\varphi(z)]\right]$$

$$=\sum_{j=k-1}^{l}\sum_{i=k-1}^{j}\frac{(-1)^j 2^k j!}{(j-i)!\,(i-k+1)!}z^{2(i-k+1)}\pi_{ik}(\psi_j), \qquad (3.46)$$

$$k=1,\ldots,l+1,$$

where π_{ik} is the composition of the operators

$$\mu_\alpha = zd/dz + (\alpha + l + 1/2),\ \alpha = i - k + 2,\ \ldots,\ i.$$

Formula (3.46) is a system of $l+1$ differential equations for the $l+1$ functions $\varphi(z)$, $\psi_1(z)$, ..., $\psi_l(z)$. Defining these functions from (3.46), from them we construct the corresponding harmonic functions uniquely, as was done in Chapter 1. The harmonic function corresponding to the function φ will be the solution of the oblique derivative problem sought.

In order that the harmonic functions constructed from the functions φ, ψ_0, ..., ψ_l be regular in the ball Σ, these functions themselves must be holomorphic in the unit disk $|z| < 1$. Thus, in the case considered the oblique derivative problem for harmonic functions which are regular in the unit ball reduces to a problem of the analytic theory of ordinary differential equations, namely to finding solutions of system (3.46) which are holomorphic in the unit disk.

The investigation of problem (3.38) simplifies essentially in the case when the polynomials p_1, p_2, p_3 are linear. Now the function $F = p_1u_x + p_2u_y + p_3u_z$ is biharmonic for any harmonic function u. Formula (3.39) assumes the form

$$p_1u_x + p_2u_y + p_3u_z = g_0 + (1 - x^2 - y^2 - z^2)g_1,$$

from which we get directly the analog of (3.45):

$$\Delta\left[\frac{p_1u_x + p_2u_y + p_3u_z}{1 - x^2 - y^2 - z^2}\right] = \Delta\left(\frac{g_0}{1 - x^2 - y^2 - z^2}\right), \quad \Delta u = 0,$$

and (3.42) has the form

$$(1-r^2)\left[(1-r^2)\Delta F + 4r\frac{\partial F}{\partial r}\right] + 2(3+r^2)F$$

$$= 4(1-r^2)\,r\frac{\partial g_0}{\partial r} + 2(3+r^2)\,g_0,$$

where $r^2 = x^2 + y^2 + z^2$.

As an example we consider the oblique derivative problem for harmonic functions which are regular in the ball Σ with the boundary condition [17]

$$xu_x + yu_y - (z-\beta)u_z = f, \quad \beta \equiv \text{const}. \qquad (3.47)$$

In this case, after substitution of the expressions, (3.41) assumes the form

$$\begin{gathered} xu_x + yu_y - (z-\beta)u_z = g_0 + (1 - x^2 - y^2 - z^2)g_1, \\ 2u_{zz} = 2r\frac{\partial g_0}{\partial r} + 3g_1. \end{gathered} \tag{3.48}$$

For the harmonic functions u, g_0, and g_1 we write the following representations:

$$u = \sum_{l=0}^{\infty} u_l(x^2+y^2, z)\, p_l(x, y), \quad g_1 = \sum_{l=0}^{\infty} v_l(x^2+y^2, z)$$
$$\times p_l(x, y), \quad g_0 = \sum_{l=0}^{\infty} h_l(x^2+y^2, z)\, p_l(x, y).$$

Substituting these expressions in (3.48) and equating coefficients of the same p_l, we find

$$2\sigma\frac{\partial u_l}{\partial \sigma} - (z-\beta)\frac{\partial u_l}{\partial z} + lu_l = h_l + (1-\sigma-z^2)\, v_l,$$
$$2\frac{\partial^2 u_l}{\partial z^2} = 4\sigma\frac{\partial v_l}{\partial \sigma} + 2z\frac{\partial v_l}{\partial z} + (3+2l)\, v_l,$$
$$l = 0, 1, \ldots.$$

We introduce the notation $\psi_l(z) = u_l(0, z)$, $\varphi_l(z) = v_l(0, z)$, $q_l(z) = h_l(0, z)$ and in the preceding equations we set $\sigma = 0$. We have

$$\begin{gathered} -(z-\beta)\,\psi_l' + l\psi_l = q_l + (1-z^2)\,\varphi_l, \\ \psi_l'' = z\varphi_l' + (l+3/2)\,\varphi_l, \quad l = 0, 1, \ldots. \end{gathered} \tag{3.49}$$

It is very easy to investigate (3.49) for $l = 0$. Eliminating the function ψ_0 from them, for the function φ_0 we get the first-order equation

$$\varphi_0' - \left[\frac{1}{z-\beta} + \frac{3\beta - z}{2(1-z\beta)}\right]\varphi_0 = \frac{q_0 - q_0'(z-\beta)}{(z-\beta)(1-z\beta)},$$

from which we find

$$\varphi_0(z) = (z-\beta)(1-z\beta)^{(1-3\beta^2)/2\beta^2}\exp\left(\frac{z}{2\beta}\right)$$
$$\times\left[C + \int \frac{q_0(t) - q_0'(t)(t-\beta)}{(t-\beta)^2}(1-t\beta)^{(\beta^2-1)/2\beta^2}\exp\left(-\frac{t}{2\beta}\right)dt\right],$$

where C is an arbitrary constant. It is clear from the formula for φ_0 that there can be a singularity of this function for $|\beta| > 1$ at the point $z = \beta^{-1}$, and for $|\beta| < 1$, at the point $z = \beta$. Obviously for $|\beta| > 1$ the function $\varphi_0(z)$ is holomorphic in the disk $|z| < 1$ if and only if $C = 0$. For $|\beta| < 1$ the function $\varphi_0(z)$ will be holomorphic in the disk $|z| < 1$ when the integrand can be expanded in a neighborhood of the point $z = \beta$ in a Laurent series which does not contain a term with $(z-\beta)^{-1}$, because after integration a logarithm appears due to this term. Expanding the integrand in a Laurent series, we find an expansion of the form

$$a_0(z-\beta)^{-2}+a_1+a_2(z-\beta)+\dots$$

After integrating and multiplying by $z-\beta$, we get that $\varphi_0(z)$ is holomorphic for $|z|<\beta$ for an arbitrary value of the constant C. Thus, for $l=0$ for all values of the parameter $\beta\neq 1$ there exists a solution of (3.49) which is holomorphic in the disk $|z|<1$. The corresponding homogeneous system ($q_0=0$) for $|\beta|>1$ has the solution $\varphi_0=0$, $\psi_0=A$, where A is an arbitrary constant, and for $|\beta|<1$ the following solution:

$$\varphi_0(z)=C(z-\beta)(1-z\beta)^{(1-3\beta^2)/2\beta^2}\exp\left(\frac{z}{2\beta}\right),$$

$$\psi_0(z)=A+\int_\beta^z(1-t\beta)^{(1-3\beta^2)/2\beta^2}\exp\left(\frac{t}{2\beta}\right)dt,$$

where A and C are arbitrary constants.

For $l>0$, as a result of eliminating ψ_l from (3.49) we get the equation

$$(1-z\beta)\varphi_l''-\left(\frac{z}{2}+l\beta+5\beta/2\right)\varphi_l'+[(l+1)/2-l^2]\varphi_l=-q_l'',$$

relating the functions φ_l and q_l. For $|\beta|<1$ the coefficients of the highest derivative of the function φ_l in the disk $|z|<1$ are different from zero, so the inhomogeneous equation is always solvable, and the corresponding homogeneous equation has two linearly independent solutions which are holomorphic in the unit disk. For $|\beta|>1$ the inhomogeneous equation also has a solution which is holomorphic in the unit disk for any right side which is holomorphic in this disk. One can find this solution directly by expanding the functions φ_l and q_l in powers of $1-z\beta$ and substituting the series into the equation. The corresponding homogeneous equation now has only one linearly independent solution which is holomorphic in the unit disk. Thus for any previously given function g_0 one can always choose a harmonic function g_1 such that the first equation of (3.48) has a solution which is harmonic in ball Σ.

We construct a harmonic function g_1 from the previously given harmonic function g_0 such that the equation

$$xu_x+yu_y-(z-\beta)u_z=g_0+(1-x^2-y^2-z^2)g_1 \tag{3.50}$$

has a solution u which is harmonic in ball Σ. The vector field $P=(x,y,\beta-z)$ goes into the tangent plane to sphere S at the set of points satisfying the equations $x^2+y^2=1-z^2$, $z=(\beta\pm\sqrt{\beta^2+8})/4$. For $|\beta|<1$ this set consists of two circles, where the half Σ^+ of ball Σ which lies in the half-space $z>\beta$ is filled by trajectories of the field P passing through points of the section of ball Σ by the plane $z=z_1=(\beta+\sqrt{\beta^2+8})/4$, and the part Σ^- of ball Σ which lies in the half-space $z<\beta$ is filled by trajectories of the field P which pass through points of the section of the ball Σ by the plane $z=z_2=(\beta-\sqrt{\beta^2+8})/4$. We introduce new independent variables

$$\xi=\rho(z-\beta),\ \eta=\varphi,\ \zeta=z, \tag{3.51}$$

where ρ, z, and φ are cylindrical coordinates $x = \rho\cos\varphi$, $y = \rho\sin\varphi$, z. The change of variables (3.51) is one-to-one both in part Σ^+ and in part Σ^- of ball Σ. In the new variables the Laplace operator assumes the form

$$\Delta u = [\xi^2(z-\beta)^{-2} + (\zeta-\beta)^2]u_{\xi\xi} + 2\xi(\zeta-\beta)u_{\xi\zeta} + u_{\zeta\zeta} + (\zeta-\beta)^2\xi^{-1}u_\xi + \xi^{-2}(\zeta-\beta)^2u_{\eta\eta},$$

and (3.50) can be rewritten as follows: $(\zeta-\beta)u_\zeta = -F(\xi, \eta, \zeta)$. Here F denotes the right side of the equation in the new independent variables.

From (3.50), written in the new variables in part Σ^+ of ball Σ we find

$$u = \chi(\xi, \eta) - \int_{z_1}^{\zeta} (t-\beta)^{-1} F(\xi, \eta, t)\, dt, \tag{3.52}$$

and in part Σ^- we have

$$u_1 = \chi_1(\xi, \eta) - \int_{z_2}^{\zeta} (t-\beta)^{-1} F(\xi, \eta, t)\, dt, \tag{3.53}$$

where χ is an arbitrary function of two variables. Since function F in (3.50) is such that this equation has a harmonic solution, taking the Laplacian from (3.52) it suffices to equate the result just for $z_1 = \zeta$ to zero. It follows from this that function χ must satisfy the equation

$$\xi^2[(z_1-\beta)^4 + \xi^2]\chi_{\xi\xi} + \xi(z_1-\beta)^4\chi_\xi + (z_1-\beta)^4\chi_{\eta\eta} = h(\xi, \eta),$$

$$h(\xi, \eta) = 2\xi^3(z_1-\beta)\frac{\partial}{\partial\xi}[F(\xi, \eta, z_1)/(z_1-\beta)] + \xi^2(z_1-\beta)^4\left[\frac{\partial}{\partial\zeta}\frac{F(\xi, \eta, \zeta)}{\zeta-\beta}\right]\Big|_{\zeta=z_1}. \tag{3.54}$$

Formula (3.54) is elliptic everywhere in the disk K_1: $\{|\xi| < (z_1-\beta)\sqrt{1-z_1^2}\}$ except for the point $\xi = 0$ at which it degenerates. The disk K_1 corresponds to the section of the ball Σ by the plane $z = z_1$, and ξ and η are polar coordinates in this section. In the disk K_1 one can represent function h by a trigonometric series

$$p(\xi, \eta) = \xi^2 \sum_{l=0}^{\infty} [a_l(\xi)\xi^l \cos l\varphi + b_l(\xi)\xi^l \sin l\varphi].$$

We shall also seek a solution of (3.54) in the form of a series

$$\chi(\xi, \eta) = \sum_{l=0}^{\infty} [u_l(\xi)\xi^l \cos l\varphi + v_l(\xi)\xi^l \sin l\varphi].$$

Substituting these series in (3.54) and equating coefficients of corresponding trigonometric functions, we get

$$[(z_1-\beta)^4 + \xi^2]\xi u_l'' + [(2l+1)(z_1-\beta)^4 + 2l\xi^2]u_l' + l(l-1)\xi u_l = \xi a_l,$$

$$[(z_1-\beta)^4 + \xi^2]\xi v_l'' + [(2l+1)(z_1-\beta)^4 + 2l\xi^2]v_l' + l(l-1)\xi v_l = \xi b_l.$$

These ordinary differential equations for all l have a unique solution which assumes preassigned value for $\xi = \xi_0$ and is bounded at the point $\xi = 0$. It follows from this that the Dirichlet problem for (3.54) in the disk K_1 with any continuously differentiable data on the boundary of K_1 is always solvable and has a unique solution. Analogously from (3.54) for the function χ_1 in the disk K_2: $\{|\xi| < (\beta - z_2)\sqrt{1 - z_2^2}\}$ one gets the following equation:

$$\xi^2 [(z_2 - \beta)^4 + \xi^2] \frac{\partial^2 \chi_1}{\partial \xi^2} + (z_2 - \beta)\xi \frac{\partial \chi_1}{\partial \xi} + (z_2 - \beta)^4 \frac{\partial^2 \chi_1}{\partial \eta^2} = h_1(\xi, \eta), \tag{3.55}$$

for which the Dirichlet problem in the disk K_2 is always solvable and has a unique solution.

For such a choice of F, χ, and χ_1, (3.52) gives a harmonic function which is regular in Σ^+, and (3.53) gives a harmonic function which is regular in Σ^-. In order that these harmonic functions make up a harmonic function which is regular in ball Σ, it is necessary and sufficient that one have

$$u|_{z=\beta} = u_1|_{z=\beta}, \quad \left.\frac{\partial u}{\partial z}\right|_{z=\beta} = \left.\frac{\partial u_1}{\partial z}\right|_{z=\beta} \tag{3.56}$$

The second condition, in the variables ξ, η, ζ, by (3.50), assumes the form $u_\xi|_{z=\beta} = u_{1\xi}|_{z=\beta}$. Now (3.56) is transformed as follows:

$$\chi(0, \eta) - \int_{z_1}^{\beta} (t - \beta)^{-1} F(0, \eta, t)\, dt = \chi_1(0, \eta) - \int_{z_2}^{\beta} (t - \beta)^{-1} F(0, \eta, t)\, dt,$$

$$\left.\frac{\partial \chi}{\partial \xi}\right|_{\xi=0} - \int_{z_1}^{\beta} (t - \beta)^{-1} \left.\frac{\partial F(\xi, \eta, t)}{\partial \xi}\right|_{\xi=0} dt = \frac{\partial \chi_1}{\partial \xi} - \int_{z_2}^{\beta} (t - \beta)^{-1} \left.\frac{\partial F(\xi, \eta, t)}{\partial \xi}\right|_{\xi=0} dt.$$

These conditions are equivalent to the fact that (3.50) has a harmonic solution which is regular in ball Σ, and does not impose any conditions on functions χ and χ_1 since $\chi(0, \eta)$ and $\chi_1(0, \eta)$ and their derivatives with respect to ξ for $\xi = 0$ vanish by virtue of the equations which these functions satisfy. Now the existence of a harmonic solution of (3.50) is guaranteed by the special choice of function F, so the compatibility conditions hold.

If $|\beta| > 1$ the vector field $P = (x, y, \beta - z)$ goes into the tangent plane to sphere S on one circle, which for $\beta > 1$ one gets as a result of intersecting sphere S with the plane $z = z_2 = (\beta - \sqrt{\beta^2 + 8})/4$, and for $\beta < -1$ with the plane $z = z_1 = (\beta + \sqrt{\beta^2 + 8})/4$. Now for $\beta < -1$ one considers (3.52) and for $\beta > 1$ one takes (3.53), and the rest of the investigation proceeds even more simply than in the case $|\beta| < 1$ since it is not necessary to "glue" two different expressions for the solution.

Thus, (3.47) is solvable for all continuously differentiable f, where there always exists a unique solution of the problem, assuming previously given continuously differentiable values on the manifold on which the vector field $P = (x, y, \beta - z)$ goes into the tangent plane to sphere S, where P is the direction in which one takes the oblique derivative for any constant $\beta \neq \pm 1$. If $|\beta| = 1$, then the vector field

P vanishes at one point $x = y = 0$, $z = \beta$ of sphere S, and in this case the oblique derivative problem is not well posed.

5. REDUCTION OF THE OBLIQUE DERIVATIVE PROBLEM TO A FREDHOLM INTEGRODIFFERENTIAL EQUATION

In the preceding chapter we found polynomials a, b, c such that for any regular harmonic function u the function $au_x + bu_y + cu_z$ is also harmonic. Here we consider a more general situation. Suppose given in a domain D of three-dimensional Euclidean space R^3 three analytic functions $a(x, y, z)$, $b(x, y, z)$, and $c(x, y, z)$, which are continuously differentiable in the closed domain $\bar{D}$; let u be a harmonic function which is regular in domain D. We find out for which a, b, and c the function $v = au_x + bu_y + cu_z$ satisfies the equation

$$\Delta v + \lambda_1 v_x + \lambda_2 v_y + \lambda_3 v_z + \lambda_4 v = 0. \tag{3.57}$$

Substituting the expression for v in terms of u in (3.57), and equating the coefficients of all derivatives of function u to zero, we get

$$\begin{gathered}\lambda_1 b + \lambda_2 a + 2(a_y + b_x) = 0, \ \lambda_1 c + \lambda_3 a + 2(a_z + c_x) = 0, \\ \lambda_2 c + \lambda_3 b + 2(b_z + c_y) = 0, \\ \lambda_1 a - \lambda_3 c + 2(a_x - c_z) = 0, \ \lambda_2 b - \lambda_3 c + 2(b_y - c_z) = 0;\end{gathered} \tag{3.58}$$

$$\begin{gathered}\Delta a + \lambda_1 a_x + \lambda_2 a_y + \lambda_3 a_z + \lambda_4 a = 0, \\ \Delta b + \lambda_1 b_x + \lambda_2 b_y + \lambda_3 b_z + \lambda_4 b = 0, \\ \Delta c + \lambda_1 c_x + \lambda_2 c_y + \lambda_3 c_z + \lambda_4 c = 0.\end{gathered} \tag{3.59}$$

From (3.58) we find

$$\begin{aligned}\lambda_1 &= -2(a^2 + b^2)^{-1}[c(a_z + c_x) + a(a_x - c_z)], \\ \lambda_2 &= -2(b^2 + c^2)^{-1}[c(b_z + c_y) + b(b_y - c_z)], \\ \lambda_3 &= 2(b^2 + c^2)^{-1}[c(b_y - c_z) - b(b_z + c_y)]\end{aligned}$$

and we get two more equations relating a, b, and c:

$$\begin{gathered}c(b_z + c_y) + b(a_x - c_z) - c^2(a^2 + b^2)^{-1}[a(a_y + b_x) - b(a_x - b_y)] \\ -ab(a^2 + b^2)^{-1}[a(a_x - b_y) + b(a_y + b_x)] = 0, \\ c(a_z + c_x) + a(a_x - c_z) - (a^2 + c^2)(a^2 + b^2)^{-1}[a(a_x - b_y) + b(a_y + b_x)] = 0.\end{gathered} \tag{3.60}$$

From (3.59) we find

$$\lambda_4 = -(a^2 + b^2)^{-1}[a\Delta a + b\Delta b + \lambda_1(aa_x + bb_x) + \lambda_2(aa_y + bb_y) + \lambda_3(aa_z + bb_z)]$$

and we get two more equations relating a, b, and c:

$$
\begin{aligned}
&a\Delta b - b\Delta a + \lambda_1(ab_x - ba_x) + \lambda_2(ab_y - ba_y) + \lambda_3(ab_z - ba_z) = 0,\\
&c\Delta a - a\Delta c + \lambda_1(ca_x - ac_x) + \lambda_2(ca_y - ac_y) + \lambda_3(ca_z - ac_z) = 0.
\end{aligned} \tag{3.61}
$$

The four equations (3.60) and (3.61) give necessary and sufficient conditions for function v to satisfy (3.57) for any harmonic u. The coefficients of (3.57) are uniquely determined by functions a, b, c and it follows from their expressions that these coefficients can have singularities.

Let us assume that $c \neq 0$ and we consider two new functions $\alpha = b/c$ and $\gamma = a/c$. From (3.60) and (3.61) one gets the following two systems of equations:

$$
\begin{aligned}
&\frac{\alpha\gamma_z}{1+\gamma^2} - \frac{\gamma\alpha_z}{1+\alpha^2} + \gamma_y - \alpha_x = \frac{1}{2}\left[\alpha_x \frac{\partial}{\partial x} \ln(1+\gamma^2) + \gamma \frac{\partial}{\partial y} \ln(1+\alpha^2)\right],\\
&\frac{\gamma_x}{1+\gamma^2} - \frac{\alpha_y}{1+\alpha^2} + \frac{1}{2}\frac{\partial}{\partial z} \ln \frac{1+\alpha^2}{1+\gamma^2} = 0;
\end{aligned} \tag{3.62}
$$

$$
\begin{aligned}
&\Delta\gamma + \frac{2\gamma_z\alpha_z}{1+\gamma^2} + \frac{4\alpha_z\alpha_y}{1+\alpha^2} + \alpha_x \frac{\partial}{\partial x} \ln(1+\gamma^2) + \alpha_y \frac{\partial}{\partial y} \ln(1+\alpha^2) - \alpha_z \frac{\partial}{\partial z}(1+\alpha^2) = 0,\\
&\Delta\gamma - \frac{2\gamma_z\gamma_x}{1+\gamma^2} - 2\frac{\alpha_z\gamma_y - \alpha_y\gamma_z}{1+\alpha^2} - \gamma_x \frac{\partial}{\partial x} \ln(1+\gamma^2)\\
&\qquad - \gamma_y \frac{\partial}{\partial y} \ln(1+\alpha^2) - \gamma_z \frac{\partial}{\partial z} \ln(1+\alpha^2) = 0,
\end{aligned} \tag{3.63}
$$

and the expressions for the coefficients of (3.57) assume the form

$$
\begin{aligned}
\lambda_1 &= -2\gamma_z(1+\gamma^2)^{-1} - \frac{\partial}{\partial x} \ln(1+\gamma^2) - 2c_x/c,\\
\lambda_2 &= -2\alpha_z(1+\alpha^2)^{-1} - \frac{\partial}{\partial y} \ln(1+\alpha^2) - 2c_y/c,\\
\lambda_3 &= -2\alpha_y(1+\alpha^2)^{-1} - \frac{\partial}{\partial z} \ln(1+\alpha^2) - 2c_z/c.
\end{aligned} \tag{3.64}
$$

The two functions α and γ satisfy the system of four equations (3.62) and (3.63), so the class of such functions is not very large, although one can give examples of functions satisfying this overdetermined system of equations. In order to include a larger class of functions a, b, c, we proceed somewhat differently.

Let us assume that functions α and γ satisfy (3.62) only. Now functions u and v are related by the equation

$$
\Delta v + \lambda_1 v_x + \lambda_2 v_y + \lambda_3 v_z = A(x, y, z)u_x + B(x, y, z)u_y + C(x, y, z)u_z, \tag{3.65}
$$

where the coefficients $\lambda_1, \lambda_2, \lambda_3$ are given by (3.64) and

$$
\begin{aligned}
&A = -(\Delta a + \lambda_1 a_x + \lambda_2 a_y + \lambda_3 a_z), \quad B = -(\Delta b + \lambda_1 b_x + \lambda_2 b_y + \lambda_3 b_z),\\
&C = -(\Delta c + \lambda_1 c_x + \lambda_2 c_y + \lambda_3 c_z).
\end{aligned}
$$

When $\lambda_1, \lambda_2, \lambda_3$ are continuous in domain D and boundary Γ of domain D is sufficiently smooth, there exists a Green's function $G(X, T)$ for domain D for the equation

$$\Delta v + \lambda_1 v_x + \lambda_2 v_y + \lambda_3 v_z = 0. \tag{3.66}$$

With the help of the Green's function from (3.65) we find

$$v = F(X) + \int_D G(X, T)\, Q(T)\, d\omega,$$

where $X = (x, y, z)$, $T = (s, t, \sigma)$, F is a solution of (3.66) which is regular in D, and Q is the right side of (3.65). Integrating by parts we get

$$v(X) = F(X) + \int_D K(X, T)\, u(T)\, d\omega. \tag{3.67}$$

Here the Fredholm kernel has the form

$$K(X, T) = \frac{\partial}{\partial s}[A(T)\, G(X, T)] + \frac{\partial}{\partial t}[B(T)\, G(X, T)] + \frac{\partial}{\partial \sigma}[c(T)\, G(X, T)].$$

We consider the oblique derivative problem for harmonic functions which are regular in domain D with boundary condition

$$au_x + bu_y + cu_z = f, \tag{3.68}$$

given on boundary Γ of domain D. We shall assume that $c \neq 0$ everywhere on Γ, and the functions $\alpha = b/c$ and $\gamma = a/c$ extend to D so that they satisfy the system of equations (3.62) in D. We extend c harmonically to domain D, and α and γ as solutions of (3.62). Obviously $c \neq 0$ everywhere on D, so the coefficients of (3.66) are continuous in D. With the help of (3.67) we find

$$au_x + bu_y + cu_z = F(X) + \int_D K(X, X_1)\, u(X_1)\, d\omega, \tag{3.69}$$

where $X = (x, y, z)$, $X_1 = (x_1, y_1, z_1)$ are points of domain D, and F is a solution of (3.66) which is regular in domain D and coincides with f on boundary Γ of domain D. Condition $c \neq 0$ in D guarantees the existence in D of two independent integrals ξ and η of the equation $aw_x + bw_y + cw_z = 0$ and the existence of a solution ζ which is holomorphic in D of the inhomogeneous equation $aw_x + bw_y + cw_z = 1$. In domain D we introduce new independent variables ξ, η, ζ and we denote by E the image of domain D under the map affecting the given change of variables. This map τ: $D \to E$ is a homeomorphism under rather broad assumptions on domain D [22]. In particular, if the trajectories of the vector field $P = (a, b, c)$ which intersect with D intersect with it along connected arcs, then map τ is a homeomorphism. In the new variables (3.69) assumes the form

$$u_\zeta = \Phi(Y) + \int_E K_1(Y, Y_1)\, u(Y_1)\, d\omega.$$

Here $Y = (\xi, \eta, \zeta)$, $Y_1 = (\xi_1, \eta_1, \zeta_1)$, Φ is the function F in the new variables, and K_1 is the kernel K in the new variables. Integrating with respect to ζ and returning to the variables x, y, z, we get

$$u(X) = h(X) + \chi(\xi(X), \eta(X)) + \int_D L(X, X_1)\, u(X_1)\, d\omega, \tag{3.70}$$

where h is a known function, uniquely determined by the function F, $\chi(\xi, \eta)$ is an arbitrary function of the variables ξ and η which is holomorphic in domain E, and L is a Fredholm kernel, uniquely determined by kernel K. The kernel L is obtained by integration with respect to point X along the trajectories of the field $P = (a, b, c)$ from kernel K, which for $X = X_1$ has a singularity of the second order, so kernel L has a singularity of the first order on the piece of the trajectory of field P issuing from point X_1. This proves that kernel L is Fredholm.

According to the Fredholm theory of equations [30], from (3.70) we find

$$u(X) = \chi_1(X) + \int_D R(X, X_1)\, \chi_1(X_1)\, d\omega + \sum_{i=1}^{m} C_1 \psi_i(X) + H(X); \qquad (3.71)$$

$$\int_D \nu_i(X)\,[h(X) + \chi_1(X)]\, d\omega = 0, \quad i = 1, \ldots, m, \qquad (3.72)$$

where $\chi_1(X) = \chi(\xi(X), \eta(X))$, and $R(X, X_1)$ is the generalized resolvent of kernel L, ψ_i, $i = 1, \ldots, m$, are linearly independent solutions of the homogeneous equation corresponding to (3.70), ν_i, $i = 1, \ldots, m$, are solutions of the adjoint homogeneous equation, C_i are arbitrary constants, and

$$H(X) = h(X) + \int_D R(X, X_1)\, d\omega$$

is a known function.

Obviously it is not the case that any function u which can be expressed by (3.71) is a solution of the original oblique derivative problem, because it is also necessary to guarantee the harmonicity of function u. We shall guarantee the harmonicity of function u by imposing additional restrictions on domain D. Let us assume that the trajectories of the vector field $P = (a, b, c)$ which intersect with domain D intersect with it along connected arcs and the vector field P goes into the tangent plane to the boundary Γ of domain D along one closed smooth line σ, through which there passes an analytic surface Σ, defined by an equation $z = \mu(x, y)$. Each trajectory of the field P which intersects with D passes through the two-dimensional manifold $M = D \cap \Sigma$ at one point. We shall call such domains D convex with respect to the vector field P, or P-convex. Obviously the manifold M projects uniquely to the plane $z = 0$ and its projection M' is homeomorphic to a disk. We choose the functions ξ and η so that

$$\xi_0(x, y) = \xi(x, y, \mu(x, y)), \quad \eta_0(x, y) = \eta(x, y, \mu(x, y))$$

effects a homeomorphism of M' to the unit disk, and we choose ζ so that $\zeta = 0$ on M. One can always choose ξ, η, ζ in this way [28], while these ξ, η, ζ affect a homeomorphism of domain D to its image in the space of variables ξ, η, ζ.

Since the function $\chi(\xi, \eta)$ is constant along trajectories of the vector field P, one can rewrite (3.72):

$$\int_M p_i(\xi, \eta)\, \chi(\xi, \eta)\, ds = q_i, \quad q_i = \int_D \nu_i(X)\, h(X)\, d\omega, \quad i = 1, \ldots, m, \qquad (3.73)$$

where p_i is a function defined on M, which at point $(\xi, \eta) \in M$ assumes a value equal to the integral of $\nu_i(X)$ over an arc lying in domain D of the trajectory of vector field P passing through this point of manifold M. The function $v = u_\zeta$ satisfies (3.65), which can be transformed as follows:

$$\Delta (au_x + bu_y + cu_z) + \lambda_1 \frac{\partial}{\partial x} (au_x + bu_y + cu_z)$$
$$+ \lambda_2 \frac{\partial}{\partial y} (au_x + bu_y + cu_z) + \lambda_3 \frac{\partial}{\partial z} (au_x + bu_y + cu_z)$$
$$- [\Delta a + \lambda_1 a_x + \lambda_2 a_y + \lambda_3 a_z] u_x$$
$$-[\Delta b + \lambda_1 b_x + \lambda_2 b_y + \lambda_3 b_z] u_y - [\Delta c + \lambda_1 c_x$$
$$+\lambda_2 c_y + \lambda_3 c_z] u_z = 0.$$

Considering the expressions for the functions $\lambda_1, \lambda_2, \lambda_3$ and the system of equations (3.62), one can reduce this equation to the form

$$a \frac{\partial}{\partial x} \Delta u + b \frac{\partial}{\partial y} \Delta u + c \frac{\partial}{\partial z} \Delta u + c \frac{\gamma_x - \gamma\gamma_z}{1 + \gamma^2} \Delta u = 0.$$

It follows from the last equation with Δu that for u to be harmonic it is necessary and sufficient that $\chi(\xi, \eta)$ in (3.70) and (3.71) be chosen so that $\Delta u = 0$ only on $\zeta = 0$. This condition leads to an integrodifferential equation for χ on manifold M.

In calculating the Laplacian of function u, defined by (3.71), difficulties can arise in calculating the second derivatives of the function

$$G(X) = \int_D R(X, X_1) \chi_1 (X_1)\, d\omega,$$

which can be written in terms of the variables ξ, η, ζ as

$$G(Y) = \int_M S_0 (Y, Y_1) \chi (Y_1)\, d\sigma,$$

where $Y = (\xi, \eta, \zeta)$, $Y_1 = (\xi_1, \eta_1)$, and S_0 is obtained from R by the change of variables and integration with respect to ζ_1 along the trajectories of field P. It follows from (3.70) that one can differentiate G under the integral sign once with respect to ξ, η, or ζ, where the integrals expressing the derivatives G_ξ, G_η, G_ζ exist in the usual sense. For G_ζ we have the following equation:

$$G_\zeta = \int_D \frac{\partial}{\partial \zeta} R (X, X_1) \chi_1 (X_1)\, d\omega.$$

Here the function R_ζ has the same singularity as the kernel of (3.70), i.e., for $X = X_1$ there is a second-order singularity. Now for the second derivative $G_{\zeta\zeta}$ we find [28]

$$G_{\zeta\zeta}|_{\zeta=0} = \int_D \frac{\partial^2}{\partial \zeta^2} R (X, X_1) \chi_1 (X_1)\, d\omega - \gamma (\xi, \eta) \chi (\xi, \eta),$$

$$\gamma = \int_S \{[(x-x_1)^2 + (y-y_1)^2 + (z-z_1)^2]\, R(X, X_1)\} \cos(r, \zeta)\, dS.$$

Let the Laplacian in the variables ξ, η, ζ have the form

$$\Delta = A_0 \frac{\partial^2}{\partial\zeta^2} + B_1 \frac{\partial^2}{\partial\xi\partial\zeta} + B_2 \frac{\partial^2}{\partial\eta\partial\zeta} + A_1 \frac{\partial^2}{\partial\xi^2} + A_2 \frac{\partial^2}{\partial\eta^2} + B_3 \frac{\partial^2}{\partial\xi\partial\eta} + D_1 \frac{\partial}{\partial\zeta} + D_2 \frac{\partial}{\partial\eta} + D_3 \frac{\partial}{\partial\xi}.$$

Calculating the trace of the expression for Δu on M and setting it equal to zero, we get for the function χ an equation on manifold M which guarantees the harmonicity of u:

$$\begin{aligned} &a_{11}(\xi, \eta)\chi_{\xi\xi} + 2a_{12}(\xi, \eta)\chi_{\xi\eta} + a_{22}(\xi, \eta)\chi_{\eta\eta} + b_1(\xi, \eta)\chi_\xi + b_2(\xi, \eta)\chi_\eta \\ &= g(\xi, \eta) + \gamma(\xi, \eta)\chi(\xi, \eta) + \sum_{i=1}^{m} C_i e_i(\xi, \eta) + \int_M S_1(\xi, \eta, s, t)\chi(s, t)\, d\sigma \\ &+ \frac{\partial}{\partial\xi}\int_M S_2(\xi, \eta, s, t)\chi(s, t)\, d\sigma + \frac{\partial}{\partial\eta}\int_M S_3(\xi, \eta, s, t)\chi(s, t)\, d\sigma \\ &+ \int_D S_4(\xi, \eta, s, t)\chi(s, t)\, d\omega, \end{aligned} \tag{3.74}$$

where the last integral is understood in the sense of the principal value, and we have introduced the notation

$$\begin{gathered} a_{11} = [\xi_x^2 + \xi_y^2 + \xi_z^2]\,|_M, \quad a_{12} = [\xi_x\eta_x + \xi_y\eta_y + \xi_z\eta_z]\,|_M, \\ a_{22} = [\eta_x^2 + \eta_y^2 + \eta_z^2]\,|_M, \\ b_1 = \Delta\xi\,|_M, \quad b_2 = \Delta\eta\,|_M, \quad g = -\Delta H\,|_M, \quad e_i = -\Delta\psi_i\,|_M, \quad i = 1, \ldots, m, \\ S_1 = \left[D_1 \frac{\partial S_0}{\partial\zeta} + D_2 \frac{\partial S_0}{\partial\eta} + D_3 \frac{\partial S_0}{\partial\xi}\right]\Big|_M, \quad S_2 = \left[B_1 \frac{\partial S_0}{\partial\zeta} + A_1 \frac{\partial S_0}{\partial\xi} \right. \\ \left. + B_3 \frac{\partial S_0}{\partial\eta}\right]\Big|_M, \quad S_3 = \left[B_2 \frac{\partial S_0}{\partial\zeta} + A_2 \frac{\partial S_0}{\partial\eta}\right]\Big|_M, \quad S_4 = \left[\frac{\partial^2}{\partial\zeta^2} R(X, X_1)\right]\Big|_{\zeta=0}. \end{gathered}$$

We consider the Dirichlet problem with data on the line σ for the differential equation

$$a_{11}\chi_{\xi\xi} + 2a_{12}\chi_{\xi\eta} + a_{22}\chi_{\eta\eta} + b_1\chi_\xi + b_2\chi_\eta = Q(\xi, \eta).$$

Here Q is a function defined on M. Since M projects single-valuedly to the plane $z = 0$, introducing the corresponding change of variables in this equation, we get

$$A_{11}(x, y)\chi_{xx} + 2A_{12}(x, y)\chi_{xy} + A_2(x, y)\chi_{yy} + B_1(x, y)\chi_x + B_2(x, y)\chi_y = \rho(x, y) \tag{3.75}$$

on the projection of manifold M to plane $z = 0$. We denote this projection by N. A solution of the Dirichlet problem for (3.75) always exists in domain N, is unique, and is given by the formula [26]

$$\chi(x, y) = q_0(x, y) + \int_N G_1(x, y, s, t)\,\rho(s, t)\,ds dt,$$

where G_1 is the Green's function of domain N for (3.75), and q_0 is a solution which is regular in domain N of the corresponding homogeneous equation, assuming the given values on the boundary of domain N. Returning in the last formula to the variables ξ and η, we get

$$\chi(\xi, \eta) = q(\xi, \eta) + \int_M G(\xi, \eta, s, t)\,Q(s, t)\,d\sigma. \tag{3.76}$$

Setting Q equal to the right side of (3.74) in (3.76), integrating by parts, and then changing the order of integration, we arrive at the relation

$$\chi(\xi, \eta) = q(\xi, \eta) + g_0(\xi, \eta) + \sum_{i=1}^{m} C_i k_i(\xi, \eta) - \int_M T(\xi, \eta, s, t)\,\chi(s, t)\,d\sigma, \tag{3.77}$$

in which we have introduced the notation

$$g_0(\xi, \eta) = \int_M G(\xi, \eta, s, t)\,g(s, t)\,d\sigma, \quad k_i(\xi, \eta)$$

$$\int_M G(\xi, \eta, s, t)\,e_i(s, t)\,d\sigma, \quad T(\xi, \eta, s, t) = \int_M G(\xi, \eta, s_1, t_1)\,[S_1(s_1, t_1, s, t)$$

$$+ \gamma(s_1, t_1)]\,d\sigma + \int_M \left[S_2(s_1, t_1, s, t)\,\frac{\partial G(\xi, \eta, s_1, t_1)}{\partial s_1} + S_3(s_1, t_1, s, t)\right.$$

$$\left.\times \frac{\partial G(\xi, \eta, s_1, t_1)}{\partial t_1}\right] d\sigma + \int_D G(\xi, \eta, s_1, t_1)\,S_4(s_1, t_1, s, t, \tau)\,d\omega.$$

It is clear from these formulas that the kernel T of (3.77) is Fredholm.

If in (3.73) at least one constant q_i is different from zero, then we choose the indexing so that $q_1 \neq 0$, and we replace these conditions by the following:

$$\begin{gathered}\int_M p_1(\xi, \eta)\,\chi(\xi, \eta)\,d\sigma = q_1, \quad \int_M P_i(\xi, \eta)\,\chi(\xi, \eta)\,d\sigma = 0,\\ P_i(\xi, \eta) = q_1 p_i(\xi, \eta) - q_i p_1(\xi, \eta), \quad i = 2, \ldots, m.\end{gathered} \tag{3.78}$$

Since (3.77) is Fredholm, if a finite number k of orthogonality conditions imposed on the function

$$R(\xi, \eta) = q(\xi, \eta) + g_0(\xi, \eta) + \sum_{i=1}^{m} C_i k_i(\xi, \eta),$$

hold, there exists a solution of this equation containing k more new constants E_1, ..., E_k linearly. Substituting the solution found in (3.78), we get m linear algebraic equations relating the $l = m + k$ constants $C_1, \ldots, C_m, E_1, \ldots, E_k$. Moreover, the constants $C_1, \ldots, C_m$ satisfy the system of k linear equations which is obtained from the orthogonality conditions imposed on function R:

$$0 = \int_M \mu_i(\xi, \eta) R(\xi, \eta)\, d\sigma = r_i + \sum_{i=1}^{m} \int_M \mu_i(\xi, \eta) k_i(\xi, \eta)\, d\sigma, \quad i = 1, \ldots, k,$$

$$r_i = \int_M \mu_i(\xi, \eta) [q(\xi, \eta) + g_0(\xi, \eta)]\, d\sigma, \tag{3.79}$$

and the μ_i are known functions. Thus, the solution of (3.77) contains l arbitrary constants, related by $l = m + k$ linear equations (3.78) and (3.79), linearly.

The linear algebraic system relating the constants $C_1, \ldots, C_m, E_1, \ldots, E_k$ is solvable if a finite number j of conditions of the form

$$q_1\lambda_{0\mu} + \sum_{i=1}^{k} \lambda_{i\mu} r_i = 0, \quad \mu = 1, \ldots, j \leqslant l \tag{3.80}$$

hold, and its solution contains j arbitrary constants linearly [20]. Considering that $\lambda_{i\mu}$ are known numbers and the function $q(\xi, \eta)$ is determined uniquely by its values $\varphi(s)$ on the curve σ, bounding manifold M, substituting the expressions for q_1 and r_i into (3.80), these conditions acquire the form

$$\int_\sigma \alpha_i(s) \varphi(s)\, ds + \int_\Gamma \beta_i(s, t) f(s, t)\, d\Gamma = 0, \quad i = 1, \ldots, j, \tag{3.81}$$

where α_i and β_i are completely definite known functions. The relations (3.81) are orthogonality conditions imposed on the vector (φ, f). The solution of (3.77) is the following:

$$\chi(\xi, \eta) = G(\xi, \eta) + \sum_{i=1}^{j} C_i \omega_i(\xi, \eta).$$

Here G and ω_i are well-defined known functions, and C_i are arbitrary constants.

Now we investigate the smoothness of the solution of the oblique derivative problem constructed. It follows from (3.69) and (3.70) that the smoothness of u is determined by the smoothness of the functions F, h, and χ. The smoothness of $F(X) = \Phi(\xi, \eta, \zeta)$ in closed domain $D \cup \Gamma$ is the same as the smoothness of function f in (3.68). Function $h + \chi$ is a solution of the equation $w_\zeta = \Phi(\xi, \eta, \zeta)$, so for the continuity of the first derivatives of h in $D \cup \Gamma$, it suffices for Φ to be continuously differentiable in $D \cup \Gamma$, so f must also be continuously differentiable on Γ. Since χ satisfies (3.75) and on σ assumes the values φ, for the continuous differentiability of χ in $D \cup \Gamma$ it suffices that $\varphi(s)$ be continuously differentiable. Thus we have

Theorem 3.12. Let domain D be convex relative to the vector field $P = (a, b, c)$, let $c \neq 0$, and let the functions $\alpha = b/c$ and $\gamma = a/c$ satisfy (3.62) everywhere in D. Then the problem of finding a harmonic function which is regular in domain D and satisfies (3.68) on boundary Γ of domain D, and assumes previously given values $\varphi(s)$ on line σ on which the vector field P goes into the tangent plane to Γ, is Fredholm. If the functions f and φ are continuously differentiable, then u is continuously differentiable in the closed domain $D \cup \Gamma$.

We assumed above that $c \neq 0$ and the functions α and γ are regular in domain D. This means that the vector field P is continuous in D and nonzero everywhere in D. Now if at least one of the functions α or γ has a singularity in D, then the vector field P can vanish in D, and the coefficients in (3.65) can have singularities in D. These singularities cause additional difficulties in studying the oblique derivative problem with boundary condition (3.68), since the investigation of the question of the existence and properties of the Green's function for (3.66) becomes more complicated.

The reduction of the oblique derivative problem made is essentially based on the system of equations (3.62). Although this system is nonlinear and complicated, it is not overdetermined, and by the Cauchy–Kowalewsky theorem its solution is determined by two holomorphic functions of two variables, so the set of its solutions is quite large. In particular, if we set $\alpha = 0$, then for γ we get the equations $\gamma_y = 0$, $\gamma_x - \gamma\gamma_z = 0$, and if we set $\gamma = 0$ then we get the equations $\alpha_x = 0$, $\alpha_y - \alpha\alpha_z = 0$. For both these systems we can write the general solution. In particular, the function $\gamma(x, z) = 2^{-1}\{x + 1 + [(x + 1)^2 + 4z]^{1/2}\}$ satisfies the first system.

In Theorem 3.12 the condition of P-convexity is imposed on the domain D, so the question arises: Do such domains exist in general? We show that for any sufficiently smooth vector field P there exist P-convex domains. For a previously given vector field $P = (a, b, c)$ we consider a sufficiently smooth function $\psi(x, y, z)$, such that

$$a\psi_x + b\psi_y + c\psi_z \geqslant \lambda > 0, \quad \operatorname{grad} \psi \neq 0.$$

We take a piece M, homeomorphic to a disk, of the level surface $\psi = c$ of function ψ, on which $\operatorname{grad} \psi \neq 0$. On M we introduce an orthogonal coordinate system ξ and η. On M we consider a twice continuously differentiable function $h(\xi, \eta)$ positive at interior points of M, equal to zero on the boundary of M, and such that $\operatorname{grad} h \neq 0$ on the boundary of M, and inside M this function has only isolated critical points. On the trajectories of field P, passing through a point of the surface M with coordinates ξ and η, we lay off an arc of length $h(\xi, \eta)$ in the positive direction of the trajectory of the field P. Making this construction for all points of surface M, we get a domain $D^+(h)$, filled by arcs of trajectories of field P. The domain $D^+(h)$ is P-convex if its closure does not contain singular points of field P, i.e., for sufficiently small function $h(\xi, \eta)$. Analogously, attaching a domain $D^-(h)$ on the other side of the piece of surface M, we get a P-convex domain $D = D^+(h) \cup D^-(h)$ with smooth boundary. In constructing the domain $D^-(h)$ instead of h one can take another function $h_1(\xi, \eta)$ having analogous properties, and in the negative direction of field P lay off segments of length h_1 on the trajectories. The domain $D^+(h) \cup D^-(h_1)$ will also be P-convex.

We note that one can generalize Theorem 3.12 to domains D, which can be represented as unions of a finite number of P-convex domains D_k, $k = 1, \ldots, N$. This generalization is effected with the help of constructions analogous to the constructions made in Section 3 of this chapter. Now the analog of (3.74) is considered on a multi-sheeted Riemann surface.

Obviously the method of investigation of the oblique derivative problem recounted in this section can hardly be applied all the time. Its applicability is restricted in the first place by the fact that the ratios of coefficients of the boundary condition must satisfy specific restrictions which follow from (3.62), so we try to find generalizations of this method. First of all, instead of the oblique derivative problem for harmonic functions, we consider the analogous problem for regular solutions of the equation

$$\Delta u + Au_x + Bu_y + Cu_z = 0 \tag{3.82}$$

with analytic coefficients. Instead of (3.58) we now get the equations

$$\begin{aligned}
b(\lambda_1 - A) + a(\lambda_2 - B) &= -2(a_y + b_x),\\
a(\lambda_1 - A) - b(\lambda_2 - B) &= 2(b_y - a_x),\\
c(\lambda_1 - A) + a(\lambda_3 - C) &= -2(a_z + c_x),\\
a(\lambda_1 - A) - c(\lambda_3 - C) &= 2(c_z - a_x),\\
c(\lambda_2 - B) + b(\lambda_3 - C) &= -2(b_z + c_y),
\end{aligned}$$

which differ from (3.58) only in that $\lambda_1, \lambda_2, \lambda_3$ are replaced by $\lambda_1 - A, \lambda_2 - B, \lambda_3 - C$. Consequently, the oblique derivative problem for solutions of (3.82) reduces to a Fredholm integrodifferential equation only when this reduction is possible for $A = B = C = 0$. Thus, the set of boundary conditions for which the oblique derivative problem reduces by the method described to a Fredholm integrodifferential equation remains rather small even for Eqs. (3.82) with lowest terms.

We consider the more general equation with analytic coefficients

$$L(u) \equiv a_{00}u_{xx} + a_{01}u_{xy} + a_{02}u_{xz} + a_{11}u_{yy} + a_{12}u_{yz} + a_{22}u_{zz} = 0, \tag{3.83}$$

where first we shall not even require ellipticity. Let the coefficients b_0, b_1, b_2 in the expression $v = b_0u_x + b_1u_y + b_2u_z$ also be analytic and satisfy the relations

$$\begin{gathered}
Ab_0 - \mu a_{00} = l(a_{00}) - 2a_{00}\frac{\partial b_0}{\partial x} - a_{01}\frac{\partial b_0}{\partial y} - a_{02}\frac{\partial b_0}{\partial z},\ Ab_1 + Bb_0 - \mu a_{01}\\
= l(a_{01}) - a_{01}\frac{\partial b_0}{\partial x} - 2a_{11}\frac{\partial b_0}{\partial y} - a_{12}\frac{\partial b_0}{\partial z} - 2a_{00}\frac{\partial b_1}{\partial x} - a_{01}\frac{\partial b_1}{\partial y} - a_{02}\frac{\partial b_1}{\partial z},\\
Ab_2 + Cb_0 - \mu a_{02} = l(a_{02}) - a_{02}\left(\frac{\partial b_0}{\partial x} + \frac{\partial b_2}{\partial z}\right)\\
- a_{12}\frac{\partial b_0}{\partial y} - 2a_{22}\frac{\partial b_0}{\partial z} - 2a_{00}\frac{\partial b_2}{\partial x} - a_{01}\frac{\partial b_2}{\partial y},\quad Bb_1 - \mu a_{11} = l(a_{11}) - a_{01}\frac{\partial b_1}{\partial x} - 2a_{11}\frac{\partial b_1}{\partial y}\\
- a_{12}\frac{\partial b_1}{\partial z},\ Bb_2 + Cb_1 - \mu a_{12} = l(a_{12}) - a_{12}\left(\frac{\partial b_1}{\partial y} + \frac{\partial b_2}{\partial z}\right) - a_{02}\frac{\partial b_1}{\partial x} - 2a_{22}\frac{\partial b_1}{\partial z} - a_{01}\frac{\partial b_2}{\partial x}\\
- 2a_{11}\frac{\partial b_2}{\partial y},\quad Cb_2 - \mu a_{22} = l(a_{22}) - a_{02}\frac{\partial b_2}{\partial x} - 2a_{22}\frac{\partial b_2}{\partial z} - a_{12}\frac{\partial b_2}{\partial y},
\end{gathered} \tag{3.84}$$

where A, B, C and μ are analytic functions and

$$l = b_0\partial/\partial x + b_1\partial/\partial y + b_2\partial/\partial z.$$

If the relations (3.84) hold, then for any solution of (3.83) one has the identity

$$\begin{gathered} L(v) + Av_x + Bv_y + Cv_z = -[L(b_0) + \lambda(b_0)]u_x \\ -[L(b_1) + \lambda(b_1)]u_y - [L(b_2) + \lambda(b_2)]u_z, \\ \lambda = A\partial/\partial x + B\partial/\partial y + C\partial/\partial z. \end{gathered} \tag{3.85}$$

To derive (3.84) we form the following two expressions:

$$\begin{aligned} L(v) &= b_0L(u_x) + b_1L(u_y) + b_2L(u_z) + u_xL(b_0) + u_yL(b_1) + u_zL(b_2) \\ &+ u_{xx}\left[2a_{00}\frac{\partial b_0}{\partial x} + a_{01}\frac{\partial b_0}{\partial y} + a_{02}\frac{\partial b_0}{\partial z}\right] + u_{xy}\left[a_{01}\frac{\partial b_0}{\partial x} + 2a_{11}\frac{\partial b_0}{\partial y} + a_{12}\frac{\partial b_0}{\partial z}\right. \\ &\left.+ 2a_{00}\frac{\partial b_1}{\partial x} + a_{01}\frac{\partial b_1}{\partial y} + a_{02}\frac{\partial b_2}{\partial z}\right] + u_{yy}\left[a_{01}\frac{\partial b_1}{\partial x} + 2a_{11}\frac{\partial b_1}{\partial y} + a_{12}\frac{\partial b_1}{dz}\right] \\ &+ u_{yz}\left[a_{02}\frac{\partial b_1}{\partial x} + a_{12}\frac{\partial b_1}{\partial y} + 2a_{22}\frac{\partial b_1}{\partial z} + a_{01}\frac{\partial b_2}{\partial x} + 2a_{11}\frac{\partial b_2}{\partial y} + a_{12}\frac{\partial b_2}{\partial z}\right] \\ &+ u_{zz}\left[a_{02}\frac{\partial b_2}{\partial x} + a_{12}\frac{\partial b_2}{\partial y} + 2a_{22}\frac{\partial b_2}{\partial z}\right], \\ lL(u)) &= b_0\, L\,(u_x) + b_1L\,(u_y)\, b_2L\,(u_z) + u_{xx}l\,(a_{00}) + u_{xy}l\,(a_{01}) + u_{xz}l\,(a_{02}) \\ &+ u_{yy}l(a_{11}) + u_{yz}l(a_{12}) + u_{zz}l(a_{22}). \end{aligned}$$

Calculating the derivatives of function v and substituting them into the right side of (3.85), we find

$$\begin{gathered} L(v) + Av_x + Bv_y + Cv_z = l(L(u)) + (h_{00} + Ab_0)u_{xx} \\ + (h_{01} + Ab_1 + Bb_0)u_{xy} + (h_{02} + Ab_2 + Cb_0)u_{xz} + (h_{11} + Bb_1)u_{yy} \\ + (h_{12} + Bb_2 + Cb_1)u_{yz} + (h_{22} + Cb_2)u_{zz} + [L(b_0) + \lambda(b_0)]u_x \\ + [L(b_1) + \lambda(b_1)]u_y + [L(b_2) + \lambda(b_2)]u_z, \end{gathered}$$

where h_{ij} denotes the right sides of the corresponding equations of (3.84) with opposite signs. This equation does not contain second derivatives of function u and turns into (3.85) if the operator

$$\begin{gathered} M(u) = (h_{00} + Ab_0)u_{xx} + (h_{01} + Ab_1 + Bb_0)u_{xy} + (h_{02} + Ab_2 + Cb_0)u_{xz} \\ + (h_{11} + Bb_1)u_{yy} + (h_{12} + Bb_2 + Cb_1)u_{yz} + (h_{22} + Cb_2)u_{zz} \end{gathered}$$

has the form $M(u) = \mu L(u)$, where μ is a fixed function, and this latter equation is equivalent to the conditions (3.84) holding.

Let the functions b_0, b_1, b_2 satisfy the inequality

$$b_0^2 + b_1^2 + b_2^2 > 0, \tag{3.86}$$

i.e., they do not vanish simultaneously anywhere in their domain D of definition. Now the first-order equation $b_0\zeta_x + b_1\zeta_y + b_2\zeta_z = 1$ has a holomorphic solution ζ and the equation $b_0\omega_x + b_1\omega_y + b_2\omega_z = 0$ has two holomorphic independent integrals ξ and η [33]. Let us assume that ζ, η, and ξ are holomorphic in the closed domain $D \cup \Gamma$. In the domain $D \cup \Gamma$ there exists an analytic function $\lambda \neq 0$ such that $\lambda b_0 = \xi_y \eta_z - \xi_z \eta_y$, $\lambda b_1 = \xi_z \eta_x - \xi_x \eta_z$, $\lambda b_2 = \xi_x \eta_y - \xi_y \eta_x$. Instead of the independent variables x, y, and z in the domain D we introduce new independent variables ξ, η, ζ. The Jacobian of this change is the following:

$$J = \frac{\partial(\xi, \eta, \zeta)}{\partial(x, y, z)} = \lambda \neq 0.$$

In the new variables (3.83) assumes the form

$$N(u) \equiv A_{00} u_{\xi\xi} + A_{01} u_{\xi\eta} + A_{02} u_{\xi\zeta} + A_{11} u_{\eta\eta} + A_{12} u_{\eta\zeta} + A_{22} u_{\zeta\zeta} + B_1 u_\xi + B_2 u_\eta + B_3 u_\zeta = 0. \tag{3.87}$$

$$A_{00} = a_{00}\xi_x^2 + a_{01}\xi_x\xi_y + a_{02}\xi_x\xi_z + a_{11}\xi_y^2 + a_{12}\xi_y\xi_z + a_{22}\xi_z^2,\ A_{01} = 2a_{00}\xi_x\eta_x + a_{01}(\xi_x\eta_y + \eta_x\xi_y) + a_{02}(\xi_x\eta_z + \eta_x\xi_z) + 2a_{11}\xi_y\eta_y + a_{12}(\xi_y\eta_z + \eta_y\xi_z) + 2a_{22}\xi_z\eta_z,$$

$$A_{02} = 2a_{00}\xi_x\zeta_x + a_{01}(\xi_x\zeta_y + \zeta_x\xi_y) + a_{02}(\xi_x\zeta_z + \zeta_x\xi_z) + 2a_{11}\xi_y\zeta_y + a_{12}(\xi_y\zeta_z + \xi_z\zeta_y) + 2a_{22}\xi_z\zeta_z,\ A_{11} = a_{00}\eta_x^2 + a_{01}\eta_x\eta_y + a_{02}\eta_x\eta_z + a_{11}\eta_y^2 + a_{12}\eta_y\eta_z + a_{22}\eta_z^2,\ A_{12} = 2a_{00}\eta_x\zeta_x + a_{01}(\eta_x\zeta_y + \zeta_x\eta_y) + a_{02}(\eta_x\zeta_z + \zeta_x\eta_z) + 2a_{11}\eta_y\zeta_y + a_{12}(\eta_y\zeta_z + \zeta_y\eta_z) + 2a_{22}\eta_z\zeta_z,\ A_{22} = a_{00}\zeta_x^2 + a_{01}\eta_x\zeta_y + a_{02}\zeta_x\zeta_z + a_{12}\zeta_y\zeta_z + a_{22}\zeta_z^2,\ B_1 = L(\xi),\ B_2 = L(\eta),\ B_3 = L(\zeta),$$

and function v goes into the function u_ζ. Obviously in the new variables the conditions (3.84) are equivalent to the equations

$$A_{00}^{-1}\frac{\partial A_{00}}{\partial \zeta} = A_{01}^{-11}\frac{\partial A_{01}}{\partial \zeta} = A_{11}^{-1}\frac{\partial A_{11}}{\partial \zeta} = \mu, \tag{3.88}$$

where we have

$$N(v) + \left(\frac{\partial A_{02}}{\partial \zeta} - \mu A_{02}\right) v_\xi + \left(\frac{\partial A_{12}}{\partial \zeta} - \mu A_{12}\right) v_\eta - \frac{\partial}{\partial \zeta} N(u) - \mu N(u) + \left(\frac{\partial A_{22}}{\partial \zeta} - \mu A_{22}\right) v_\zeta - \left(\frac{\partial B_1}{\partial \zeta} - \mu B_1\right) u_\xi - \left(\frac{\partial B_2}{\partial \zeta} - \mu B_2\right) u_\eta - \left(\frac{\partial B_3}{\partial \zeta} - \mu B_3\right) u_\zeta = 0.$$

If function u satisfies the equation $N(u) = 0$, then this equation assumes the form

$$N(v) + \left(\frac{\partial A_{02}}{\partial \zeta} - \mu A_{02}\right) v_\xi + \left(\frac{\partial A_{12}}{\partial \zeta} - \mu A_{12}\right) v_\eta + \left(\frac{\partial A_{22}}{\partial \zeta} - \mu A_{22}\right) v_\zeta = \left(\mu B_1 - \frac{\partial B_1}{\partial \zeta}\right) u_\xi + \left(\mu B_2 -- \frac{\partial B_2}{\partial \zeta}\right) u_\eta + \left(\mu B_3 - \frac{\partial B_3}{\partial \zeta}\right) u_\zeta. \tag{3.89}$$

From (3.88) it follows easily that

$$A_{00} = H(\xi, \eta, \zeta)\alpha(\xi, \eta),\ A_{01} = H(\xi, \eta, \zeta)\beta(\xi, \eta),$$

$$A_{11} = H(\xi, \eta, \zeta)\,\gamma(\xi, \eta),\ H = \exp\left\{\int \mu(\xi, \eta, t)\,dt\right\}.$$

If (3.83) is elliptic in domain D, then (3.87) is also elliptic in domain D_1, to which D is mapped by the change of variables x, y, z to ξ, η, ζ. By the ellipticity of operator N the function H does not vanish at any point of domain D_1, because otherwise the characteristic form $\Omega(\lambda_1, \lambda_2, \lambda_3)$ of operator N for $\lambda_3 = 0$ would vanish for any λ_1 and λ_2 at those points of domain D_1 at which $H = 0$, which contradicts the ellipticity of (3.87). Dividing (3.87) by H, we reduce it to the form

$$N_0(u) \equiv a_{00}(\xi, \eta)u_{\xi\xi} + a_{01}(\xi, \eta)u_{\xi\eta} + A_{02}(\xi, \eta, \zeta)u_{\xi\zeta} + a_{11}(\xi, \eta)u_{\eta\eta} + A_{12}(\xi, \eta, \zeta)u_{\eta\zeta} + A_{22}(\xi, \eta, \zeta)u_{\zeta\zeta} + \alpha u_\xi + \beta u_\eta + \gamma u_\zeta = 0, \tag{3.90}$$

where all the coefficients are arbitrary functions of their arguments such that the form $\Omega_0(\lambda_1, \lambda_2, \lambda_3) = a_{00}\lambda_1^2 + a_{01}\lambda_1\lambda_2 + A_{02}\lambda_1\lambda_3 + a_{11}\lambda_2^2 + A_{12}\lambda_2\lambda_3 + A_{22}\lambda_3^2$ has the same sign at all points of domain D_1 for any $\lambda_1, \lambda_2, \lambda_3, \lambda_1^2 + \lambda_2^2 + \lambda_3^2 \neq 0$.

If the vector field $P = (b_0, b_1, b_2)$ which figures in the boundary condition of the oblique derivative problem

$$b_0u_x + b_1u_y + b_2u_z = f \tag{3.91}$$

for (3.83) can be extended twice continuously differentiably from boundary Γ of domain D to domain D so that (3.86) holds, then there exists an infinite set of equations of the form (3.90), for which (3.84) is valid. For such equations, (3.89) assumes the form

$$N(v) = \frac{\partial A_{02}}{\partial \zeta} v_\xi + \frac{\partial A_{12}}{\partial \zeta} v_\eta + \frac{\partial A_{22}}{\partial \zeta} v_\zeta = 0, \tag{3.92}$$

if one sets $\alpha = \beta = \gamma = 0$.

Now suppose at the point $Y \in D$ we have $b_0 = b_1 = b_2 = 0$, while at this point all roots of the polynomial

$$Q(\lambda) = \begin{vmatrix} \partial b_0/\partial x - \lambda & \partial b_0/\partial y & \partial b_0/\partial z \\ \partial b_1/\partial x & \partial b_1/\partial y - \lambda & \partial b_1/\partial z \\ \partial b_2/\partial x & \partial b_2/\partial y & \partial b_2/\partial z - \lambda \end{vmatrix}$$

form are real and different from zero. In this case, by a change of variables which is one-to-one in a neighborhood of point Y, the functions b_i can be reduced to the norm $b_0 = \delta_1\xi$, $b_1 = \delta_2\eta$, $b_3 = \delta_3\zeta$, $\delta_i = \operatorname{sign}\lambda_i$, $i = 1, 2, 3$ [33]. We again denote the variables ξ, η, ζ by x, y, z. All the λ_i can have the same sign or two have the same and the third the opposite sign, so without loss of generality one can consider two versions: $b_0 = x$, $b_1 = y$, $b_2 = z$, and $b_0 = x$, $b_1 = y$, $b_2 = -z$, to which all other cases are easily reduced. In the first case, along with function u, function $v = xu_x + yu_y + zu_z$ is also harmonic, and in the second the function $v = xu_x + yu_y - zu_z$ is related to the function u by

$$L(v) = \left(x\frac{\partial}{\partial x} + y\frac{\partial}{\partial y} - z\frac{\partial}{\partial z}\right) L(u) - 2L(u),$$

where L is the degenerate elliptic operator

$$L(u) = (x^2 + y^2)^2(u_{xx} + u_{yy}) + u_{zz}.$$

Consequently, for any solution of the equation $L(u) = 0$ we have $L(v) = 0$. In this case the operator L is degenerate; however, the Dirichlet problem is well-posed for the equation $L(u) = 0$, and the maximum principle is valid.

Definition 3.4. The second-order operator L and first-order operator l

$$L = \sum_{i,j=0}^{m} a_{ij}(X)\frac{\partial^2}{\partial x_i \partial x_j}, \quad l = \sum_{i=0}^{m} b_i(X)\frac{\partial}{\partial x_i}, \quad X = (x_0, \ldots, x_m),$$

will be said to commute up to lowest terms if there exist functions A_i, $i = 0, \ldots, m$, B_i, $i = 0, \ldots, m$, and C such that for any twice differentiable function u one has the identity

$$L(l(u)) + \sum_{i=1}^{m} A(X)\frac{\partial}{\partial x_i} l(u) = l(L(u)) + CL(u) + \sum_{i=0}^{m} B_i(X)\frac{\partial u}{\partial x_i}. \tag{3.93}$$

If operators L and l commute up to lowest terms and the function u satisfies the equation

$$L(l(u)) + \sum_{i=0}^{m} A_i(X)\frac{\partial}{\partial x_i} l(u) = \sum_{i=0}^{m} B_i(X)\frac{\partial u}{\partial x_i}, \tag{3.94}$$

then it follows from (3.93) that in this case

$$l(L(u)) + C(X)L(u) = 0. \tag{3.95}$$

Consequently, for any solution of (3.94) we have

$$L(u) = w, \quad l(w) + C(X)w = 0, \tag{3.96}$$

and it follows from this that in order that the solution u of (3.94) also be a solution of the equation $L(u) = 0$, it is necessary and sufficient that the equation $L(u) = 0$ holds only on a two-dimensional surface M, which is not tangent to a characteristic of the operator l at any of its points, i.e., on a surface which can support the initial data for the second equation (3.96).

Definition 3.5. Let domain D and the first-order operator l defined in it satisfy the following conditions:

1) in the domain D there exists an m-dimensional connected analytic set M such that domain D is filled by characteristics of the operator l, passing through points of M;

2) M is an m-dimensional real-analytic manifold at all of its points at which

$$\sum_{i=0}^{m} [b_i(X)]^2 > 0;$$

3) each characteristic of the operator l, which intersects D generally, intersects D along a connected set, and M in a unique point.

Then we shall call domain D convex with respect to the operator l or l-convex.

If in addition to conditions 1–3 the following condition also holds: 4) the set of points N of boundary Γ of domain D, at which the characteristics of operator l are tangent to Γ, consists of a finite number of disjoint manifolds of dimension $m-1$, then we shall call domain D strongly convex with respect to operator l, or strongly l-convex.

Suppose given in a strongly l-convex domain D along with the first-order operator l a uniformly elliptic operator L which commutes up to smallest terms with operator l. For simplicity we shall assume that the coefficients of both operators l and L are analytic in the closed domain $\bar{D}$. For solutions of the equation $L(u) = 0$ which are regular in domain D we consider the oblique derivative problem with the boundary condition $l(u) = f$ on Γ, where f is a continuously differentiable function defined on Γ. We restrict ourselves to a three-dimensional domain D whose boundary has continuous principal curvatures. By (3.94) we have

$$L(v) + A_1(X)v_x + A_2(X)v_y + A_3(X)v_z = B_1(X)u_x + B_2(X)u_y + B_3(X)u_z, \quad (3.97)$$

where $X = (x, y, z)$, $l(u) = au_x + bu_y + cu_z$, and A_i, B_i, $i = 1, 2, 3$ are known functions. Further, this problem can be investigated in exactly the same way that the problem (3.68) was investigated for Laplace's equation.

With the help of the Green's function and integration by parts from (3.97), we find

$$v(X) = F(X) + \int_D K_1(X, Q)\, v(Q)\, d\omega + \int_D K_2(X, Q)\, u(Q)\, d\omega. \quad (3.98)$$

Here K_1 and K_2 are well-defined kernels with weak singularities, and F is a solution of the equation $L(v) = 0$ which is regular in domain D and which, on boundary Γ of domain D, satisfies condition $v = f$. By virtue of the uniqueness of the solution of the Dirichlet problem for the homogeneous equation corresponding to (3.97), from (3.98) one can express the function v as follows [30]:

$$v(X) = h(X) + \int_D L(X, Q)\, u(Q)\, d\omega, \quad (3.99)$$

where L is a new Fredholm kernel which can be expressed in terms of the kernel K_2 and the resolvent R of the kernel K_1, and

$$h(X) = F(X) + \int_D R(X, Q)\, F(Q)\, d\omega$$

is a known function. Substituting the expression for v in terms of u into (3.99), we arrive at the integrodifferential equation

$$au_x + bu_y + cu_z = h(X) + \int_D L(X, Q)\, u(Q)\, d\omega, \tag{3.100}$$

which is equivalent to (3.95), where in our case $C \equiv 0$. Although (3.100) is not equivalent to the original oblique derivative problem, all solutions of this problem are included among the solutions of (3.100).

From the conditions imposed on domain D it follows that there exist unique solutions of the following Cauchy problems:

$$aH_x + bH_y + cH_z = h, \quad H|_M = 0,$$
$$aP_x + bP_y + cP_z = L, \quad P|_M = 0,$$

where M is the surface which occurs in the definition of the l-convexity of domain D. Integrating (3.100) along characteristics of operator l, we get

$$u(X) = H(X) + g(X) + \int_D P(X, A)\, u(A)\, d\omega. \tag{3.101}$$

Here g is a general solution of the equation $l(g) = 0$. Formula (3.101) is a Fredholm equation with respect to function u, so we have [30]

$$u(X) = G(X) + g(X) + \sum_{i=1}^{m} C_i \psi_i(X) + \int_D R(X, A)\, g(A)\, d\omega, \tag{3.102}$$

$$\int_D \chi_i(X)\,[H(X) + g(X)]\, d\omega = 0, \quad i = 1, \ldots, k, \tag{3.103}$$

where R is the generalized resolvent of kernel P, G, ψ_i, χ_i are known functions, and C_i are arbitrary constants. Since the kernel P vanishes on the surface M, $R = 0$ when $X \in M$. Function g is constant along characteristics of operator l. Now it remains to choose function g so that $L(u) = 0$ for function u from (3.102), and for this it suffices that $L(u) = 0$ only on surface M. Calculating $L(u)$ and equating the result to zero on M, as in the investigation of the problem (3.68) for the Laplace equation, we get an integrodifferential equation for g on surface M, which can be investigated in exactly the same way that (3.74) was investigated. Analogously to Theorem 3.12, one proves the following:

Theorem 3.13. If domain D is strongly l-convex and the operators L and l commute up to smallest terms, then the problem $L(u) = 0$ in D, $l(u) = f$ on Γ, $u = \varphi$ on $\Gamma \cap M$, where f and φ are arbitrary continuously differentiable functions, can always be solved and the solution is unique.

In investigating the oblique derivative problem in this section, the investigation of (3.100) is the essential difficulty even in the case where the operators L and l commute and (3.100) does not contain terms with integral. This is well illustrated by the example $L = \Delta$, $l = y\partial/\partial x - x\partial/\partial y$. For harmonic functions which are regular in the ball Σ: $\{(x - x_0)^2 + (y - y_0)^2 + z^2 < R^2\}$, $x_0^2 + y_0^2 > R^2$, we consider the oblique derivative problem with the boundary condition given on sphere S bounding Σ

$$yu_x - xu_y = f. \tag{3.104}$$

This problem can always be solved. A solution of the corresponding homogeneous problem is any harmonic function which is symmetric with respect to the Oz axis and regular in the domain $D = \{(\sqrt{x^2+y^2} - \sqrt{x_0^2+y_0^2})^2 + z^2 < R^2\}$, formed by rotating ball Σ about the Oz axis. If instead of ball Σ we take the domain E: $\{(\sqrt{x^2+y^2} - \rho_0)^2 + z^2 < (R + rx\sqrt{x^2+y^2})^2\}$, $\rho_0 > R > r > 0$, then any harmonic function which is symmetric with respect to the Oz axis and regular in the domain E_1: $\{(\sqrt{x^2+y^2} - \rho_0)^2 + z^2 < (R + r)^2\}$ will be a solution of the homogeneous problem corresponding to (3.104). Now the inhomogeneous problem (3.104) is solvable in domain E only for those f which are the values on the boundary of harmonic functions w, which are regular in domain E, and which satisfy

$$\int_q w\,ds = 0, \quad q: \{x^2 + y^2 = \rho^2, z = z_0\}, \quad (\rho - \rho_0)^2 + z_0^2 \leqslant (R - r)^2.$$

All such circles q are closed characteristics of the operator l, lying entirely in domain E. Obviously in the latter case the l-convexity condition is violated, i.e., the domain is not convex with respect to operator l.

Remark. Let domain D be strongly l-convex, and the operators L and l commute up to lowest terms; let $B(\epsilon)$ be the set of points of domain D whose distance from boundary Γ of domain D does not exceed $\epsilon > 0$. Obviously for any second-order elliptic operator M with sufficiently smooth coefficients coinciding with L on $B(\epsilon)$ for any fixed arbitrarily small $\epsilon > 0$, the assertion of Theorem 3.13 remains valid since any solution of the problem $L(u) = 0$ in D, $l(u) = f$ on Γ is a solution of the problem $M(u) = 0$ in D, $l(u) = f$ on Γ, and conversely.

6. BOUNDARY PROBLEM FOR A SYSTEM OF HARMONIC FUNCTIONS

The following problem for a system of harmonic functions is very similar to the oblique derivative problem: Find n functions $u_1, \ldots, u_n$ which are regular in domain D of three-dimensional space, continuously differentiable in closed domain $\bar{D}_n$, and which, on boundary Γ of domain D, satisfy the conditions

$$\sum_{i=1}^{n} \left[a_{ij}(X)\frac{\partial u_i}{\partial x} + b_{ij}(X)\frac{\partial u_i}{\partial y} + c_{ij}(X)\frac{\partial u_i}{\partial z}\right] = f_j(X), \; j = 1, \ldots, n, \tag{3.105}$$

where $X = (x, y, z)$ and a_{ij}, b_{ij}, c_{ij}, and f_j are Hölder continuous functions given on Γ. For $n = 1$ this problem becomes the oblique derivative problem. To investigate the problem (3.105) one can adapt many methods used in studying the oblique derivative problem.

In some cases one can reduce problem (3.105) to the investigation of a system of Fredholm equations. Let us assume that in domain D there exist analytic functions A_{ij}, B_{ij}, C_{ij} such that on Γ one has

$$a_{ij} = A_{ij}, \; b_{ij} = B_{ij}, \; c_{ij} = C_{ij}, \quad i, j = 1, \ldots, n.$$

In domain D we consider the functions

$$v_j(X) = \sum_{i=1}^{n} \left[A_{ij}(X) \frac{\partial u_i}{\partial x} + B_{ij}(X) \frac{\partial u_i}{\partial y} + C_{ij}(X) \frac{\partial u_i}{\partial z} \right],$$
$$j = 1, \ldots, n.$$

From them we form the expressions

$$L_i = \Delta v_j + \sum_{i=1}^{n} \left[\alpha_{ij} \frac{\partial v_i}{\partial x} + \beta_{ij} \frac{\partial v_i}{\partial y} + \gamma_{ij} \frac{\partial v_i}{\partial z} \right].$$

Since all the u_i are harmonic, we find

$$L_j = \sum_{i=1}^{n} \left[P_{ij} \frac{\partial u_i}{\partial x} + Q_{ij} \frac{\partial u_i}{\partial y} + R_{ij} \frac{\partial u_i}{\partial z} \right] + \sum_{i=1}^{n} \left[\Omega_{ij}^{(1)} \frac{\partial^2 u_i}{\partial x^2} + \Omega_{ij}^{(2)} \frac{\partial^2 u_i}{\partial x \partial y} + \right.$$
$$\left. + \Omega_{ij}^{(3)} \frac{\partial^2 u_i}{\partial x \partial z} + \Omega_{ij}^{(4)} \frac{\partial^2 u_i}{\partial y^2} + \Omega_{ij}^{(5)} \frac{\partial^2 u_i}{\partial y \partial z} \right],$$

where we have introduced the notation

$$\Omega_{ij}^{(1)} = 2\left(\frac{\partial A_{ij}}{\partial x} - \frac{\partial C_{ij}}{\partial z} \right) + \sum_{k=1}^{n} (\alpha_{kj} A_{ik} - \gamma_{kj} C_{ik}),$$

$$\Omega_{ij}^{(2)} = 2\left(\frac{\partial A_{ij}}{\partial y} + \frac{\partial B_{ij}}{\partial x} \right) + \sum_{k=1}^{n} (\alpha_{kj} B_{ik} + \beta_{kj} A_{ik}),$$

$$\Omega_{ij}^{(3)} = 2\left(\frac{\partial A_{ij}}{\partial z} + \frac{\partial C_{ij}}{\partial x} \right) + \sum_{k=1}^{n} (\alpha_{kj} C_{ik} + \gamma_{kj} A_{ik}),$$

$$\Omega_{ij}^{(4)} = 2\left(\frac{\partial B_{ij}}{\partial y} - \frac{\partial C_{ij}}{\partial z} \right) + \sum_{k=1}^{n} (\beta_{ki} B_{ik} - \gamma_{kj} C_{ik}),$$

$$\Omega_{ij}^{(5)} = 2\left(\frac{\partial B_{ij}}{\partial z} + \frac{\partial C_{ij}}{\partial y} \right) + \sum_{k=1}^{n} (\gamma_{kj} B_{ik} + \beta_{kj} C_{ik}),$$

and P_{ij}, Q_{ij}, R_{ij} can also be expressed in a known way in terms of α_{ij}, β_{ij}, γ_{ij}, A_{ij}, B_{ij}, C_{ij}. In order that the expressions L_j not contain the second derivatives of the function u, it is necessary and sufficient that the following conditions hold:

$$\Omega_{ij}^{(l)} = 0, \quad i, j = 1, \ldots, n, \; l = 1, 2, 3, 4, 5. \tag{3.106}$$

These equations are a system of $5n^2$ equations with respect to the $6n^2$ functions A_{ij}, B_{ij}, C_{ij}, α_{ij}, β_{ij}, γ_{ij}. From any $3n^2$ equations of the system (3.106) one can express the functions α_{ij}, β_{ij}, γ_{ij} in terms of A_{ij}, B_{ij}, C_{ij} and their first derivatives. Substituting these expressions in the remaining $2n^2$ equations of (3.106), we get a system of $2n^2$ nonlinear differential equations for the $3n^2$ functions A_{ij}, B_{ij}, C_{ij}.

We consider the matrices $A = \|A_{ij}\|$, $B = \|B_{ij}\|$, $C = \|C_{ij}\|$, and we shall assume that at least one of them is nondegenerate. To be definite, say C is nondegenerate; then (3.105) can be rewritten as follows:

$$\frac{\partial u_j}{\partial z} + \sum_{i=1}^{n} \left[p_{ij}(X) \frac{\partial u_i}{\partial x_i} + q_{ij}(X) \frac{\partial u_i}{\partial y} \right] = g_j(X),$$

where the matrices $P = \|p_{ij}\|$, $Q = \|q_{ij}\|$ have the form $P = AC^{-1}$, $Q = BC^{-1}$, and the functions g_j are linear combinations of the functions f_j with unknown coefficients. Consequently, without loss of generality one can assume the matrix C is the identity. In the equations relating A_{ij}, B_{ij}, C_{ij}, we set $C_{ij} = 0$, $i \neq j$, $C_{ii} = 1$; then we get a system of $2n^2$ equations for $2n^2$ unknown functions.

Considering (3.106), we get the relations

$$\Delta v_j + \sum_{i=1}^{n} \left[\alpha_{ij} \frac{\partial v_i}{\partial x} + \beta_{ij} \frac{\partial v_i}{\partial y} + \gamma_{ij} \frac{\partial v_i}{\partial z} \right] = \sum_{i=1}^{n} \left[P_{ij}(X) \frac{\partial u_i}{\partial x} + Q_{ij}(X) \frac{\partial u_i}{\partial y} + R_{ij}(X) \frac{\partial u_i}{\partial z} \right], \quad j = 1, \ldots, n. \tag{3.107}$$

From these equations, just as in investigating the oblique derivative problem, we find

$$\sum_{i=1}^{n} A_{ij} \frac{\partial u_i}{\partial x} + B_{ij} \frac{\partial u_i}{\partial y} + C_{ij} \frac{\partial u_i}{\partial z} \Big] = g_j(X) + \int_D \sum_{i=1}^{n} K_{ij}(X, Y) u_i(Y) \, dY, \tag{3.108}$$

$$j = 1, \ldots, n,$$

where the g_j are harmonic functions which are regular in domain D, which coincide with f_j on Γ, and K_{ij} are well-defined Fredholm kernels. On the basis of the system of integral equations (3.108), problem (3.105) can be investigated just as the oblique derivative problem was with the help of (3.69).

The most essential point in the reduction given of (3.105) to the system of equations (3.108) is the study of the system of equations arising from the conditions (3.106). If we set $\alpha_{ij} = \beta_{ij} = \gamma_{ij} = 0$, then these conditions split into n^2 systems of the form

$$A_x - C_z = 0, \; A_y + B_x = 0, \; A_z + C_x = 0, \; B_y - C_z = 0, \; B_z + C_y = 0, \tag{3.109}$$

where now in (3.107) the first derivatives of the functions v_j do not occur. The functions v_j will be harmonic for any harmonic u_i if and only if each triple of functions A_{ij}, B_{ij}, C_{ij} is a linear solution of the system (3.109). If these functions are such that $v(X)$ is a regular harmonic function for any harmonic function $u(X)$, then problem (3.105) reduces to finding harmonic solutions of the first-order system of linear equations

$$\sum_{i=1}^{n}\left[A_{ij}(X)\frac{\partial u_i}{\partial x}+B_{ij}(X)\frac{\partial u_i}{\partial y}+C_{ij}(X)\frac{\partial u_i}{\partial z}\right]=F_j(X),\quad j=1,\ldots,n, \tag{3.110}$$

where F_j is a harmonic function which is regular in domain D and coincides with f_j on boundary Γ of domain D. In particular, this holds when all the coefficients a_{ij}, b_{ij}, c_{ij} in (3.105) are constant.

To investigate problem (3.105) one can generalize the method of Bouligand–Giraud of reduction to Fredholm equations. We illustrate this with the case $j = 2$. Now we seek two harmonic functions u and v, which are regular in domain D and satisfy on boundary Γ of the domain the conditions

$$a_j u_x + b_j u_y + c_j u_z + \alpha_j v_x + \beta_j v_y + \gamma_j v_z = f_j,\quad j=1,\ 2, \tag{3.111}$$

where a_j, b_j, c_j, α_j, β_j, γ_j, f_j are functions given on Γ. To study this problem it is necessary to construct the matrix analog of the function $\omega(X, A)$ considered in Section 4 of Chapter 2. At first we shall assume the coefficients of the left side of boundary condition (3.111) are constant. We shall seek a solution of the problem in the form

$$\begin{aligned} u(X) &= \int_\Gamma [G_0(X,A)\,\mu_1(A) + G_1(X,A)\,\mu_2(A)]\,d\sigma,\\ v(X) &= \int_\Gamma [H_0(X,A)\,\mu_1(A) + H_1(X,A)\,\mu_2(A)]\,d\sigma. \end{aligned} \tag{3.112}$$

Here $X = (x, y, z)$ is a point of domain D, $A = (a, b, c)$ is an arbitrary point, μ_1 and μ_2 are unknown functions, and G_0, G_1, H_0, H_1 are harmonic solutions of the equation

$$\begin{aligned} a_1G_{0x} + b_1G_{0y} + c_1G_{0z} + \alpha_1H_{0x} + \beta_1H_{0y} + \gamma_1H_{0z} &= \frac{\partial}{\partial n}(r^{-1}),\\ a_2G_{0x} + b_2G_{0y} + c_2G_{0z} + \alpha_2H_{0x} + \beta_2H_{0y} + \gamma_2H_{0z} &= 0,\\ a_1G_{1x} + b_1G_{1y} + c_1G_{1z} + \alpha_1H_{1x} + \beta_1H_{1y} + \gamma_1H_{1z} &= 0,\\ a_2G_{1x} + b_2G_{1y} + c_2G_{1z} + \alpha_2H_{1x} + \beta_2H_{1y} + \gamma_2H_{1z} &= \frac{\partial}{\partial n}(r^{-1}), \end{aligned} \tag{3.113}$$

which are regular in domain D, where in these equations X varies over the whole space and the point A over the closure of the complement of domain D, and the normal derivative to Γ is taken with respect to the coordinates of point A. For now we set aside the question of conditions on domain D which guarantee the existence of the solutions needed of system (3.113). Assuming that there exist functions G_i and H_i, $i = 0, 1$, we substitute expressions (3.112) and (3.111) and we let the point $X = (x, y, z)$ of domain D tend to boundary Γ of domain D. By the equations of (3.113) and the properties of the potential of a double layer, we get

$$\begin{aligned} \mu_1(X) + \frac{1}{2\pi}\int_\Gamma \frac{\partial}{\partial n}(r^{-1})\,\mu_1(A)\,d\sigma &= \frac{1}{2\pi} f_1(X),\\ \mu_2(X) + \frac{1}{2\pi}\int_\Gamma \frac{\partial}{\partial n}(r^{-1})\,\mu_2(A)\,d\sigma &= \frac{1}{2\pi} f_2(X). \end{aligned} \tag{3.114}$$

To each solution μ_1, μ_2 of the system of integral equations (3.114), the formulas (3.112) associate a solution of problem (3.111), but it is unclear whether all solutions of (3.111) can be represented in the form (3.112), so here, as in the oblique derivative problem, the important point is the investigation of the equivalence of the given reduction to problem (3.111).

In the case where the coefficients of the left sides of the boundary conditions (3.111) are variable, we shall define the functions G_i and H_i, $i = 0, 1$, from the equations

$$\begin{gathered}
a_1(A)\,G_{0x} + b_1(A)\,G_{0y} + c_1(A)\,G_{0z} + \alpha_1(A)\,H_{0x} + \beta_1(A)\,H_{0y} \\
+ \gamma_1(A)\,H_{0z} = \frac{\partial}{\partial n}(r^{-1}), \\
a_2(A)G_{0x} + b_2(A)G_{0y} + c_2(A)G_{0z} + \alpha_2(A)H_{0x} + \beta_2(A)H_{0y} + \gamma_2(A)H_{0z} = 0; \\
a_1(A)G_{1x} + b_1(A)G_{1y} + c_1(A)G_{1z} + \alpha_1(A)H_{1x} + \beta_1(A)H_{1y} + \gamma_1(A)H_{1z} = 0, \\
a_2(A)\,G_{1x} + b_2(A)\,G_{1y} + c_2(A)\,G_{1z} + \alpha_2(A)\,H_{1x} \\
+ \beta_2(A)\,H_{1y} + \gamma_2(A)\,H_{1z} = \frac{\partial}{\partial n}(r^{-1}).
\end{gathered} \tag{3.115}$$

Now instead of (3.114) one gets the following system of integral equations:

$$\begin{gathered}
\mu_1(X) + \frac{1}{2\pi}\int_\Gamma \frac{\partial}{\partial n}(r^{-1})\,\mu_1(A)\,d\sigma + \int_\Gamma [L_{11}(X, A)\,\mu_1(A) + L_{12}(X, A)\,\mu_2(A)]\,d\sigma = \frac{1}{2\pi}f_1(X), \\
\mu_2(X) + \frac{1}{2\pi}\int_\Gamma \frac{\partial}{\partial n}(r^{-1})\,\mu_2(A)\,d\sigma + \int_\Gamma [L_{21}(X, A)\,\mu_1(A) \\
+ L_{22}(X, A)\,\mu_2(A)]\,d\sigma = \frac{1}{2\pi}f_2(X),
\end{gathered} \tag{3.116}$$

where L_{ij}, $i, j = 1, 2$, are unknown kernels. In general these kernels are not necessarily Fredholm, but if they are Fredholm, then, with the help of system (3.116), one can construct some solutions of problem (3.111). It is by no means simpler to establish that the L_{ij} are Fredholm than it was for the kernels which arise in investigating the oblique derivative problem.

In investigating problem (3.111), aside from clarifying whether the kernels L_{ij} in (3.116) are Fredholm and establishing the equivalence of these equations to the problem, there arise complications in the construction of the kernels in the integral representations (3.112), i.e., in seeking solutions of systems (3.113) and (3.114). Obviously the Fredholm character of L_{ij} depends essentially on the properties of the functions G_i and H_i, $i = 0, 1$, i.e., in the end, on the structure of the solutions of systems (3.113) and (3.115). By the second equation of system (3.113) one can set $G_0(X, A) = \alpha_2\Omega_x + \beta_2\Omega_y + \gamma_2\Omega_z$, $H_0(X, A) = -a_2\Omega_x - b_2\Omega_y - c_2\Omega_z$, where $\Omega(X, A)$ is a function which is harmonic in the variable $X = (x, y, z)$. Substituting these expressions in the first equation of (3.113), we get for the function Ω the second-order equation

$$(a_1\alpha_2 - \alpha_1 a_2)\,\Omega_{xx} + (a_1\beta_2 + b_1\alpha_2 - b_2\alpha_1 - a_2\beta_1)\,\Omega_{xy} + \tag{3.117}$$

$$+ (a_1\gamma_2 + c_1\alpha_2 - \alpha_1 c_2 - \gamma_1 a_2)\,\Omega_{xz} + (b_1\beta_2 - \beta_1 b_2)\,\Omega_{yy}$$
$$+ (c_1\beta_2 + b_1\gamma_2 - \gamma_1 b_2 - \beta_1 c_2)\,\Omega_{yz} + (c_1\gamma_2 - \gamma_1 c_2)\,\Omega_{zz} = \frac{\partial}{\partial n}(r^{-1}).$$

Now to construct kernels G_0 and H_0 it suffices to find a particular harmonic solution of (3.117), which is regular in domain D. Analogously, from the third equation of (3.113) we find $G_1(X, A) = \alpha_1 E_x + \beta_1 E_y + \gamma_1 E_z$, $H_1(X, A) = -a_1 E_x - b_1 E_y - c_1 E_z$, where $E(X, A)$ is an arbitrary function of $X = (x, y, z)$, which is harmonic in domain D. Substituting these expressions for G_1 and H_1 in terms of E into the fourth equation of (3.113), we get for $E(X, A)$ an equation which differs from (3.117) only in that the signs of all the coefficients of the left side of the equation are replaced by the opposite ones. Consequently, we can set $E(X, A) = -\Omega(X, A)$ and, finally we get

$$G_0(X, A) = \alpha_2\Omega_x + \beta_2\Omega_y + \gamma_2\Omega_z, \quad H_0(X, A) = -a_2\Omega_x - b_2\Omega_y - c_2\Omega_z,$$
$$G_1(X, A) = -\alpha_1\Omega_x - \beta_1\Omega_y - \gamma_1\Omega_z, \quad H_1(X, A) = a_1\Omega_x + b_1\Omega_y + c_1\Omega_z,$$

where $\Omega(X, A)$ is a particular harmonic solution of (3.117), which is regular in domain D.

In the general case the problem of constructing kernels in (3.112) reduces to the construction of particular harmonic solutions of (3.117), whose right side has singularities on the boundary of the domain. In some cases one can construct such solutions say, for example, satisfying the conditions

$$a_1\beta_2 + b_1\alpha_2 - b_2\alpha_1 - a_2\beta_1 = 0, \quad a_1\gamma_2 + c_1\alpha_2 - \alpha_1 c_2 - \gamma_1 a_2 = 0,$$
$$c_1\beta_2 + b_1\gamma_2 - \gamma_1 b_2 - \beta_1 c_2 = 0, \quad a_1\alpha_2 - \alpha_1 a_2 = b_1\beta_2 - \beta_1 b_2,$$
$$c_1\gamma_2 - \gamma_1 c_2 - b_1\beta_2 + \beta_1 b_2 = \lambda \neq 0.$$

Then, using the harmonicity of Ω we reduce (3.117) to the form

$$\lambda\Omega_{zz} = \frac{\partial}{\partial n}(r^{-1}),$$

from which it follows that one can set

$$\Omega = \lambda\frac{\partial}{\partial n}\{(z - c)\ln[z - c \pm [(x - a)^2 + (y - b)^2$$
$$+ (z - c)^2]^{1/2}] \mp [(x - a)^2 + (y - b)^2 + (z - c)^2]^{1/2}\},$$

where $A = (a, b, c)$ and the plus or minus sign is chosen so that the singularities of the logarithm lie outside domain D in which function Ω is considered; the derivative is taken along the normal relative to the coordinates of point A. Obviously here it is necessary to require the convexity of domain D with respect to the direction of the Oz axis. If in boundary condition (3.111) we have $b_1 = b_2 = \alpha_1 = 1$, $\alpha_2 = 1$, and all other coefficients are zero, then one can set

$$G_0(X, A) = 2^{-1} \ln \{y - b \pm [(x-a)^2 + (y-b)^2 + (z-c)^2]^{1/2}\},$$
$$H_0(X, A) = 0,\ G_1(X, A) = 2^{-1} \ln \{y - b \pm [(x-a)^2 + (y-b)^2 + (z-c^2)]^{1/2}\},$$
$$H_1(X, A) = 2^{-1} \ln \{x - a \pm [(x-a)^2 + (y-b)^2 + (z-c)^2]^{1/2}\}.$$

Chapter 4

SYSTEMS OF PARTIAL DIFFERENTIAL EQUATIONS RELATED TO MULTIDIMENSIONAL GENERALIZATIONS OF THE CAUCHY–RIEMANN SYSTEM

The oblique derivative problem for harmonic functions of two independent variables is closely connected with the Riemann–Hilbert problem: Find a solution which is regular in the domain D of the Cauchy–Riemann system $u_x - v_y = 0$, $u_y + v_x = 0$, which on boundary Γ of domain D satisfies the condition $au + bv = f$, where a, b, and f are given functions. If we set $u = \varphi_y$, $v = \varphi_x$, then we get the oblique derivative problem with boundary condition $a\varphi_y + b\varphi_x = f$ for a harmonic function φ which is regular in domain D. One can investigate the Riemann–Hilbert problem directly, without reducing it to the oblique derivative problem, with the help of the apparatus of the theory of functions of a complex variable [13]. Essentially, this reduction to a problem of the theory of functions also gives an apparatus for investigating the oblique derivative problem for harmonic functions of two independent variables. Multidimensional generalizations of the Cauchy–Riemann system are also known [60]. In the three-dimensional case the analog of the Cauchy–Riemann system has the form

$$\begin{aligned} &u_x + v_y + w_z = 0, \quad s_x + w_y - v_z = 0, \\ &s_y + u_z - w_x = 0, \quad s_z - u_y + v_x = 0, \end{aligned} \tag{4.1}$$

and in four-dimensional space

$$\begin{aligned} &s_t - u_x - v_y - w_z = 0, \quad u_t + s_x + w_y - v_z = 0, \\ &v_t - w_x + s_y + u_z = 0, \quad w_t + v_x - u_y + s_z = 0. \end{aligned} \tag{4.2}$$

In this chapter we shall consider the analog of the Riemann–Hilbert problem for system (4.1) and we shall investigate some second-order systems connected with systems (4.1) and (4.2).

1. ANALOG OF THE RIEMANN–HILBERT PROBLEM

The analog of the Riemann–Hilbert problem for system (4.1) can be posed as follows [7]: Find a solution $U = (s, u, v, w)$ of system (4.1) which is regular in the domain and, on boundary Γ of domain D, satisfies the conditions

$$a_i s + b_i u + c_i v + d_i w = f_i, \quad i = 1, 2, \tag{4.3}$$

where a_i, b_i, c_i, d_i, f_i are given functions on Γ. This problem also reduces to the oblique derivative problem for harmonic functions.

We construct a general solution of system (4.1). From this system we easily get that all components of any of its solutions are harmonic functions. For example, the harmonicity of function s can be obtained by differentiating the second equation of system (4.1) with respect to x, the third with respect to y, and the fourth with respect to z. After addition of the results the harmonicity of u, v, and w is established analogously. We take two arbitrary harmonic functions s and u and construct harmonic functions σ and ω satisfying the equations $\sigma_x = s$, $\omega_x = u$. Obviously these functions are determined up to arbitrary harmonic functions of two variables y and z. From the third and fourth equations of system (4.1) we find

$$w = \sigma_y + \omega_z + \psi(y, z), \quad v = -\sigma_z + \omega_y + \varphi(y, z),$$

where φ and ψ are arbitrary harmonic functions of two independent variables. We substitute the expressions for s, u, v, and w into the first two equations of (4.1):

$$\omega_{xx} - \sigma_{zy} + \omega_{yy} + \varphi_y + \sigma_{zy} + \omega_{zz} + \psi_z = 0,$$
$$\sigma_{xx} + \sigma_{yy} + \omega_{zy} + \psi_y + \sigma_{zz} - \omega_{zy} - \varphi_z = 0.$$

By the harmonicity of the functions σ and ω these equations assume the form $\varphi_y + \psi_z = 0$, $\psi_z - \varphi_z = 0$, from which it follows that $\varphi = \chi_y$, $\psi = \chi_z$, $\Delta\chi = 0$. Now replacing ω by $\omega + \chi$, without loss of generality we can write

$$s = \sigma_x, \quad u = \omega_x, \quad v = \omega_y - \sigma_z, \quad w = \omega_z + \sigma_y. \tag{4.4}$$

The relations of (4.4) express a solution of (4.1) in terms of two arbitrary harmonic functions σ and ω.

By the representation (4.4) of solutions of (4.1) from conditions (4.3) we get

$$a_i\sigma_x + d_i\sigma_y - c_i\sigma_z + b_i\omega_x + c_i\omega_y + d_i\omega_z = f_i, \quad i = 1, 2. \tag{4.5}$$

These conditions are conditions of the type of the oblique derivative problem for two harmonic functions. Taking representations of σ and ω as potentials of a simple layer, one can reduce problem (4.5) in the usual way to a system of singular integral equations [41]. In general in considering problem (4.5) the same difficulties arise as in the oblique derivative problem for harmonic functions. Problem (4.3) has also been little investigated, since there is no analog here of the apparatus of the theory of functions of a complex variable [7].

One can study special cases of problem (4.5) separately by the same methods with the help of which the oblique derivative problem for harmonic functions was investigated in Chapters 2 and 3. In particular, one can consider problem (4.5) for harmonic functions with polynomial coefficients, regular in the ball, or for harmonic functions with constant coefficients $a_i, b_i, c_i, d_i, i = 1, 2$, regular in an arbitrary domain. If $a_1 = b_2 = 1$ and all the remaining coefficients are zero, then boundary conditions (4.5) assume the form

$$\sigma_x = f_1, \quad \omega_x = f_2 \text{ on } \Gamma, \tag{4.6}$$

i.e., problem (4.5) splits into two oblique derivative problems for harmonic functions σ and ω. As was already established in Chapter 2, solutions of (4.6) which are regular in domain D have the form $\sigma = \tau + \varphi(y, z)$, $\omega = \Omega + \psi(y, z)$, where τ and Ω are harmonic functions which are regular in domain D, which are uniquely defined by functions f_1 and f_2, and φ and ψ are arbitrary harmonic functions of two variables which are regular in domain D. Further we find $v = \Omega_y - \tau_z + \psi_y - \varphi_z$, $w = \Omega_z + \tau_y + \psi_z + \varphi_y$. We set $\alpha = \psi_y - \varphi_z$, $\beta = \psi_z + \varphi_y$. Obviously $\alpha_z - \beta_y = 0$, $\alpha_y + \beta_z = 0$, so there exists a harmonic function χ such that $\alpha = \chi_y$, $\beta = \chi_z$. Thus we have

$$s = \tau_x, \quad u = \Omega_x, \quad v = \Omega_y - \tau_z + \chi_y, \quad w = \Omega_z + \tau_y + \chi_z. \tag{4.7}$$

In the case where domain D is convex, for functions τ and Ω one can construct integral representations in terms of functions f_1 and f_2, as was done in Section 4 of Chapter 2.

If domain D is the half-space H: $\{x > 0\}$, then in the class of functions σ and ω which tend to zero at infinity, one can explicitly write [19]:

$$\sigma = \frac{1}{2\pi} \int\limits_{-\infty}^{+\infty}\!\!\int \frac{f_1(a, b)\, da db}{\sqrt{x^2 + (y-a)^2 + (z-b)^2}},$$

$$\omega = \frac{1}{2\pi} \int\limits_{-\infty}^{+\infty}\!\!\int \frac{f_2(a, b)\, da db}{\sqrt{x^2 + (y-a)^2 + (z-b)^2}},$$

where it is assumed that functions f_1 and f_2 also tend to zero at infinity. In the formulas of (4.7) now $\chi \equiv 0$, since this harmonic function is regular in the whole plane and tends to zero at infinity. Consequently, in this case problem (4.5) is always solvable in the class of functions tending to zero at infinity, and has a unique solution, so for $a_1 = b_2 = 1$, $b_1 = c_1 = d_1 = a_2 = c_2 = d_2 = 0$ in the half-space H, and problem (4.3) for system (4.1) is always solvable and has a unique solution.

If the components of the vector $P = (p_1, p_2, p_3, p_4)$ satisfy system (4.1), then the components of the vector $Q = GP$, where G is the constant matrix

$$G = \left\| \begin{matrix} g_1 & g_2 & g_3 & g_4 \\ -g_2 & g_1 & -g_4 & g_3 \\ -g_3 & g_4 & g_1 & -g_2 \\ -g_4 & -g_3 & g_2 & g_1 \end{matrix} \right\|,$$

and P is considered as column-vector, also satisfy (4.1) [7]. Obviously $q_1 = g_1p_1 + g_2p_2 + g_3p_3 + g_4p_4$, $q_2 = -g_2p_1 + g_1p_2 - g_4p_3 + g_3p_4$, $q_3 = -g_3p_1 + g_4p_2 + g_1p_3 - g_2p_4$, $q_4 = -g_4p_1 - g_3p_2 + g_2p_3 + g_1p_4$. If in the boundary condition (4.3) the coefficients are constant and $a_2 = -b_1$, $b_2 = a_1$, $c_2 = -d_1$, $d_2 = c_1$, then setting $a_1 = g_1$, $b_1 = g_2$, $c_1 = g_3$, $d_1 = g_4$ and considering, instead of the solution U of (4.1) sought, $V = GU$, for it we get problem (4.3), in which $a_1 = b_2 = 1$, and all the other coefficients are zero. This problem is always solvable and all its solutions are given by (4.7); the solution U of the original problem can now be expressed by the formula $U = G^{-1}V$, where

$$G^{-1} = \frac{1}{g_1^2 + g_2^2 + g_3^2 + g_4^2} \times \begin{Vmatrix} g_1 & -g_2 & -g_3 & -g_4 \\ g_2 & g_1 & g_4 & -g_3 \\ g_3 & -g_4 & g_1 & g_2 \\ g_4 & g_3 & -g_2 & g_1 \end{Vmatrix}.$$

For problem (4.3) with constant coefficients this lets us write a solution explicitly, and for a wide class of boundary conditions with variable coefficients one can reduce the problem to Fredholm equations similarly to the way this was done in Chapter 3 in investigating the oblique derivative problem.

Let the vector $P = (p_1, p_2, p_3, p_4)$ satisfy the system of equations

$$\begin{aligned} &\frac{\partial p_2}{\partial x} + \frac{\partial p_3}{\partial y} + \frac{\partial p_4}{\partial z} = 0, \quad \frac{\partial p_1}{\partial x} - \frac{\partial p_3}{\partial z} + \frac{\partial p_4}{\partial y} = 0, \\ &\frac{\partial p_1}{\partial y} + \frac{\partial p_2}{\partial z} - \frac{\partial p_4}{\partial x} = 0, \quad \frac{\partial p_1}{\partial z} - \frac{\partial p_2}{\partial y} + \frac{\partial p_3}{\partial x} = 0, \end{aligned} \tag{4.8}$$

in addition let a, b, c, and d be arbitrary sufficiently smooth functions; then the vector $Q = (q_1, q_2, q_3, q_4)$ with components $q_1 = ap_1 + bp_2 + cp_3 + dp_4$, $q_2 = -bp_1 + ap_2 - dp_3 + cp_4$, $q_3 = -cp_1 + dp_2 + ap_3 - bp_4$, $q_4 = -dp_1 - cp_2 + bp_3 + ap_4$ satisfies the relations

$$\begin{aligned} \frac{\partial q_2}{\partial x} + \frac{\partial q_3}{\partial y} + \frac{\partial q_4}{\partial z} &= -p_1(b_x + c_y + d_z) + p_2(a_x + d_y - c_z) \\ &\quad + p_3(-d_x + a_y + b_z) + p_4(c_x - b_y + a_z), \\ \frac{\partial q_1}{\partial y} + \frac{\partial q_2}{\partial z} - \frac{\partial q_4}{\partial x} &= p_1(a_y - b_z + d_x) + p_2(b_y + a_z + c_x) \\ &\quad + p_3(c_y - d_z - b_x) + p_4(d_y + c_z - a_x), \\ \frac{\partial q_1}{\partial x} - \frac{\partial q_3}{\partial z} + \frac{\partial q_4}{\partial y} &= p_1(a_x + c_z - d_y) + p_2(b_x - d_z - c_y) \\ &\quad + p_3(c_x - a_z + b_y) + p_4(d_x + b_z + a_y), \\ \frac{\partial q_1}{\partial z} - \frac{\partial q_2}{\partial y} + \frac{\partial q_3}{\partial x} &= p_1(a_z + b_y - c_x) + p_2(b_z - a_y + d_x) \\ &\quad + p_3(c_z + d_y + a_x) + p_4(d_z - c_y - b_x). \end{aligned} \tag{4.9}$$

Now if instead of Q we take the vector $V = (v_1, v_2, v_3, v_4)$ with components $v_1 = ap_1 + bp_2 + cp_3 + dp_4$, $v_2 = bp_1 - ap_2 + dp_3 - cp_4$, $v_3 = cp_1 - dp_2 - ap_3 + bp_4$, $v_4 = dp_1 + cp_2 - bp_3 - ap_4$, then instead of (4.9) we get

$$\begin{aligned}
\frac{\partial v_2}{\partial x} + \frac{\partial v_3}{\partial y} + \frac{\partial v_4}{\partial z} &= p_1 (b_x + c_y + d_z) + p_2 (c_z - a_x - d_y) \\
&\quad + p_3 (d_x - a_y - b_z) + p_4 (b_y - c_x - a_z), \\
\frac{\partial v_1}{\partial x} + \frac{\partial v_3}{\partial z} - \frac{\partial v_4}{\partial y} &= p_1 (a_x + c_z - d_y) + p_2 (b_x - d_z - c_y) \\
&\quad + p_3 (c_x - a_z + b_y) + p_4 (d_x - b_z + a_y), \\
\frac{\partial v_1}{\partial y} - \frac{\partial v_2}{\partial z} + \frac{\partial v_4}{\partial x} &= p_1 (a_y - b_z + d_x) + p_2 (b_y + a_z + c_x) \\
&\quad + p_3 (c_y - d_z - b_x) + p_4 (d_y + c_z - a_x), \\
\frac{\partial v_1}{\partial z} + \frac{\partial v_2}{\partial y} - \frac{\partial v_3}{\partial x} &= p_1 (a_z + b_y - c_x) + p_2 (b_z - a_y + d_x) \\
&\quad + p_3 (c_z + d_y + a_x) + p_4 (d_z - c_y - b_x).
\end{aligned} \tag{4.10}$$

Now if instead of the vector P we take a vector $U = (u_1, u_2, u_3, u_4)$ satisfying the system

$$\begin{aligned}
&\frac{\partial u_2}{\partial x} + \frac{\partial u_3}{\partial y} + \frac{\partial u_4}{\partial z} = 0, \quad \frac{\partial u_1}{\partial x} + \frac{\partial u_3}{\partial z} - \frac{\partial u_4}{\partial y} = 0, \\
&\frac{\partial u_1}{\partial y} - \frac{\partial u_2}{\partial z} + \frac{\partial u_4}{\partial x} = 0, \quad \frac{\partial u_1}{\partial z} + \frac{\partial u_2}{\partial y} - \frac{\partial u_3}{\partial x} = 0,
\end{aligned} \tag{4.11}$$

adjoint to system (4.8), the vector $W = (w_1, w_2, w_3, w_4)$ with components $w_1 = au_1 + bu_2 + cu_3 + du_4$, $w_2 = bu_1 - au_2 - du_3 + cu_4$, $w_3 = cu_1 + du_2 - au_3 - bu_4$, $w_4 = du_1 - cu_2 + bu_3 - au_4$ satisfies the relations

$$\begin{aligned}
\frac{\partial w_2}{\partial x} + \frac{\partial w_3}{\partial y} + \frac{\partial w_4}{\partial z} &= u_1 (b_x + c_y + d_z) + u_2 (d_y - a_x - c_z) \\
&\quad + u_3 (b_z - d_x - a_y) + u_4 (c_x - b_y - a_z), \\
\frac{\partial w_1}{\partial x} + \frac{\partial w_3}{\partial z} - \frac{\partial w_4}{\partial y} &= u_1 (a_x + c_z - d_y) + u_2 (b_x + d_z + c_y) \\
&\quad + u_3 (c_x - a_z - b_y) + u_4 (d_x - b_z + a_y), \\
\frac{\partial w_1}{\partial y} - \frac{\partial w_2}{\partial z} + \frac{\partial w_4}{\partial x} &= u_1 (a_y - b_z + d_x) + u_2 (b_y + a_z - c_x) \\
&\quad + u_3 (c_y + d_z + b_x) + u_4 (d_y - c_z - a_x), \\
\frac{\partial w_1}{\partial z} + \frac{\partial w_2}{\partial y} - \frac{\partial w_3}{\partial x} &= u_1 (a_z + b_y - c_x) + u_2 (b_z - a_y - d_x) \\
&\quad + u_3 (c_z - d_y + a_x) + u_4 (d_z + c_y + b_x).
\end{aligned} \tag{4.12}$$

If everywhere in domain D one has

$$a^2 + b^2 + c^2 + d^2 > 0, \tag{4.13}$$

then one can represent (4.9), (4.10), and (4.12) as a system of equations just with respect to the components of the vectors Q, V, and W, respectively, since if this condition holds the components of the vectors P and U can be expressed uniquely in terms of the components of the vectors Q, V, and W, respectively.

One can use (4.9) and (4.10) for reducing boundary problem (4.3) with special boundary conditions for the system (4.8) to a system of Fredholm integral equations of the third kind, and analogously apply (4.12) in investigating boundary conditions for system (4.11). We consider in more detail the following boundary problem: find a solution $U = (u_1, u_2, u_3, u_4)$ of system (4.11) which is regular in domain D and satisfies on boundary Γ of domain D the conditions

$$\begin{aligned} &au_1 + bu_2 + cu_3 + du_4 = h_1, \\ &du_1 - cu_2 + bu_3 - au_4 = h_2, \end{aligned} \tag{4.14}$$

where a, b, c, d, h_1, h_2 are functions given on Γ, where (4.13) holds everywhere on Γ. We shall assume the functions a, b, c, d, h_1, h_2 to be twice continuously differentiable. Since (4.13) holds on boundary Γ of domain D, one can continuously extend the four-component vector $A = (a, b, c, d)$ from Γ to the whole domain D, so that (4.13) holds throughout D [2]. We shall assume that such an extension of the twice continuously differentiable functions a, b, c, d to D has been made, and we shall denote the extended functions by the same letters. Along with U we consider a vector W, connected with U by (4.12).

To investigate problem (4.14), we need some representations of solutions of the inhomogeneous system

$$\begin{gathered} \frac{\partial u_2}{\partial x} + \frac{\partial u_3}{\partial y} + \frac{\partial u_4}{\partial z} = f_1(X), \quad \frac{\partial u_1}{\partial x} + \frac{\partial u_3}{\partial z} - \frac{\partial u_4}{\partial x} = f_2(X), \\ \frac{\partial u_1}{\partial y} - \frac{\partial u_2}{\partial z} + \frac{\partial u_4}{\partial x} = f_3(X), \\ \frac{\partial u_1}{\partial z} + \frac{\partial u_2}{\partial y} - \frac{\partial u_3}{\partial x} = f_4(X), \end{gathered} \tag{4.15}$$

where $X = (x, y, z)$. It is easy to verify that the general solution of this system has the form

$$\begin{aligned} u_1(X) &= \omega_1(X) + \int_D \{G_x f_2(X_0) - G_y f_3(X_0) + G_z f_4(X_0)\}\, d\Omega, \\ u_2(X) &= \omega_2(X) + \int_D \{G_x f_1(X_0) - G_z f_3(X_0) + G_y f_4(X_0)\}\, d\Omega, \\ u_3(X) &= \omega_3(X) + \int_D \{G_y f_1(X_0) + G_z f_2(X_0) - G_x f_4(X_0)\}\, d\Omega, \\ u_4(X) &= \omega_4(X) + \int_D \{G_z f_1(X_0) - G_y f_2(X_0) + G_x f_3(X_0)\}\, d\Omega. \end{aligned} \tag{4.16}$$

Here $(\omega_1, \omega_2, \omega_3, \omega_4)$ is the general solution of homogeneous system (4.11) corresponding to (4.15), and $G(X, X_0)$ is the Green's function of domain D of the Dirichlet problem for Laplace's equation.

However, in investigating the boundary problems of interest to us, it is more convenient to use a different representation of the solutions of system (4.15), which we now proceed to derive. We construct harmonic functions $g_i(X, X_0)$, $i = 1, 2, 3$, which are regular in domain D, depending on a parametric point X_0, and which on boundary Γ of domain D satisfy the conditions

$$g_1|_\Gamma = G_x|_\Gamma, \quad g_2|_\Gamma = G_y|_\Gamma, \quad g_3|_\Gamma = G_z|_\Gamma. \tag{4.17}$$

Let α, β, γ be the direction cosines of the normal to Γ at the point $X \in \Gamma$. Since $G(X, X_0) = 0$ for $X \in \Gamma$, one can rewrite (4.17) as follows:

$$g_1 = \alpha \frac{dG}{dn}\bigg|_\Gamma, \quad g_2 = \beta \frac{dG}{dn}\bigg|_\Gamma, \quad g_3 = \gamma \frac{dG}{dn}\bigg|_\Gamma, \tag{4.18}$$

where dG/dn is the derivative of the function G with respect to the normal to Γ. When point X_0 lies inside D, the function $dG/dn|_\Gamma$ is continuous, so the functions $g_i(X, X_0)$, $i = 1, 2, 3$, are also continuous; now if $X_0 \in \Gamma$, then as $X \to X_0$ inside domain D, these functions can have singularities of order no higher than two with respect to the distance between points X and X_0, i.e., the same singularities as dG/dn. We consider the following harmonic functions which are regular in domain D:

$$Q_1(X) = -\int_D \{g_1(X, X_0) f_2(X_0) + g_2(X, X_0) f_3(X_0)$$
$$+ g_3(X, X_0) f_4(X_0)\}\, d\Omega, \quad Q_4(X) = -\int_D \{g_3(X, X_0) f_1(X_0)$$
$$- g_2(X, X_0) f_2(X_0) + g_1(X, X_0) f_4(X_0)\}\, d\Omega$$

and we try to choose harmonic functions $Q_2(X)$ and $Q_3(X)$ which are regular in domain D, such that the quadruple of functions $Q_i(X)$, $i = 1, 2, 3, 4$, satisfies (4.11).

In order that the functions Q_2 and Q_3 be simpler to construct, we impose additional restrictions on domain D. Let us assume that domain D intersects the plane $z = 0$ in a simply connected planar domain Δ with piecewise-smooth boundary γ, and any line parallel to the Oz axis and passing through a point of Δ intersects D in a connected segment, and any such line passing through any point of the complement of Δ in the plane $z = 0$ does not intersect domain D at all. We shall call domains satisfying these conditions convex with respect to the direction of the Oz axis. In such domains, from the second and third equations of system (4.11) we find

$$Q_2(X) = \int_0^z \left(\frac{\partial Q_1}{\partial y} + \frac{\partial Q_4}{\partial x}\right) dz + A(x, y),$$

$$Q_3(X) = \int_0^z \left(\frac{\partial Q_4}{\partial y} - \frac{\partial Q_1}{\partial x}\right) dz + B(x, y),$$

where A and B are arbitrary functions of the variables x and y. Substituting these expressions for Q_2 and Q_3 in the first and fourth equations of (4.11), we get for A and B the inhomogeneous Cauchy–Riemann system

$$A_x + B_y = -\left.\frac{\partial Q_4}{\partial z}\right|_{z=0}, \quad A_y - B_x = -\left.\frac{\partial Q_1}{\partial z}\right|_{z=0},$$

from which it follows that one can set [10]

$$A(x, y) = \frac{1}{\pi}\int_{\Delta} \rho^{-2}\left\{(\eta - y)\left.\frac{\partial Q_1(\xi, \eta, z)}{\partial z}\right|_{z=0} + (\xi - x)\left.\frac{\partial Q_4(\xi, \eta, z)}{\partial z}\right|_{z=0}\right\} d\xi d\eta,$$

$$B(x, y) = \frac{1}{\pi}\int_{\Delta} \rho^{-1}\left\{(x - \xi)\left.\frac{\partial Q_1(\xi, \eta, z)}{\partial z}\right|_{z=0} - (\eta - y)\left.\frac{\partial Q_4(\xi, \eta, z)}{\partial z}\right|_{z=0}\right\} d\xi d\eta,$$

where $\rho^2 = (x - \xi)^2 + (y - \eta)^2$, and $d\xi d\eta$ is the area element.

Along with the representation (4.16) of the general solution of (4.15) one can also use the following:

$$\begin{aligned}
u_1(X) &= \omega_1(X) + Q_1(X) + \int_D \{G_x f_2(X_0) + G_y f_3(X_0) + G_z f_4(X_0)\}\, d\Omega, \\
u_2(X) &= \omega_2(X) + Q_2(X) + \int_D \{G_x f_1(X_0) - G_z f_3(X_0) + G_y f_4(X_0)\}\, d\Omega, \\
u_3(X) &= \omega_3(X) + Q_3(X) + \int_D \{G_y f_1(X_0) + G_z f_2(X_0) - G_x f_4(X_0)\}\, d\Omega, \\
u_4(X) &= \omega_4(X) + Q_4(X) + \int_D \{G_z f_1(X_0) - G_y f_2(X_0) + G_x f_3(X_0)\}\, d\Omega.
\end{aligned} \tag{4.19}$$

Here $(\omega_1, \omega_2, \omega_3, \omega_4)$ is the general solution of the homogeneous system (4.11) corresponding to (4.15). This representation is more convenient for investigating boundary problems, since on boundary Γ of domain D in (4.19) we have $u_1(X) = \omega_1(X)$, $u_4(X) = \omega_4(X)$, so the problem of finding a solution (u_1, u_2, u_3, u_4) of system (4.15), which is regular in the domain and, on the boundary Γ of domain D, satisfies the conditions

$$u_1(X)|_\Gamma = \varphi_1, \quad u_2(X)|_\Gamma = \varphi_2, \tag{4.20}$$

where φ_1 and φ_2 are functions given on Γ, reduces with the help of the representation (4.19) of solutions of system (4.15) to the investigation of problem (4.20) for homogeneous system (4.11).

Problem (4.20) for homogeneous system (4.11) is solved in the following way. We construct harmonic functions u_1 and u_4 which are regular in domain D, satisfying (4.20). Further we set

$$u_2(X) = \int_0^z \left(\frac{\partial u_1}{\partial y} + \frac{\partial u_4}{\partial x}\right) dz + p(x, y),$$

$$u_3(X) = \int_0^z \left(\frac{\partial u_4}{\partial y} - \frac{\partial u_1}{\partial x}\right) dz + q(x, y),$$

assuming that domain D is convex with respect to the direction of the Oz axis. The four functions u_1, u_2, u_3, u_4 satisfy the second and third equations of (4.11). Substituting these functions into the first and fourth equations of the system, we get that in order that these functions satisfy (4.11), it is necessary and sufficient that

$$p_x + q_y = -\left.\frac{\partial u_4}{\partial z}\right|_{z=0}, \quad p_y - q_x = -\left.\frac{\partial u_1}{\partial z}\right|_{z=0}. \tag{4.21}$$

These conditions are an inhomogeneous Cauchy–Riemann system in the section of domain D by the plane $z = 0$ for the functions p and q of two variables x and y. A solution of (4.21) always exists and is defined up to a pair of conjugate harmonic functions of two variables which are regular in domain Δ [10]. Hence problem (4.20) for system (4.11) is always solvable, and its solution depends linearly on a pair of conjugate harmonic functions of the variables x and y, regular in domain Δ.

Now we turn to the investigation of problem (4.14) for system (4.11). Along with the solution $U = (u_1, u_2, u_3, u_4)$ sought, we consider a vector $W = (w_1, w_2, w_3, w_4)$ with components $w_1 = au_1 + bu_2 + cu_3 + du_4$, $w_2 = bu_1 - au_2 - du_3 + cu_4$, $w_3 = cu_1 + du_2 - au_3 - bu_4$, $w_4 = du_1 - cu_2 + bu_3 - au_4$. The vectors U and W are related by (4.12), which we shall consider as an inhomogeneous system (4.15), setting, in (4.15) and (4.19),

$$f_1(X) = (b_x + c_y + d_z)u_1 + (d_y - c_z - a_x)u_2 + (b_z - a_y - d_x)u_3 + (c_x - b_y - a_z)u_4,$$
$$f_2(X) = (a_x + c_z - d_y)u_1 + (b_x + d_z + c_y)u_2$$
$$+ (c_x - a_z - b_y)u_3 + (d_x - b_z + a_y)u_4,$$
$$f_3(X) = (a_y - b_z + d_x)u_1 + (b_y + a_z - c_x)u_2$$
$$+ (c_y + d_z + b_x)u_3 + (d_y - c_z - a_x)u_4,$$
$$f_4(X) = (a_z + b_y - c_x)u_1 + (b_z - a_y - d_x)u_2$$
$$+ (c_z - d_y + a_x)u_3 + (d_z + c_y + b_x)u_4.$$

We introduce the following notation:

$$A = \begin{Vmatrix} b_x + c_y + d_z & d_y - a_x - c_z & b_z - d_x - a_y & c_x - b_y - a_z \\ a_x + c_z - d_y & b_x + c_y + d_z & c_x - a_z - b_y & d_x - b_z + a_y \\ a_y - b_z + d_x & b_y + a_z - c_y & c_y + d_z + b_x & d_y - c_z - a_x \\ a_z + b_y - c_x & b_z - a_y - d_x & c_z - d_y + a_x & c_y + b_x + d_z \end{Vmatrix},$$

$$G = \begin{Vmatrix} 0 & G_x & G_y & G_z \\ G_x & 0 & -G_z & G_y \\ G_y & G_z & 0 & -G_x \\ G_z & -G_y & G_x & 0 \end{Vmatrix}, \quad E = \begin{Vmatrix} 0 & g_1 & g_2 & g_3 \\ g_1 & 0 & -g_3 & g_2 \\ g_2 & g_3 & 0 & -g_1 \\ g_3 & -g_2 & g_1 & 0 \end{Vmatrix}.$$

The elements $K_{ij}(X, X_0)$ of the matrix $K = G \times A$ are Fredholm kernels, and relations (4.19) can be written in vector form as follows:

$$W(X) = \Omega(X) + Q(X) + \int_D K(X, X_0)\, u(X_0)\, d\Sigma, \tag{4.22}$$

where $\Omega = (\omega_1, \omega_2, \omega_3, \omega_4)$, $Q = (Q_1, Q_2, Q_3, Q_4)$.

In the expression for the components of the vector Q there occur functions f_1, ..., f_4, so now this vector depends on u_1, ..., u_4, so we consider its structure in more detail. In the expressions for the functions $Q_1(X)$ and $Q_4(X)$ there occur $L_{ij}(X, X_0)$, $i = 1, 4$, $j = 1, 2, 3, 4$, the elements of the matrix $L = H \times A$. When point X lies in domain D, these elements L_{ij} have as many derivatives with respect to the coordinates of point X_0 as the functions a, b, c, and d do, and they are analytic in the coordinates of point X. If $X \in \Gamma$, then as $X_0 \to X$ inside D, the functions L_{ij} can have singularities of the same character as the singularities of the normal derivative of the Green's function on the boundary of the domain. With the help of L_{ij} we construct the functions

$$M_i(X, X_0) = -\int_0^{z_0} \left\{ \frac{\partial L_{1i}(X, x_0, y_0, \xi)}{\partial x_0} + \frac{\partial L_{4i}(X, x_0, y_0, \xi)}{\partial y_0} \right\} d\xi,$$

$$N_i(X, X_0) = -\int_0^{z_0} \left\{ \frac{\partial L_{4i}(X, x_0, y_0, \xi)}{\partial y_0} - \frac{\partial L_{1i}(X, x_0, y_0, \xi)}{\partial x_0} \right\} d\xi,$$

$$X_0 = (x_0, y_0, z_0), \quad X = (x, y, z).$$

When points X and X_0 lie in domain D, the functions M_i and N_i, $i = 1, 2, 3, 4$, are smooth, while if $X \in \Gamma$ and X_0 tends to X inside D, then these functions can have singularities like those of the kernels considered in the last section of Chapter 1. In order to avoid difficulties caused by these singularities, let us assume that the functions a, b, c, and d are extended from boundary Γ to the whole domain D so that, in addition to (4.13), one also has

$$\begin{gathered}
\gamma d_z + \beta d_y + \alpha d_x + \alpha a_y - \beta a_x + \gamma b_x - \alpha b_z + \gamma c_y - \beta c_z = 0, \\
\gamma c_z + \beta c_y + \alpha c_x - \alpha b_y + \beta b_x + \gamma a_x - \alpha a_z - \gamma d_y + \beta d_z = 0, \\
\gamma b_z + \beta b_y + \alpha b_x + \alpha c_y - \beta c_x - \gamma a_y + \beta a_z - \gamma d_x + \alpha d_z = 0, \\
\gamma a_z + \beta a_y + \alpha a_x - \alpha d_y + \beta d_x + \gamma b_y - \beta b_z - \gamma c_x + \alpha c_z = 0
\end{gathered} \tag{4.23}$$

on surface Γ, where α, β, γ are the direction cosines of the normal to Γ. These conditions do not restrict the generality. It follows from (4.23) that $L_{ij}(X, X_0) = 0$ when $X \in D$ and $X_0 \in \Gamma$, so now the functions M_i and N_i have only integrable singularities, and the kernels

$$T_i(x_0, y_0, X) = \frac{1}{\pi} \int_\Delta \left\{ (\eta - y) \left. \frac{\partial L_{1i}(\xi, \eta, z_0, X)}{\partial z_0} \right|_{z_0=0} + (\xi - x_0) \left. \frac{\partial L_{4i}(\xi, \eta, z_0, X)}{\partial z_0} \right|_{z_0=0} \right\} \frac{\partial\xi\, \partial\eta}{(\xi - x)^2 + (\eta - y)^2},$$

$$S_i(x_0, y_0, X) = -\frac{1}{\pi}\int\limits_{\Delta}\left\{(\xi - x_0)\frac{\partial L_{1i}(\xi, \eta, z_0, X)}{\partial z_0}\bigg|_{z_0=0}\right.$$
$$\left. - (\eta - y_0)\frac{\partial L_{4i}(\xi, \eta, z_0, X)}{\partial z_0}\bigg|_{z_0=0}\right\}\frac{d\xi d\eta}{(\xi - x_0)^2 + (\eta - y_0)^2}$$

are Fredholm.

With the help of the notation introduced, we rewrite (4.22) as follows:

$$w_1(X) = \omega_1(X) - \sum_{i=1}^{4}\int\limits_D L_{1i}(X, X_0)\, u_i(X_0)\, d\Omega + \sum_{i=1}^{4}\int\limits_D K_{1i}(X, X_0)\, u_i(X_0)\, d\Omega,$$
$$w_2(X) = \omega_2(X) + \sum_{i=1}^{4}\int\limits_D M_i(X, X_0)\, u_i(X_0)\, d\Omega + \sum_{i=1}^{4}\int\limits_D T_i(x, y, X_0)$$
$$\times\, u(X_0)\, d\Omega + \sum_{i=1}^{4}\int\limits_D K_{2i}(X, X_0)\, u_i(X_0)\, d\Omega, \tag{4.24}$$
$$w_3(X) = \omega_3(X) + \sum_{i=1}^{4}\int\limits_D N_i(X, X_0)\, u_i(X_0)\, d\Omega + \sum_{i=1}^{4}\int\limits_D S_i(x, y, X_0)\, u_i(X_0)\, d\Omega$$
$$+ \sum_{i=1}^{4}\int\limits_D K_{3i}(X, X_0)\, u_i(X_0)\, d\Omega,$$
$$w_4(X) = \omega_4(X) - \sum_{i=1}^{4}\int\limits_D L_{4i}(X, X_0)\, u_i(X_0)\, d\Omega + \sum_{i=1}^{4} K_{4i}(X, X_0)\, u_i(X_0)\, d\Omega,$$

where $(\omega_1, \omega_2, \omega_3, \omega_4)$ is a solution of the homogeneous system (4.11), satisfying the conditions $\omega_1|_\Gamma = h_1$, $\omega_2|_\Gamma = h_2$. This solution is defined up to a pair of conjugate harmonic functions of two independent variables, and (4.24) is a Fredholm system of equations for the functions $u_1, \ldots, u_4$. Thus problem (4.14) for system (4.11) reduces to the investigation of a Fredholm system of equations in which there figure as known functions $\omega_1, \ldots, \omega_4$, depending linearly on a pair of conjugate harmonic functions of two independent variables. Consequently, problem (4.14) for system (4.11) differs only in being Fredholm from problem (4.20) for the same system and, in particular, problem (4.14) is solvable provided no more than a finite number of orthogonality conditions imposed on the functions h_1 and h_2 hold; in the solution of the homogeneous problem corresponding to (4.14) there occurs the same indefiniteness as in the solution of the homogeneous problem corresponding to (4.20).

When the domain D is the half-space $z > 0$, in (4.19) one can write the Green's function explicitly as well as the functions Q_1, Q_2, Q_3, Q_4. Now these formulas assume the form

$$u_1(X) = \omega_1(X) + \int\limits_D \{G_x f_2(X_0) + G_y f_3(X_0) + [G_z + H_z]\, f_4(X_0)\}\, d\Omega,$$
$$u_2(X) = \omega_2(X) + \int\limits_D \{[G_x + H_x]\, f_1(X_0) - G_z f_3(X_0) + [G_y + H_y]\, f_4(X_0)\}\, d\Omega, \tag{4.25}$$
$$u_3(X) = \omega_3(X) + \int\limits_D \{[G_y + H_y]\, f_1(X_0) + G_z f_2(X_0) - [G_x + H_x]\, f_4(X_0)\}\, d\Omega,$$

$$u_4(X) = \omega_4(X) + \int_D \{[G_z + H_z] f_1(X_0) - G_y f_2(X_0) + G_x f_3(X_0)\} \, d\Omega,$$

where G is the Green's function and H is the Neumann function of the half-space $z > 0$, i.e.,

$$G = \frac{1}{4\pi} \{[(x - x_0)^2 + (y - y_0)^2 + (z - z_0)^2]^{-1/2} - [(x - x_0)^2 + (y - y_0)^2 + (z + z_0)^2]^{-1/2}\},$$

$$H = \frac{1}{2\pi} [(x - x_0)^2 + (y - y_0)^2 + (z - z_0)^2]^{-1/2}.$$

Exactly as in deriving the system of equations (4.22), substituting into (4.25) the expressions for the functions $f_1, \ldots, f_4$; we get a system of Fredholm equations

$$W(X) = F(X) + \int_D E(X, X_0) U(X_0) \, d\Omega, \tag{4.26}$$

where E is a known matrix, $F = (\omega_1, \omega_2, \omega_3, \omega_4)$, where the vector F is a solution of problem (4.20) in the half-space $z > 0$ for system (4.11). In the class of functions tending to zero at infinity this problem is always solvable and has a unique solution, so F is a completely definite vector-function. From the reduction of problem (4.14) for system (4.11) in the half-space to a system of Fredholm integral equations (4.26) it follows that

Theorem 4.1. If domain D is the half-space $z > 0$ and the coefficients of boundary condition (4.14) satisfy the inequality (4.13), then in the class of solutions which tend to zero at infinity the problem (4.14) for system (4.11) is always Fredholm.

2. GENERALIZATIONS OF A HOLOMORPHIC VECTOR

System (4.1) is a natural three-dimensional analog of the Cauchy–Riemann system and its solutions are sometimes called holomorphic vectors. System (4.1) is also sometimes called a Moisil–Theodoresco system [7]. We consider the system

$$\begin{aligned} &u_x + v_y + w_z = 0, \quad v_z - w_y = 0, \\ &u_z - w_x = 0, \quad u_y - v_x = 0, \end{aligned} \tag{4.27}$$

which the gradient of a harmonic function of three independent variables satisfies. One gets a Moisil–Theodoresco system from (4.27) by introducing a fourth unknown function s as follows:

$$\begin{aligned} &u_x + v_y + w_z = 0, \quad s_x - v_z + w_y = 0, \\ &s_y + u_z - w_x = 0, \quad s_z - u_y + v_x = 0. \end{aligned} \tag{4.28}$$

One can make an analogous construction, starting from an arbitrary second-order elliptic equation with three independent variables. For simplicity we restrict ourselves to the equation

$$AU_{zz} + aU_{xx} + 2bU_{xy} + cU_{yy} = 0 \tag{4.29}$$

with analytic coefficients, satisfying the inequalities

$$A > 0, \ ac - b^2 > 0. \tag{4.30}$$

The gradient $U = (u, v, w)$ of any solution of (4.29) satisfies the system

$$\begin{gathered} Aw_z + au_x + bu_y + bv_x + cv_y = 0, \\ v_z - w_y = 0, \ u_z - w_x = 0, \ u_y - v_x = 0, \end{gathered} \tag{4.31}$$

which one can consider a generalization of (4.27). Just as in (4.27), we introduce into (4.31) a fourth unknown function s. We get the system

$$\begin{gathered} Aw_z + au_x + bu_y + bv_x + cv_y = 0, \\ s_x - v_z + w_y = 0, \ s_y + u_z - w_x = 0, \\ s_z - u_y + v_x = 0, \end{gathered} \tag{4.32}$$

whose characteristic determinant has the form $\Delta = -(\lambda_1^2 + \lambda_2^2 + \lambda_3^2)[a\lambda_1^2 + 2b\lambda_1\lambda_2 + c\lambda_2^2 + A\lambda_3^2]$. By (4.30) the system (4.32) is elliptic.

System (4.32) differs from (4.28) only in the first equation, and any solution of it of the form $(0, u, v, w)$ is the gradient of a solution of (4.29). One can consider this system as one possible generalization of the Moisil–Theodoresco system. Just as for (4.28), a consequence of (4.32) is the equation $\Delta s = 0$ for function s.

We construct the general solution of (4.32). We set $w = \omega_z$ and we shall assume the functions s and ω unknown in the system. From the second and third equations of the system we find

$$u = \omega_x - \int_0^z s_y dz + \varphi(x, y), \quad v = \omega_y + \int_0^z s_x dz + \psi(x, y).$$

Substituting these expressions into the first and fourth equations of system (4.32), we get

$$\begin{gathered} A\omega_{zz} + a\omega_{xx} + 2b\omega_{xy} + c\omega_{yy} \\ + (c - a)\int_0^z s_{xy} dz + b\int_0^z (s_{xx} - s_{yy})\, dz + a\varphi_x + c\psi_y + b(\varphi_y + \psi_x) = 0, \\ s_z + \int_0^z (s_{xx} + s_{yy})\, dz - \varphi_y + \psi_x = 0. \end{gathered}$$

Considering the harmonicity of function s, we transform the second equation by integration by parts into the condition

$$\psi_x - \varphi_y = s_z|_{z=0}. \tag{4.33}$$

We set $s = \sigma_z$, where σ is a harmonic function, and

$$\varphi = -\sigma_y|_{z=0} + p(x, y), \; \psi = \sigma_x|_{z=0} + q(x, y).$$

By the harmonicity of σ, (4.33) assumes the form $q_x - p_y = 0$, and for ω we get the equation

$$A\omega_{zz} + a\omega_{xx} + 2b\omega_{xy} + c\omega_{yy}$$
$$= (a-c)\sigma_{xy} + b(\sigma_{yy} - \sigma_{xx}) + ap_x + cq_y + b(p_y + q_x).$$

It follows from the relation $q_x - p_y = 0$ that there exists a function of two variables $\chi(x, y)$ such that $q = \chi_y$, $p = \chi_x$, and the last equation assumes the form

$$A\omega_{zz} + a(\omega_{xx} + \chi_{xx}) + 2b(\omega_{xy} + \chi_{xy}) + c(\omega_{yy} + \chi_{yy}) =$$
$$= (a-c)\sigma_{xy} + b(\sigma_{yy} - \sigma_{xx}).$$

We again denote the function $\omega + \chi(x, y)$ by ω and then the preceding relation can be rewritten as follows:

$$A\omega_{zz} + a\omega_{xx} + 2b\omega_{xy} + c\omega_{yy} = (a-c)\sigma_{xy} + b(\sigma_{yy} - \sigma_{xx}). \tag{4.34}$$

The expressions for u and v assume the form $u = \omega_x - \sigma_y$, $v = \omega_y + \sigma_x$. For solutions of (4.32) one gets the following representation:

$$s = \sigma_z, \quad u = \omega_x - \sigma_y, \quad v = \omega_y + \sigma_x, \quad w = \omega_z. \tag{4.35}$$

where σ is an arbitrary harmonic function, and ω is an arbitrary solution of (4.34). This representation of solutions of (4.32) generalizes the representation (4.4) of solutions of system (4.1).

The situation particularly simplifies when $a = c$ and $b = 0$. Now (4.32) has the form

$$\begin{gathered} Aw_z + a(u_x + v_y) = 0, \; s_x - v_z + w_y = 0, \\ s_y + u_z - w_x = 0, \; s_z - u_y + v_x = 0, \end{gathered} \tag{4.36}$$

and the functions ω and σ figuring in (4.35) satisfy the equations

$$A\omega_{zz} + a(\omega_{xx} + \omega_{yy}) = 0, \quad \Delta\sigma = 0, \tag{4.37}$$

i.e., in this case the system of equations for ω and σ splits into separate equations for ω and σ.

Representation (4.35) of solutions of (4.32) lets us reduce the Riemann–Hilbert problem with the condition (4.3) to the oblique derivative problem for a system of two second-order equations. However, even the simplest problem with boundary condition

$$s|_\Gamma = h_1, \quad w|_\Gamma = h_2 \tag{4.38}$$

here can frequently not be carried through to a conclusion [48]. One can investigate the problem (4.38) completely in the case $a = c$, $b = 0$, since now it splits into two oblique derivative problems, which is easy to study.

Although system (4.32) has much in common with system (4.1), in contrast with (4.1) the characteristic determinant of system (4.32) splits into two different irreducible factors of degree two. We show that one can modify this system so that the characteristic determinant of the modified system becomes the square of an irreducible polynomial of second degree and that the structural properties of the solutions are preserved. Instead of (4.32) we consider the system

$$\begin{aligned} &Aw_z + au_x + bu_y + bv_x + cv_y = 0,\\ &\alpha_1 s_x + \beta_1 s_y + \gamma_1 s_z - v_z + w_y = 0,\\ &\alpha_2 s_x + \beta_2 s_y + \gamma_2 s_z + u_z - w_x = 0,\\ &\alpha_3 s_x + \beta_3 s_y + \gamma_3 s_z - u_y + v_x = 0. \end{aligned} \tag{4.39}$$

A consequence of this system is the equation

$$\frac{\partial}{\partial x}(\alpha_1 s_x + \beta_1 s_y + \gamma_1 s_z) + \frac{\partial}{\partial y}(\alpha_2 s_x + \beta_2 s_y + \gamma_2 s_z) + \frac{\partial}{\partial z}(\alpha_3 s_x + \beta_3 s_y + \gamma_3 s_z) = 0,$$

whose characteristic form is $\delta_1 = \alpha_1\lambda_1{}^2 + (\beta_1 + \alpha_2)\lambda_1\lambda_2 + (\gamma_1 + \alpha_2)\lambda_1\lambda_3 + \beta_2\lambda_2{}^2 + (\gamma_2 + \beta_2)\lambda_2\lambda_3 + \gamma_3\lambda_3{}^2$, and the characteristic determinant of system (4.39) is equal to $\delta_1\delta_2$, where $\delta_2 = a\lambda_1{}^2 + 2b\lambda_1\lambda_2 + c\lambda_2{}^2 + A\lambda_3{}^2$. In order that $\delta_1 = \delta_2$, it suffices to set $\alpha_1 = a$, $2b = \beta_1 + \alpha_2$, $\beta_2 = c$, $\gamma_3 = A$, $\gamma_1 + \alpha_3 = 0$, $\gamma_2 + \beta_3 = 0$ and, in particular, one can take $\beta_1 = \alpha_2 = b$, $\gamma_1 = \alpha_3 = \gamma_2 = \beta_3 = 0$. In this case (4.30) assumes the form

$$\begin{aligned} &Aw_z + au_x + bu_y + bv_x + cv_y = 0,\\ &as_x + bs_y - v_z + w_y = 0,\\ &bs_x + cs_y + u_z - w_x = 0,\\ &As_z - u_y + v_x = 0, \end{aligned} \tag{4.40}$$

and for s we get the equation

$$\frac{\partial}{\partial x}(as_x + bs_y) + \frac{\partial}{\partial y}(bs_x + cs_y) + \frac{\partial}{\partial z}(As_z) = 0. \tag{4.41}$$

In many respects the properties of (4.40) are analogous to the properties of (4.32), and the characteristic determinant of (4.40) has the form $\Delta = (a\lambda_1{}^2 + 2b\lambda_1\lambda_2 + c\lambda_2{}^2 + A\lambda_3{}^2)^2$.

For (4.40) one can generalize the representation of solutions (4.35). Let $w = \omega_z$, and s be a solution of (4.41). From the third and second equations of (4.40) we find

$$u = \omega_x - \int_0^z (bs_x + cs_y)\,dz + \varphi(x, y),$$

$$v = \omega_y + \int_0^z (as_x + bs_y)\,dz + \psi(x, y).$$

Substituting these expressions for u and v in the last equation of (4.40), and using (4.41), we get

$$\psi_x - \varphi_y = -As_z|_{z=0}. \tag{4.42}$$

Analogously, from the first equation of (4.40) we have

$$\begin{aligned} &A\omega_{zz} + a\omega_{xx} + 2b\omega_{xy} + c\omega_{yy} \\ &= a\int_0^z \frac{\partial}{\partial x}(bs_x + cs_y)\,dz - b\left[2\int_0^z \frac{\partial}{\partial x}(as_x + bs_y)\,dz + As_z\right] \\ &\quad + b\,[(As_z)|_{z=0}] - a\varphi_x - b\varphi_y - b\psi_x - c\psi_y\,. \end{aligned} \tag{4.43}$$

We set $\psi = \chi_x + p_y$, $\varphi = -\chi_y + p_x$; then from (4.42) for χ we get

$$\Delta\chi = -As_z|_{z=0}, \tag{4.44}$$

and in (4.43) we replace ω by $\omega + p$, where $\omega + p$ is again denoted by ω; then this equation has the form

$$\begin{aligned} &A\omega_{zz} + a\omega_{xx} + 2b\omega_{xy} + c\omega_{yy} \\ &= a\int_0^z \frac{\partial}{\partial x}(bs_x + cs_y)\,dz - b\left[2\int_0^z \frac{\partial}{\partial x}(as_x + bs_y)\,dz + As_z\right] \\ &\quad + b\,(As_z)|_{z=0} + (c - a)\,\chi_{xy} + b\,(\chi_{yy} - \chi_{xx}). \end{aligned} \tag{4.45}$$

The general solution of (4.40) now has the form

$$s, \quad u = \omega_x - \int_0^z (bs_x + cs_y)\,dz - \chi_y,$$

$$v = \omega_y + \int_0^z (as_x + bs_y)\,dz + \chi_x, \quad w = \omega_z,$$

where s is a solution of (4.41), χ is a solution of (4.44), and ω is a solution of (4.45).

System (4.40) is obtained as a generalization of a system equivalent to (4.29), which one can consider the canonical form of a second-order elliptic equation. We consider the more general second-order equation

$$a_1\omega_{xx} + (b_1 + a_2)\omega_{xy} + (e_1 + a_3)\omega_{xz} + b_2\omega_{yy} + (c_2 + b_3)\omega_{yz} + c_3\omega_{zz} = 0. \qquad (4.46)$$

Now, analogously, instead of (4.40) we get the system

$$\begin{aligned} &a_1u_x + b_1u_y + c_1u_z + a_2v_x + b_2v_y + cv_z + a_3w_x + b_3w_y + c_3w_z = 0,\\ &a_1s_x + b_1s_y + c_1s_z - v_z + w_y = 0,\\ &a_2s_x + b_2s_y + c_2s_z + u_z - w_x = 0,\\ &a_3s_x + b_3s_y + c_3s_z - u_y + v_x = 0 \end{aligned} \qquad (4.47)$$

with characteristic determinant $\Delta = [a_1\lambda_1^2 + (b_1 + a_2)\lambda_1\lambda_2 + (c_1 + a_3)\lambda_1\lambda_3 + b_2\lambda_2^2 + (c_2 + b_3)\lambda_2\lambda_3 + c_3\lambda_3^2]$. From this point of view one can consider the system the greatest generalization of the Moisil–Theodoresco system.

In the first equation of (4.1), we change the sign to the opposite, and assuming that s, u, v, and w are functions of the variables t, x, y, and z, in the first equations of the system obtained we add s_t on the left, in the second u_t, in the third v_t, and in the fourth w_t. As a result of these transformations one gets a system (4.2) which occurs in the theory of a holomorphic quaternion and which one can also consider a four-dimensional generalization of the Cauchy–Riemann system [60]. With the help of the analogous procedure, one can associate with (4.32) the following system of four equations for four unknown functions:

$$\begin{aligned} &s_t - au_x - bu_y - bv_x - cv_y - Aw_z = 0,\\ &u_t + s_x - v_z + w_y = 0,\\ &v_t + s_y + u_z - w_x = 0,\\ &w_t + s_z - u_y + v_x = 0, \end{aligned} \qquad (4.48)$$

with characteristic determinant $\Delta = (\lambda_0^2 + \lambda_1^2 + \lambda_2^2 + \lambda_3^2)(\lambda_0^2 + a\lambda_1^2 + 2b\lambda_1\lambda_2 + c\lambda_2^2 + A\lambda_3^2)$. Consequently the system (4.48) is elliptic if and only if

$$U_{tt} + aU_{xx} + 2bU_{xy} + cU_{yy} + AU_{zz} = 0 \qquad (4.49)$$

is elliptic. If we take any solution U of (4.49) and we set $s = -U_t$, $u = U_x$, $v = U_y$, $w = U_z$, then we get a solution of (4.48), so it is natural to consider (4.48) as a generalization of (4.2).

Analogously, starting from (4.40) we arrive at the four-dimensional system

$$\begin{aligned} &s_t - au_x - bu_y - bv_x - cv_y - Aw_z = 0,\\ &u_t + as_x + bs_y - v_z + w_y = 0,\\ &v_t + bs_x + cs_y + u_z - w_x = 0,\\ &w_t + As_z - u_y + v_x = 0 \end{aligned} \qquad (4.50)$$

with characteristic determinant $\Delta = \lambda_0^2[\lambda_1^2 + \lambda_2^2 + \lambda_3^2 + (a\lambda_1 + b\lambda_2)^2 + (b\lambda_1 + c\lambda_2)^2 + A\lambda_3^2] + (a\lambda_1^2 + 2b\lambda_1\lambda_2 + c\lambda_2^2 + A\lambda_3^2)^2$. However, many vectors of the form $(-U_t, U_x, U_y, U_z)$, where U is a solution of (4.49), do not satisfy (4.50). Because of this fact, (4.48) is a more natural generalization of (4.2) than (4.50). Obviously the systems (4.48) and (4.50) can be deformed by a continuous deformation in the class of elliptic systems to system (4.2).

The systems considered below are very similar in their properties to (4.50). We introduce operators D^* and $\bar{D}^*$, which act on vector functions $V = (s, u, v, w)$, as will be defined below. By $A_1, A_2, A_3, B_1, B_2, C$ we denote linear homogeneous operators relative to the functions on which they act, which contain differentiations only with respect to x, y, and z; further, we set

$$D^*V = \begin{Vmatrix} s_t + A_1(u) + A_2(v) + A_3(w) \\ -A_1(s) + u_t + B_1(v) + B_2(w) \\ -A_2(s) - B_1(u) + v_t + C(w) \\ -A_3(s) - B_2(u) - C(v) + w_t \end{Vmatrix},$$

$$\bar{D}^*V = \begin{Vmatrix} s_t - A_1(u) - A_2(v) - A_3(w) \\ A_1(s) + u_t - B_1(v) - B_2(w) \\ A_2(s) + B_1(u) + v_t - C(w) \\ A_3(s) + B_2(u) + C(v) + w_t \end{Vmatrix}.$$

The systems of equations $D^*V = 0$ and $\bar{D}^*V = 0$ can be considered generalizations of (4.50). We denote by a_1, a_2, a_3, b_1, b_2, and c the symbols of the operators $A_1, A_2, A_3, B_1, B_2, C$, respectively. These symbols are homogeneous linear functions of the variables ξ_1, ξ_2, ξ_3, and the symbol of the operator $\partial/\partial t$ is equal to ξ_0. With the operators D^* and $\bar{D}^*$ are associated the symbolic matrices

$$\begin{Vmatrix} \xi_0 & a_1(\xi) & a_2(\xi) & a_3(\xi) \\ -a_1(\xi) & \xi_0 & b_1(\xi) & b_2(\xi) \\ -a_2(\xi) & -b_1(\xi) & \xi_0 & c(\xi) \\ -a_2(\xi) & -b_2(\xi) & -c(\xi) & \xi_0 \end{Vmatrix}, \quad \begin{Vmatrix} \xi_0 & -a_1(\xi) & -a_2(\xi) & -a_3(\xi) \\ a_1(\xi) & \xi_0 & -b_1(\xi) & -b_2(\xi) \\ a_2(\xi) & b_1(\xi) & \xi_0 & -c(\xi) \\ a_3(\xi) & b_2(\xi) & c(\xi) & \xi_0 \end{Vmatrix},$$

denoted by M and $\bar{M}$, respectively. Here $\xi = (\xi_1, \xi_2, \xi_3)$. These matrices are the transposes of one another, so their determinants coincide and are given by the formula $\Delta = \xi_0^2(\xi_0^2 + b_1^2 + b_2^2 + c^2) + (a_1c - a_2b_2 + a_3b_1)^2$. It follows from this expression for the characteristic determinant that a necessary and sufficient condition for the ellipticity of the systems $D^*V = 0$ and $\bar{D}^*W = 0$ is the positive or negative definiteness of the quadratic form $a_1c - a_2b_2 + a_3b_1$ in the variables ξ_1, ξ_2, ξ_3, which is equivalent with the ellipticity of the equation

$$A_1C(s) - A_2B_2(s) + A_3B_1(s) = 0,$$

which is a second-order partial differential equation in the variables x, y, and z. System $\bar{D}^*V = 0$ can be considered a vast generalization of system (4.2), and many of the properties should extend to it in the case where $\bar{D}^*V = 0$ is elliptic. These systems are especially similar to one another when the coefficients of $\bar{D}^*V = 0$ are constant.

Let the operators D^* and $\bar{D}^*$ act on vectors V, independently of the variable t. On such vectors the following operators are generated by these operators:

$$D_0V = \left\| \begin{matrix} A_1(u) + A_2(v) + A_3(w) \\ A_1(s) + B_1(v) - B_2(w) \\ A_2(s) - B_1(u) + C(w) \\ A_3(s) + B_2(u) - C(v) \end{matrix} \right\|,$$

$$\bar{D}_0V = \left\| \begin{matrix} A_1(u) + A_2(v) + A_3(w) \\ A_1(s) - B_1(v) + B_2(w) \\ A_2(s) + B_1(u) - C(w) \\ A_3(s) - B_2(u) + C(v) \end{matrix} \right\|$$

respectively. The systems $D_0V = 0$ and $\bar{D}_0V = 0$ are elliptic if and only if the equation

$$A_1C(s) + A_2B_2(s) + A_3B_1(s) = 0$$

is elliptic. When the system $\bar{D}_0V = 0$ is elliptic, it can also be considered a generalization of the Moisil–Theodoresco system and with the help of the operators D_0 and $\bar{D}_0$ one can construct a generalization of the theory of the holomorphic vector.

If besides choosing vectors V independent of t we also assume that the first component s of vector V is identically zero, then the system $\bar{D}_0V = 0$ assumes the form

$$A_1(u) + A_2(v) + A_3(w) = 0, \quad B_2(w) - B_1(v) = 0,$$
$$B_1(u) - C(w) = 0, \quad C(v) - B_2(u) = 0.$$

It can be considered a generalization of the system which the gradient of a harmonic function of three independent variables satisfies.

When the coefficients of all the operators $A_1, A_2, A_3, B_1, B_2, C$ are constant, and from the system $\bar{D}_0V = 0$, just as from the Moisil–Theodoresco system, one got Laplace's equation for each component, we have the equations $A_1C(s) + A_2B_2(s) + A_3B_1(s) = 0$, $A_1C(w) + A_2B_2(w) + A_3B_1(w) = 0$ and the same kind of equations for v and u. If the coefficients are variable, then the similar equations contain terms depending on the other components of vector V. The presence of such consequences of a first-order system lets us construct representations of solutions of this system in terms of a solution of a second-order equation, analogously to the representation of solutions of the Moisil–Theodoresco system in terms of harmonic functions. In particular, if $B_1 = \partial/\partial z$, $B_2 = \partial/\partial y$, $C = \partial/\partial x$, then (4.47) can be written in the form $\bar{D}_0V = 0$, where

$$A_i = a_i\,\partial/\partial x + b_i\,\partial/\partial y + c_i\,\partial/\partial z, \quad i = 1, 2, 3.$$

We construct the general solution of (4.2). For this we introduce independent complex variables $\xi_1 = x + it$, $\xi_2 = y + iz$ and complex unknown functions $p = s + iu$, $q = w + iv$. Now system (4.2) can be rewritten as follows:

$$\frac{\partial p}{\partial \zeta_1} + \frac{\partial q}{\partial \bar{\zeta}_2} = 0, \quad \frac{\partial p}{\partial \zeta_2} - \frac{\partial q}{\partial \bar{\zeta}_1} = 0. \tag{4.51}$$

The general solution of the second equation of this system is given by

$$p = \frac{\partial \varphi}{\partial \bar{\zeta}_1}, \quad q = \frac{\partial \varphi}{\partial \zeta_2}, \tag{4.52}$$

where $\varphi(\zeta_1, \bar{\zeta}_1, \zeta_2, \bar{\zeta}_2)$ is an analytic function of all four of its arguments. Substituting expressions (4.52) into the first equation of (4.51), we get for the complex function φ the Laplace equation

$$4^{-1}\Delta\varphi \equiv \frac{\partial^2 \varphi}{\partial \zeta_1 \partial \bar{\zeta}_1} + \frac{\partial^2 \varphi}{\partial \zeta_2 \partial \bar{\zeta}_2} = 0.$$

Consequently, in order that (4.52) give a general solution of (4.51), the function φ must have the form $\varphi = \sigma(t, x, y, z) + i\omega(t, x, y, z)$, where σ and ω are arbitrary real harmonic functions. One can rewrite (4.51) in real form as follows:

$$s = \sigma_x - \omega_t, \quad u = \omega_x + \sigma_t, \quad v = \omega_y - \sigma_z, \quad w = \sigma_y + \omega_z, \tag{4.53}$$

where σ and ω are arbitrary harmonic functions. If σ and ω do not depend on t, then (4.55) turns into (4.4), and solutions of (4.2) which are independent of t satisfy (4.1.).

With the help of representations (4.53) of solutions of (4.2), one can reduce the Riemann–Hilbert problem (4.3) for this system to the problem of finding two harmonic functions σ and ω, which are regular in domain D and satisfy on boundary Γ of domain D the conditions

$$b_i\sigma_t + a_i\sigma_x + d_i\sigma_y - c_i\sigma_z - a_i\omega_t + b_i\omega_x + c_i\omega_y + d_i\omega_z = f_i, \quad i = 1, 2. \tag{4.54}$$

It is known that the Riemann–Hilbert problem for (4.2) in any domain D does not satisfy the Lopatinskii condition for any coefficients, so this problem must fail to be Noetherian [11, 37]. Analogous complications must arise in the investigation of problem (4.54) also. We consider some examples. Let domain D be the half-space H: $\{t > 0\}$, and let conditions (4.3) have the form $s|_{t=0} = f_1$, $u|_{t=0} = f_2$. Now conditions (4.54) are the following:

$$(\sigma_x - \omega_t)|_{t=0} = f_1, \quad (\omega_x + \sigma_t)|_{t=0} = f_2.$$

We construct harmonic functions F_1 and F_2 which are regular in the half-space H and coincide, for $t = 0$, with f_1 and f_2, respectively. Now the functions σ and ω are determined from the inhomogeneous Cauchy–Riemann system

$$\sigma_x - \omega_t = F_1, \quad \omega_x + \sigma_t = F_2 \tag{4.55}$$

with respect to the variables t and x. Since σ and ω are regular for all y and z, these functions are determined from (4.55) up to the gradient of a harmonic function $\chi(x, t)$ of two variables, which is regular in the half-space $t > 0$. However, in order that the functions v and w be continuous in the closed half-space H, the function σ and ω must have derivatives with respect to y and z, which are continuous up to the boundary, and for this it is necessary that the functions f_1 and f_2 be differentiable in y and z, since as a result of solving system (4.55) the smoothness in these variables is not increased up to the boundary. It is also obvious that functions (4.53) do not depend on the harmonic function of two variables $\chi(x, t)$; in this case problem (4.3) is always solvable and has a unique solution, but increased smoothness of the boundary data is required. Now if on the boundary of the half-space H the values of v and w are given, then a solution of problem (4.3) is determined up to a harmonic function of two variables $\chi(x, t)$, which is regular in the half-space H, which enters into s and u. When u and v are given on the boundary, problem (4.3) reduces to the search for harmonic solutions of the equations

$$\omega_x + \sigma_t = F_1, \quad \omega_y - \sigma_z = F_2$$

with harmonic right sides. For all these examples, loss of smoothness is typical, i.e., the continuity of f_1 and f_2 is insufficient for the solvability of the problem, and it is necessary to require some smoothness of these functions.

3. SECOND-ORDER SYSTEMS OF EQUATIONS

The Noetherian property of classical boundary problems can fail for elliptic systems of partial differential equations. The first example of such systems was the system of A. V. Bitsadze, which in complex notation has the form [24]

$$w_{\bar{z}\bar{z}} = 0, \quad w = u + iv, \quad z = x + iy. \tag{4.56}$$

For system (4.56), the Noetherian property of the Dirichlet problem in a disk and in a half-plane fails. It is closely connected with the Cauchy–Riemann system, so it is natural to try to construct a multidimensional generalization of (4.56) according to the analogous principle, starting from multidimensional generalizations of the Cauchy–Riemann system [8, 60]. Since (4.1) is the three-dimensional analog of the Cauchy–Riemann system, we try, with its help, to construct the three-dimensional analog of (4.56). We denote by D the operator which associates with the vector $U = (s, u, v, w)$ the left sides of (4.1), and we consider the system of equations $D^2U = 0$. It is easy to verify by direct calculation that this system has the form

$$\begin{gathered}
\Delta s = 0, \\
-\Delta u + 2\frac{\partial}{\partial x}(u_x + v_y + w_z) = 0, \\
-\Delta v + 2\frac{\partial}{\partial y}(u_x + v_y + w_z) = 0, \\
-\Delta w + 2\frac{\partial}{\partial z}(u_x + v_y + w_z) = 0.
\end{gathered}$$

In it one can throw out the first equation, and as the three-dimensional analog of (4.56) take

$$
\begin{aligned}
-\Delta u + 2\frac{\partial}{\partial x}(u_x + v_y + w_z) &= 0,\\
-\Delta v + 2\frac{\partial}{\partial y}(u_x + v_y + w_z) &= 0,\\
-\Delta w + 2\frac{\partial}{\partial z}(u_x + v_y + w_z) &= 0.
\end{aligned}
\tag{4.57}
$$

One can generalize this system to a space of any dimension

$$
-\Delta u_j + 2\frac{\partial}{\partial x_j}\sum_{i=1}^{n}\frac{\partial u_i}{\partial x_i} = 0, \quad j = 1, \ldots, n.
\tag{4.58}
$$

Hence we shall not consider (4.57) separately, but consider (4.58) below.

One can consider (4.2) as the four-dimensional analog of the Cauchy–Riemann system. We denote by D the operator associating $U = (s, u, v, w)$ with the left sides of (4.2), and by $\bar{D}$ the operator associating with the vector the left sides of the equations of the system

$$
\begin{aligned}
&s_t + u_x + v_y + w_z = 0, \; u_t - s_x - w_y + v_z = 0,\\
&v_t + w_x - s_y - u_z = 0, \; w_t - v_x + u_y - s_z = 0.
\end{aligned}
\tag{4.59}
$$

Now the systems $D^2U = 0$ and $\bar{D}^2U = 0$ have the form

$$
\begin{aligned}
-\Delta s + 2\frac{\partial}{\partial t}(s_t - u_x - v_y - w_z) &= 0,\\
-\Delta u + 2\frac{\partial}{\partial t}(u_t + s_x + w_y - v_z) &= 0,\\
-\Delta v + 2\frac{\partial}{\partial t}(v_t - w_x + s_y + u_z) &= 0,\\
-\Delta w + 2\frac{\partial}{\partial t}(w_t + v_x - u_y + s_z) &= 0;
\end{aligned}
\tag{4.60}
$$

$$
\begin{aligned}
-\Delta s + 2\frac{\partial}{\partial t}(s_t + u_x + v_y + w_z) &= 0,\\
-\Delta u + 2\frac{\partial}{\partial t}(u_t - s_x - w_y + v_z) &= 0,\\
-\Delta v + 2\frac{\partial}{\partial t}(v_t + w_x - s_y - u_z) &= 0,\\
-\Delta w + 2\frac{\partial}{\partial t}(w_t - v_x + u_y - s_z) &= 0
\end{aligned}
\tag{4.61}
$$

respectively. One can consider system (4.61) the four-dimensional analog of (4.56). By direct calculation one verifies the identities $D\bar{D} = \bar{D}D = \Delta$, where Δ is the Laplace operator, from which it follows that $D^2\bar{D}^2 = \bar{D}^2D^2 = \Delta^2$. From this we get that all components of any solution of (4.60) or of (4.61) are biharmonic functions. This fact lets us write the representation of all solutions of (4.61) in terms of harmonic functions.

We consider solutions of (4.61) which are regular in the half-space H: $\{t > 0\}$. Since all components of such solutions are biharmonic in H, one can write

$$s = \varphi_1 + t\psi_1, \quad u = \varphi_2 + t\psi_2, \quad v = \varphi_3 + t\psi_3, \quad w = \varphi_4 + t\psi_4, \tag{4.62}$$

where φ_i and ψ_i, $i = 1, 2, 3, 4$, are harmonic functions which are regular in the half-space H. Substituting these representations of the components of a solution in (4.61), and considering the harmonicity of the functions φ_i and ψ_i, we get that without loss of generality one can assume the vectors $\Phi = (\varphi_1, \varphi_2, \varphi_3, \varphi_4)$ and $\Psi = (\psi_1, \psi_2, \psi_3, \psi_4)$ are solutions of (4.59), i.e., the general solution of (4.61) has the form

$$U = \Phi + t\Psi, \quad \bar{D}\Phi = 0, \quad \bar{D}\Psi = 0, \tag{4.63}$$

$U = (s, u, v, w)$. Obviously (4.63) generalizes the representation of the solutions of (4.56) in terms of holomorphic functions. It is easy to show that one can represent the general solution of (4.59) in terms of two harmonic functions σ and ω as follows:

$$s = \omega_t + \sigma_x, \quad u = \omega_x - \sigma_t, \quad v = \omega_y - \sigma_z, \quad w = \sigma_y + \omega_z. \tag{4.64}$$

It follows from (4.63) that the homogeneous Dirichlet problem for solutions of (4.61) which are regular in the half-space H has an infinite set of linearly independent solutions of the form

$$s = t(\omega_t + \sigma_x), \quad u = t(\omega_x - \sigma_t), \quad v = t(\omega_y - \sigma_z), \quad w = t(\sigma_y + \omega_z), \tag{4.65}$$

where σ and ω are arbitrary harmonic functions which are regular in the half-space H and continuously differentiable in the closed half-space $\bar{H}$: $\{t \geq 0\}$. Obviously the representation (4.63) of solutions of (4.61) holds not only in the half-space H, but is suitable for solutions of (4.61) which are regular in any domain D, but Φ and Ψ must be solutions of (4.59) which are regular in domain D. Thus, any solution of (4.61) which is regular in domain D can be expressed with the help of (4.63) and (4.64) in terms of four arbitrary harmonic functions which are regular in domain D, and the Dirichlet problem for (4.61) reduces to a problem of the type of the oblique derivative problem for a quadruple of harmonic functions.

We show that in the half-space H, for the solvability of the Dirichlet problem

$$s|_{t=0} = f_1, \quad u|_{t=0} = f_2, \quad v|_{t=0} = f_3, \quad w|_{t=0} = f_4$$

it is necessary to impose on the data of the problem an infinite set of orthogonality conditions. By virtue of representation (4.63) and the harmonicity of the components of the vector Φ we can write

$$\varphi_i(t, x, y, z) = \int_T M(t, x-\alpha, y-\beta, z-\gamma) f_i(\alpha, \beta, \gamma)\, d\omega,$$

$$i = 1, 2, 3, 4,$$

$$M = 2^{-1}\pi^{-2} t [t^2 + (x-\alpha)^2 + (y-\beta)^2 + (z-\gamma)^2]^{-2},$$

where T is the hyperplane $t = 0$, and $d\omega$ is the volume element of this hyperplane. Since the vector Φ is a solution of (4.59), for all $t > 0$ one must have

$$\begin{aligned}
&\int_T (M_t f_1 + M_x f_2 + M_y f_3 + M_z f_4)\, d\omega = 0,\\
&\int_T (M_t f_2 - M_x f_1 - M_y f_4 + M_z f_3)\, d\omega = 0,\\
&\int_T (M_t f_3 + M_x f_4 - M_y f_1 - M_z f_2)\, d\omega = 0,\\
&\int_T (M_t f_4 - M_x f_3 + M_y f_2 - M_z f_1)\, d\omega = 0.
\end{aligned} \tag{4.66}$$

Due to the uniqueness of the solution of the Dirichlet problem for Laplace's equation, (4.66) will hold if it holds for some fixed $t_0 > 0$. Setting $t = t_0 > 0$ from (4.66) we get orthogonality conditions of the form

$$\int_T \sum_{i=1}^{4} K_{ij}(X, A)\, f_i(A)\, d_A\omega = 0, \quad j = 1, 2, 3, 4, \tag{4.67}$$

where $X = (x, y, z)$, $A = (\alpha, \beta. \gamma)$. In (4.67) the functions K_{ij} depend analytically on X, so the left sides of these conditions are analytic functions of the variables x, y, z. In order that an analytic function be identically equal to zero, it is necessary and sufficient that all its derivatives vanish along with it itself at some fixed point X_0 of its domain of definition [47]. In (4.67) we set $X_0 = 0$, equating all the derivatives of the left sides of (4.67) to zero at the point O, and from (4.67) we get a countable set of orthogonality conditions on the functions f_1, f_2, f_3, and f_4.

In all the equations of (4.59) except the first, we change the signs to the opposites

$$\begin{aligned}
&s_t + u_x + v_y + w_z = 0, \quad -u_t + s_x + w_y - v_z = 0,\\
&-v_t - w_x + s_y + u_z = 0, \quad -w_t + v_x - u_y + s_z = 0,
\end{aligned} \tag{4.68}$$

and we denote by $\overline{D}_1$ the operator which associates with the vector $U = (s, u, v, w)$ the left sides of (4.68). One easily verifies by direct calculation that the system $\overline{D}_1{}^2 U = 0$ splits into the Laplace equation $\Delta s = 0$ for function s and the system of three equations

$$\begin{aligned}
&-\Delta u + 2\frac{\partial}{\partial x}(u_x + v_y + w_z) + 2\frac{\partial}{\partial t}(u_t + v_z - w_y) = 0,\\
&-\Delta v + 2\frac{\partial}{\partial y}(u_x + v_y + w_z) + 2\frac{\partial}{\partial t}(v_t + w_x - u_z) = 0,\\
&-\Delta w + 2\frac{\partial}{\partial z}(u_x + v_y + w_z) + 2\frac{\partial}{\partial t}(w_t - v_x + u_y) = 0
\end{aligned} \tag{4.69}$$

for the functions u, v, w. System (4.69) was first considered by Shevchenko [42]. For functions which do not depend on t, (4.68) becomes the Moisil–Theodoresco system, and (4.69) becomes (4.57). We denote the left sides of the equations of (4.69) by q_1, q_2, q_3; then by direct calculation one gets the identities

$$-\Delta q_1 + 2\frac{\partial}{\partial x}\left(\frac{\partial q_1}{\partial x} + \frac{\partial q_2}{\partial y} + \frac{\partial q_3}{\partial z}\right) + 2\frac{\partial}{\partial t}\left(\frac{\partial q_1}{\partial t} - \frac{\partial q_2}{\partial z} + \frac{\partial q_3}{\partial y}\right) = \Delta^2 u,$$

$$-\Delta q_2 + 2\frac{\partial}{\partial y}\left(\frac{\partial q_1}{\partial x} + \frac{\partial q_1}{\partial y} + \frac{\partial q_3}{\partial z}\right) + 2\frac{\partial}{\partial t}\left(\frac{\partial q_2}{\partial t} + \frac{\partial q_1}{\partial z} - \frac{\partial q_3}{\partial x}\right) = \Delta^2 v,$$

$$-\Delta q_3 + 2\frac{\partial}{\partial z}\left(\frac{\partial q_1}{\partial x} + \frac{\partial q_2}{\partial y} + \frac{\partial q_3}{\partial z}\right) + 2\frac{\partial}{\partial t}\left(\frac{\partial q_3}{\partial t} - \frac{\partial q_1}{\partial y} + \frac{\partial q_2}{\partial x}\right) = \Delta^2 w,$$

from which it follows that all components of any solution of (4.69) are biharmonic functions. Using representation (4.62) of biharmonic functions, the general solution of (4.69) can be written as follows:

$$U = \Phi + t\Psi, \; U = (u, v, w), \quad \Phi = (\varphi_1, \varphi_2, \varphi_3), \; \Psi = (\psi_1, \psi_2, \psi_3), \tag{4.70}$$

and Φ and Ψ satisfy the first-order system

$$u_x - v_y' + w_z = 0, \; u_t + v_z - w_y = 0, \quad v_t - u_z + w_x = 0, \; w_t + u_y - v_x = 0, \tag{4.71}$$

which is obtained from (4.59) for $s = 0$, so by (4.64) its general solution can be written in the form

$$u = \chi_{xx} + \chi_{tt}, \; v = \chi_{yx} + \chi_{zt}, \; w = \chi_{xz} - \chi_{ty},$$

where χ is an arbitrary harmonic function of four independent variables.

It follows from representation (4.70) of solutions of (4.69) that the homogeneous Dirichlet problem for system (4.69) in the half-space H has an infinite set of linearly independent solutions $u = t(\chi_{xx} + \chi_{tt})$, $v = t(\chi_{xy} + \chi_{zt})$, $w = t(\chi_{xz} + \chi_{ty})$, depending on one arbitrary harmonic function which is regular in H. Now the inhomogeneous Dirichlet problem

$$u|_{t=0} = f_1, \; v|_{t=0} = f_2, \; w|_{t=0} = f_3$$

is solvable provided an infinite set of orthogonality conditions hold. For example, by the first equation of (4.71) the functions f_1, f_2, f_3 must satisfy the condition $f_{1x} + f_{2y} + f_{3z} = 0$.

By the same method used to construct (4.61) and (4.69) one can construct a large class of elliptic systems of second-order equations for which the Noetherian property of the Dirichlet problem fails [39]. In the space R^{n+1} of variables t, x_1, ..., x_m suppose given an elliptic system of m first-order equations for m unknown functions u_1, ..., u_m of the form

$$A_0 u_t + \sum_{i=1}^{n} A_i u_{x_i} = 0, \tag{4.72}$$

where A_i, $i = 0, \ldots, n$, are square matrices, and $U = (u_1, \ldots, u_m)$ is a column-vector. By the ellipticity of (4.72), $\det A_0 \neq 0$, so multiplying (4.72) by the matrix A_0^{-1}, we rewrite it as follows:

$$u_t + \sum_{i=1}^{n} B_i u_{x_i} = 0, \quad B_i = A_0^{-1} A_i. \tag{4.73}$$

We denote by D the operator which associates with the column-vector U the column-vector consisting of the left sides of the equations of (4.73). One can consider the system of second-order equations

$$D^2 U \equiv U_{tt} + \frac{\partial}{\partial t} \sum_{i=1}^{n} B_i U_{x_i} + \sum_{i=1}^{n} B_i U_{x_{i_t}} + \sum_{i=1}^{n} B_i \frac{\partial}{\partial x_j} \sum_{j=1}^{n} B_j U_{x_j} = 0 \tag{4.74}$$

the analog of (4.61). One can represent any solution of (4.74) in the form

$$U = V + tW, \quad DV = 0, \quad DW = 0. \tag{4.75}$$

The fact that (4.75) satisfies (4.74) is verified directly, but $D^2U = D(DU) = D[DV + D(tW)] = D[W + tDW] = 0$, since $D(tW) = WDt + tDW = W + tDW$. We show that all solutions of (4.74) have the form (4.75). System $D^2U = 0$ is equivalent to $DU = W$, $DW = 0$. Obviously tW is a particular solution of the inhomogeneous system $DU = tW$, so $U = V + tW$, $DV = 0$ is the general solution of (4.74). It is also obvious that (4.74) is always elliptic. When the coefficients of (4.73) are constant, one can write (4.74) as follows:

$$u_{tt} + 2 \sum_{i=1}^{n} B_i u_{x_i t} + \sum_{i,j=1}^{n} B_i B_j u_{x_i x_j} = 0. \tag{4.76}$$

It follows from the representation (4.75) of the solutions of (4.74) that the homogeneous Dirichlet problem in the half-space H: $\{t > 0\}$ for (4.76) has an infinite set of linearly independent solutions of the form

$$U = tW, \quad DW = 0, \tag{4.77}$$

where W is a vector-function which is regular in the half-space H.

We return to the investigation of the system (4.58):

$$-\Delta u_j + 2 \frac{\partial}{\partial x_j} \sum_{i=1}^{n} \frac{\partial u_i}{\partial x_i} = 0, \quad j = 1, \ldots, n. \tag{4.78}$$

We differentiate its first equation with respect to x_1, the second with respect to x_2, the jth with respect to x_j, the last with respect to x_n, and then we add the results of the differentiations to get the identity

$$\Delta\left(\sum_{i=1}^{n}\frac{\partial u_i}{\partial x_i}\right)=0, \tag{4.79}$$

from which it follows that $\Delta^2 u_j = 0$, $j = 1, \ldots, n$, i.e., all components of any solution of (4.78) are biharmonic functions. Consequently, all components of any solution of (4.78) which is regular in the half-space H: $\{x_n > 0\}$ can be represented in the form

$$u_j = \varphi_j + x_n\psi_j,\quad j = 1,\ \ldots,\ n, \tag{4.80}$$

and the components of a solution which is regular in the ball Σ: $\{x_1^2 + \ldots + x_n^2 < R^2\}$ in the form

$$u_j = \varphi_j + (x_1^2 + \ldots + x_n^2)\,\psi_j,\quad j = 1, \ldots, n, \tag{4.81}$$

where φ_j and ψ_j are harmonic functions which are regular in the half-space H or in the ball Σ, respectively.

With the help of (4.80) we find

$$\sum_{i=1}^{n}\frac{\partial u_i}{\partial x_i}=\sum_{i=1}^{n}\frac{\partial \varphi_i}{\partial x_i}+\psi_n+x_n\sum_{i=1}^{n}\frac{\partial \psi_i}{\partial x_i}.$$

Obviously this function will be a harmonic function which is regular for $x_n > 0$ only if the condition

$$\sum_{i=1}^{n}\frac{\partial \psi_i}{\partial x_i}=0 \tag{4.82}$$

holds. Substituting (4.80) into (4.78), considering (4.82), we get

$$-2\frac{\partial \psi_j}{\partial x_n}+2\frac{\partial}{\partial x_j}\left(\sum_{i=1}^{n}\frac{\partial \varphi_i}{\partial x_i}+\psi_n\right)=0,\quad j=1,\ldots,n.$$

It follows from this equation that

$$\psi_i=\frac{\partial \psi}{\partial x_j},\quad \psi=\sum_{i=1}^{n}\frac{\partial \varphi_i}{\partial x_i}+\psi_n,\quad j=1,\ldots,n-1,$$

$$\frac{\partial}{\partial x_n}\left(\sum_{i=1}^{n}\frac{\partial \varphi_i}{\partial x_i}\right)=0.$$

In the class of harmonic functions which are regular in the half-space H, from the last relation we have

$$\sum_{i=1}^{n}\frac{\partial \varphi_i}{\partial x_i}=0. \tag{4.83}$$

From the $n-1$ preceding equations, considering (4.82) and (4.83) we get that $\psi_i = \partial\psi/\partial x_j$, $j = 1, \ldots, n$, where ψ is an arbitrary harmonic function which is regular in the half-space H. Consequently, (4.80) assumes the form

$$u_j=\varphi_j+x_n\frac{\partial \psi}{\partial x_j},\quad j=1,\ldots,n, \tag{4.84}$$

where $\varphi_1, \ldots, \varphi_n, \psi$ are arbitrary harmonic functions which are regular in the half-space H, satisfying (4.83). Analogously, for solutions of (4.78) which are regular in ball Σ, from (4.81) we find

$$u_j=\varphi_j+\frac{1}{n-2}(x_1^2+\ldots+x_n^2)\frac{\partial}{\partial x_j}\sum_{i=1}^{n}\frac{\partial \varphi_i}{\partial x_i}, \tag{4.85}$$

$\varphi_1, \ldots, \varphi_n$ being arbitrary harmonic functions which are regular in ball Σ. Obviously one can rewrite representation (4.85) of solutions of (4.78) which are regular in ball Σ as follows:

$$u_j=\chi_j+\frac{1}{n-2}(x_1^2+\ldots+x_n^2-R^2)\frac{\partial}{\partial x_j}\sum_{i=1}^{n}\frac{\partial \chi_i}{\partial x_i}, \tag{4.86}$$
$$j=1,\ldots,n,$$

setting $\chi_j=\varphi_j+R^2\dfrac{\partial}{\partial x_j}\sum\limits_{i=1}^{n}\dfrac{\partial\varphi_i}{\partial x_i}$.

For (4.78) we consider the Dirichlet problem in the following formulation: Find a solution $u_1, \ldots, u_n$ of (4.78) which is regular in domain D, continuous in the closed domain $D \cup \Gamma$, and on boundary Γ of domain D satisfies

$$u_j|_\Gamma=f_j,\ j=1,\ldots,n, \tag{4.87}$$

where f_j are continuous functions, given on Γ. With the help of representations (4.84) or (4.85) of solutions of (4.78), one can reduce this Dirichlet problem to a problem of the type of the oblique derivative problem for harmonic functions. It follows from (4.84) that the homogeneous Dirichlet problem in the half-space H for (4.78) has an infinite set of linearly independent solutions of the form

$$u_j = x_n \partial\psi/\partial x_j, \; j = 1, \ldots, n, \tag{4.88}$$

depending on an arbitrary harmonic function ψ which is regular in the half-space H. From (4.84) and (4.87) we have $\varphi_j|_\Gamma = f_j$, $j = 1, \ldots, n$, Γ: $\{x_n = 0\}$, and one can write down the function explicitly [19]:

$$\varphi_j(X) = \frac{1}{\sigma_{n-1}} \int_\Gamma \frac{x_n f_j(a_1, \ldots, a_{n-1})\, da_1 \ldots da_{n-1}}{[(x_1 - a_1)^2 + \ldots + (x_{n-1} - a_{n-1})^2 + x_n^2]^{n/2}},$$

where σ_{n-1} is the area of the unit n-dimensional sphere. By (4.83) the inhomogeneous Dirichlet problem in the half-space H is solvable if and only if the following condition holds:

$$\int_\Gamma \sum_{j=1}^{n} \frac{\partial}{\partial x_j} \frac{x_n f_j(a_1, \ldots, a_{n-1})\, da_1 \ldots da_{n-1}}{[(x_1 - a_1)^2 + \ldots + (x_{n-1} - a_{n-1})^2 + x^2]^{n/2}} = 0. \tag{4.89}$$

In exactly the same way as in considering the conditions (4.66) one can show that (4.89) is equivalent to a countable set of orthogonality conditions. It follows from (4.86) that a solution of the Dirichlet problem (4.87) for the system in ball Σ has the form

$$u_j(X) = F_j(X) + \frac{1}{n-2}(x_1^2 + \ldots + x_n^2 - R^2)\frac{\partial}{\partial x_j} \sum_{i=1}^{n} \frac{\partial F_i}{\partial x_i}, \tag{4.90}$$
$$j = 1, \ldots, n, \; X = (x_1, \ldots, x_n),$$

where F_j is a harmonic function which is regular in ball Σ, satisfying the condition $F_j = f_j$ on sphere S: $\{x_1^2 + \ldots + x_n^2 = R^2\}$. In (4.90) the second derivatives of the functions F_j figure, so the continuity of F_j in the closed ball $\Sigma \cup S$ is insufficient for the solvability of problem (4.87) and it is necessary to require increased smoothness of the boundary data f_j. In particular, it suffices to require $f_j \in C^2$, $j = 1, \ldots, n$, i.e., the second derivatives of these functions must exist.

In the half-space H the following boundary problem for (4.78) is well posed: Find a solution of (4.78) which is regular in H and satisfies

$$u_j|_\Gamma = f_j, \; j = 1, \ldots, n-1, \; \left.\frac{\partial u_n}{\partial x_n}\right|_\Gamma = f_n, \tag{4.91}$$

and has the smoothness in the half-space $H \cup \Gamma$ which is needed for these conditions to make sense, i.e., $u_1, \ldots, u_{n-1}$ are continuous, and u_n is continuously differentiable in $H \cup \Gamma$. From representation (4.84) and conditions (4.91), we get

$$\varphi_j|_\Gamma = f_j, \; j = 1, \ldots, n-1, \; \left.\left[\frac{\partial \varphi_n}{\partial x_n} + \frac{\partial \psi}{\partial x_n}\right]\right|_\Gamma = f_n,$$

but with the help of (4.83) we can write

$$\frac{\partial \psi}{\partial x_n} = f_n + \sum_{i=1}^{n-1} \frac{\partial f_i}{\partial x_i} = g(x_1, \ldots, x_{n-1}).$$

Thus, in order to find a harmonic function ψ, it suffices to solve the Neumann problem in the half-space H. We have [19]

$$\psi(X) = \frac{1}{\sigma_{n-1}} \int \frac{g(a_1, \ldots, a_{n-1})\, da_1 \ldots da_{n-1}}{[(x_1 - a_1)^2 + \ldots + (x_{n-1} - a_{n-1})^2 + x_n^2]^{(n-2)/2}},$$

and the functions φ_j, $j = 1, \ldots, n$, are given by the formula

$$\varphi_j(X) = \frac{1}{\sigma_{n-1}} \int \frac{x_n f_j(a_1, \ldots, a_{n-1})\, da_1 \ldots da_{n-1}}{[(x_1 - a_1)^2 + \ldots + (x_{n-1} - a_{i-1})^2 + x_n^2]^{n/2}}.$$

Defining the functions φ_j with the help of (4.83), we find

$$\left.\frac{\partial \varphi_n}{\partial x_n}\right|_\Gamma = -\sum_{i=1}^{n-1} \frac{\partial f_i}{\partial x_i} = h(x_1, \ldots, x_{n-1}),$$

from which we get the following explicit formula for φ_n:

$$\varphi_n(X) = \frac{1}{\sigma_{n-1}} \int \frac{h(a_1, \ldots, a_{n-1})\, da_1 \ldots da_{n-1}}{[(x_1 - a_1)^2 + \ldots + (x_{n-1} - a_{n-1})^2 + x_n^2]^{(n-2)/2}}.$$

Obviously the functions $\varphi_1, \ldots, \varphi_n$ satisfy (4.83), where $f_1, \ldots, f_{n-1}$ in (4.91) must be differentiable, and one gets that the solution of problem (4.91) for system (4.78) is differentiable in the closed half-space $H \cup \Gamma$.

For an arbitrary domain D one can formulate problem (4.91) as follows: Find a solution of (4.78) which is regular in domain D and on the boundary Γ of domain D satisfies the conditions

$$u_j = f_j,\ j = 1, \ldots, n-1,\ \frac{\partial u_n}{\partial \nu} = f_n, \tag{4.92}$$

where ν is the normal to boundary Γ of domain D. We consider problem (4.92) in ball Σ. From (4.86) we have

$$\chi_j|_S = f_j,\ j = 1, \ldots, n-1, \quad \frac{\partial \chi_n}{\partial r} + \frac{2R}{n-2} \frac{\partial}{\partial x_n} \sum_{i=1}^{n} \frac{\partial \chi_i}{\partial x_i} = f_n.$$

Solving the Dirichlet problem for Laplace's equation, we find functions $\chi_1, \ldots, \chi_{n-1}$ which are expressed by Poisson's formula [19]. Analogously we construct a harmonic function F_n which is regular in ball Σ and satisfies the condition $F_n|_S = f_n$. Obviously the equation

$$\frac{r}{R}\frac{\partial \chi_n}{\partial r} + \frac{2R}{n-2}\frac{\partial}{\partial x_n}\sum_{i=1}^{n}\frac{\partial \chi_i}{\partial x_i} = F_n,$$

holds, from which we get the equation for defining the function χ_n,

$$r\frac{\partial \chi_n}{\partial r} + \frac{2R^2}{n-2}\frac{\partial^2 \chi_n}{\partial x_n^2} = G, \quad G = -\frac{2R^2}{n-2}\sum_{i=1}^{n-2}\frac{\partial \chi_i}{\partial x_i} + RF_n, \tag{4.93}$$

where G is a known harmonic function. As noted in Chapter 1, one can represent the functions χ_n and G by series

$$\chi_n = \sum_{l=0}^{\infty} A_l\,(x_1^2 + \ldots + x_{n-1}^2, x_n)\, p_l\,(x_1, \ldots, x_{n-1}),$$

$$G_n = \sum_{l=0}^{\infty} B_l\,(x_1^2 + \ldots + x_{n-1}^2, x_n)\, p_l\,(x_1, \ldots, x_{n-1}).$$

Here the p_l are homogeneous harmonic polynomials, and the functions A_l and B_l are uniquely determined by the functions of a complex variable x_n

$$\psi_l(x_n) = A_l(0, x_n), \; g_l(x_n) = B_l(0, x_n),$$

which are analytic in the disk $|x_n| < R$. Substituting the expressions for χ_n and G in (4.93), equating coefficients of identical p_l, and setting $x_1{}^2 + \ldots + x_{n-1}{}^2 = 0$, we get an equation relating the functions ψ_l and g_l:

$$\frac{2R^2}{n-2}\psi_l'' + x_n\psi_l' + l\psi_l = g_l, \; l = 0, 1, \ldots$$

From these equations ψ_l can always be determined by functions g_l, since the coefficient of the highest derivative is nonzero throughout the disk $|x_n| < R$. Consequently, problem (4.92) is always solvable in ball Σ.

With solutions of the homogeneous problem corresponding to problem (4.92) for system (4.78) are associated one-to-one solutions of the homogeneous equation

$$\frac{2R^2}{n-2}\psi_l'' + x_n\psi_l' + l\psi_l = 0, \; l = 0, 2, \ldots \tag{4.94}$$

Equation (4.94) has two linearly independent solutions:

$$a_l\,(x_n) = \Phi\left(1 + \frac{l}{2}, \frac{1}{2}; -\frac{R^2 x_n}{n-2}\right),$$

$$b_l\,(x_n) = x_n\Phi\left(\frac{l}{2} + \frac{3}{2}, \frac{3}{2}; -\frac{R^2 x_n}{n-2}\right),$$

where $\Phi(\alpha, \gamma; z)$ is the degenerate hypergeometric function

$$\Phi(\alpha, \gamma; z) = \sum_{k=0}^{\infty} \frac{\Gamma(\alpha+k)\,\Gamma(\gamma)}{\Gamma(\alpha)\,\Gamma(\gamma+k)} \frac{z^k}{k!}.$$

With the help of (1.48), with the functions a_l and b_l are associated harmonic functions $v_l = A_l^{(1)}(x_1^2 + \ldots + x_{n-1}^2, x_n)p_l(x_1, \ldots, x_{n-1})$ and $w_l = A_l^{(2)}(x_1^2 + \ldots + x_{n-1}^2, x_n)p_l(x_1, \ldots, x_{n-1})$, which generate linearly independent solutions of the homogeneous problem

$$u_j(X) = \frac{1}{n-2}(x_1^2 + \ldots + x_n^2 - R^2)v_l(X),$$

$$j = 1, \ldots n-1,$$

$$u_n(X) = v_l(X) + \frac{1}{n-2}(x_1^2 + \ldots + x_n^2 - R^2)\frac{\partial^2 v_l}{\partial x_n^2},$$

$$u_j(X) = \frac{1}{n-2}(x_1^2 + \ldots + x_n^2 - R^2)\,w_l(X),\ j = 1, \ldots, n-1,$$

$$u_n(X) = w_l(X) + \frac{1}{n-2}(x_1^2 + \ldots + x_n^2 - R^2)\frac{\partial^2 w_l}{\partial x_n^2}.$$

Thus, the homogeneous problem corresponding to problem (4.92) for system (4.78) has a countable set of linearly independent solutions. For the solvability of (4.92) it suffices that the functions $f_1, \ldots, f_{n-1}$ have second derivatives.

All the systems of second-order partial differential equations considered in this section are not strongly elliptic [12], so the loss of the Noetherian character of the classical boundary problems becomes possible for such systems. We note that in four-dimensional space we have two examples of systems (4.61) and (4.78) for which the Noetherian property of the Dirichlet problem in the half-space fails. However, the set of nontrivial solutions of the homogeneous Dirichlet problem for system (4.61) is larger than for system (4.78), since it depends on two arbitrary harmonic functions, and the set of solutions of the homogeneous Dirichlet problem for system (4.78) depends only on one arbitrary harmonic function.

4. ELLIPTIC SYSTEMS DEPENDING ON A PARAMETER

In a number of problems, in particular in the study of the homotopy classification of elliptic systems, systems depending on certain parameters are of interest. One can consider the system

$$-\Delta u_j + \lambda \frac{\partial}{\partial x_j} \sum_{i=1}^{n} \frac{\partial u_i}{\partial x_i} = 0,\ j = 1, \ldots, n, \tag{4.95}$$

where λ is a real parameter, as a family of systems depending on the parameter λ. The eigenvalues of the characteristic matrix of this system have the form $\mu_j = -(\xi_1^2 + \ldots + \xi_n^2)$, $j = 1, \ldots, n-1$, $\mu_n = (\lambda - 1)(\xi_1^2 + \ldots + \xi_n^2)$. Consequently, for $\lambda < 1$

all the eigenvalues of the characteristic matrix of system (4.95) have the same sign, so the system is strongly elliptic for $\lambda < 1$. For $\lambda = 1$, system (4.95) degenerates, and for $\lambda > 1$ it is again elliptic, but no longer strongly elliptic, since μ_n has the sign plus, and all the other μ_j minus, so the characteristic matrix of this system is not positive or negative definite. Obviously, for $\lambda > 1$ the system (4.95) is homotopic to (4.78), where the continuous change of parameter λ effects this homotopy.

We construct a general solution of system (4.95) in some domain D with sufficiently smooth boundary Γ. We rewrite the system:

$$\Delta u_j = \lambda \frac{\partial \chi}{\partial x_j},\ j = 1, \ldots, n,\ \chi = \sum_{i=1}^{n} \frac{\partial u_i}{\partial x_i}.$$

It follows from these equations that one can represent u_j as follows: $u_j = \varphi_j + \lambda \partial\omega/\partial x_j$, where φ_j are harmonic functions which are regular in domain D, and ω is a particular solution of the equation $\Delta\omega = \chi$; in particular we set

$$\omega(X) = \int_D G(X, Y)\chi(Y)\, d_y V,$$

G being the Green's function of the Dirichlet problem for domain D. Now we have

$$u_j(X) = \varphi_j(X) + \lambda \frac{\partial}{\partial x_j} \int_D G(X, Y)\chi(Y)\, d_y V,\ j = 1, \ldots, n,$$

but the equation

$$\chi = \sum_{i=1}^{n} \frac{\partial u_i}{\partial x_i} = \sum_{i=1}^{n} \frac{\partial \varphi_i}{\partial x_i} + \lambda\Delta \int_D G(X, Y)\chi(Y)\, d_y V = \sum_{i=1}^{n} \frac{\partial \varphi_i}{\partial x_i} + \lambda\chi,$$

$$\chi = \frac{1}{1-\lambda} \sum_{i=1}^{n} \frac{\partial \varphi_i}{\partial x_i},\ \lambda \neq 1,$$

must hold, i.e., χ is a harmonic function, so

$$u_j(X) = \varphi_j(X) + \frac{\lambda}{1-\lambda} \frac{\partial}{\partial x_j} \int_D G(X, Y) \sum_{i=1}^{n} \frac{\partial \varphi_i}{\partial y_j}\, d_y V, \qquad (4.96)$$
$$\lambda \neq 1,\ j = 1, \ldots, n.$$

This is also a representation of the solutions of system (4.95) for $\lambda \neq 1$ in terms of harmonic functions in domain D. For $\lambda = 1$, instead of the connection between χ and the harmonic functions φ_j, $j = 1, \ldots, n$, we get $\Sigma_{i=0}{}^{n}(\partial\varphi_i/\partial x_i) = 0$, and χ remains an arbitrary sufficiently smooth function, so ω is also an arbitrary function. Instead of (4.96) we have

$$u_j(X) = \varphi_j(X) + \frac{\partial \omega}{\partial x_j}, \; j = 1, \ldots, n, \; \sum_{i=1}^{n} \frac{\partial \varphi_i}{\partial x_i} = 0, \; \Delta, \varphi = 0, \tag{4.97}$$

and ω is an arbitrary function which is three times differentiable in domain D. One can use representation (4.96) of solutions of system (4.95) to reduce the Dirichlet problem for this system to singular integral equations.

In domains of special forms, for example in the half-space and in the ball, one can give other representations of solutions, generalizing (4.84) and (4.86). Just as for system (4.78), one can show that all solutions of system (4.95) which are regular in the half-space H: $\{x_n > 0\}$ have the form

$$u_j = \varphi_j + x_n \frac{\partial \psi}{\partial x_j}, \; j = 1, \ldots, n, \quad \frac{\partial \psi}{\partial x_n} = \frac{\lambda}{2-\lambda} \sum_{i=1}^{n} \frac{\partial \varphi_i}{\partial x_i}, \tag{4.98}$$

where φ_j, $j = 1, \ldots, n$, are arbitrary harmonic functions which are regular in the half-space H. It follows from this representation of solutions that for $\lambda \neq 2$ the Dirichlet problem in the half-space H for the elliptic system (4.95) is solvable for any continuously differentiable boundary conditions, while the solution is always unique. Now for known functions $\varphi_1, \ldots, \varphi_n$ the function ψ is uniquely determined. With the help of representation (4.98) of solutions of system (4.95) one can investigate the Dirichlet problem for this system in the layer $E(h)$: $\{0 < x_n < h\}$ also. Let the conditions of the Dirichlet problem have the form

$$u_j |_{x_n=0} = f_j, \; u_j |_{x_n=h} = g_j, \; j = 1, \ldots, n.$$

From (4.98) we have the equations

$$\varphi_j |_{x_n=0} = f_j, \; \varphi_j |_{x_n=0} + h \frac{\partial \psi}{\partial x_j} \Big|_{x_n=h} = g_j, \tag{4.99}$$

and by (4.98), with the help of (4.99), we find

$$\begin{gathered} \left[\frac{2-\lambda}{\lambda} \frac{\partial \varphi}{\partial x_n} - \frac{\partial \varphi}{\partial x_n}\right]\Big|_{x_n=0} = F, \; \varphi_n |_{x_n=0} = f_n, \\ \left[\varphi_n + h \frac{\partial \psi}{\partial x_n}\right]\Big|_{x_n=h} = g, \\ \frac{\partial}{\partial x_x}\left[\frac{2-\lambda}{\lambda} \psi - h \frac{\partial \psi}{\partial x_n} - \varphi_n\right]\Big|_{x_n=h} = G, \end{gathered} \tag{4.100}$$

where F and G are known functions which are given by the formulas

$$F = \sum_{i=1}^{n-1} \frac{\partial f_i}{\partial x_i}, \; G = \sum_{i=1}^{n-1} \frac{\partial g_i}{\partial x_i}.$$

Solving problem (4.100) for the pair of harmonic functions φ_n and ψ, which are regular in domain $E(h)$, for the known function ψ from (4.99) for the definition of $\varphi_1, \ldots, \varphi_{n-1}$ we get an ordinary Dirichlet problem in layer $E(h)$.

We investigate problem (4.100) with the help of the Fourier transform with respect to the variables $x_1, \ldots, x_{n-1}$. We set

$$\Psi = A(\sigma)\exp(|\sigma|x_n) + B(\sigma)\exp(-|\sigma|x_n),$$
$$\Phi = C(\sigma)\exp(|\sigma|x_n) + D(\sigma)\exp(-|\sigma|x_n).$$

Here $\sigma = (\sigma_1, \ldots, \sigma_{n-1})$, $|\delta|^2 = \delta_1{}^2 + \ldots + \delta_{n-1}{}^2$, and Ψ and Φ are Fourier transforms of the functions ψ and φ_n, respectively. From (4.100) we find

$$(2-\lambda)\lambda^{-1}|\sigma|(A-B) - (C-D)|\sigma| = \tilde{F}, \tag{4.101}$$
$$A + B = \tilde{f}_n,\ h|\sigma|[A\exp(|\sigma|h) - B\exp(-|\sigma|h)] + C\exp(-|\sigma|h)$$
$$+ D\exp(-|\sigma|h) = \tilde{g}_n,\ A|\sigma|[(2-\lambda)\lambda^{-1} - h|\sigma|]\exp(|\sigma|h)$$
$$- B|\sigma|[(2-\lambda)\lambda^{-1} - h|\sigma|]\exp(-|\sigma|h) - C|\sigma|\exp(|\sigma|h)$$
$$+ D|\sigma|\exp(-|\sigma|h) = \tilde{G}.$$

The tilde here denotes the Fourier transform of the corresponding functions. The determinant of system (4.101) has the form

$$\Delta(\sigma) = -2(2-\lambda)\lambda^{-1}|\sigma|^2[2 + \sinh(|\sigma|h)].$$

This determinant vanishes identically for $\lambda = 2$ but for $\lambda \neq 2$ only for $\sigma = 0$ due to the factor σ^2, to which there corresponds the Laplace operator, so for $\lambda \neq 2$ the Dirichlet problem in layer $E(h)$ is always solvable and has a unique solution for any continuously differentiable boundary data. For $\lambda = 2$ the corresponding homogeneous problem has an infinite set of linearly independent solutions, but for the inhomogeneous problem to be solvable it is necessary to impose an infinite set of orthogonality conditions on the boundary data. This follows from the fact that for $\lambda = 2$ the left sides of the equations of (4.101) are connected by one linear relation and three of the four equations are linearly independent.

For solutions of system (4.95) which are regular in the ball Σ: $\{x_1{}^2 + \ldots + x_n{}^2 < R^2\}$ one can generalize the representation of solutions (4.86). We shall construct solutions of system (4.95) in the form

$$u_j(X) = \chi_j(X) + \frac{1}{n-2}(x_1^2 + \ldots + x_n^2 - R^2)\frac{\partial\psi}{\partial x_j}, \tag{4.102}$$
$$j = 1, \ldots, n,$$

where $\chi_1, \ldots, \chi_n, \psi$ are harmonic functions which are regular in ball Σ. Substituting (4.102) into (4.95), as in the derivation of (4.86) we get the relation

$$\frac{2(\lambda-2)}{n-2}\sum_{i=1}^{n} x_i\frac{\partial\psi}{\partial x_i} - 2\psi = -\lambda\sum_{i=1}^{n}\frac{\partial\chi_i}{\partial x_i}. \tag{4.103}$$

As in deriving (4.90), from the boundary conditions for the Dirichlet problem for system (4.95), by virtue of (4.102) we get a Dirichlet problem for harmonic functions $\chi_j(X)$ which are regular in ball Σ with the boundary condition $\chi_j = f_j$ on sphere S, bounding ball Σ. We denote such a harmonic function by $F_j(X)$, and with the help of $F_j(X)$, $j = 1, \ldots, n$, we form the function

$$G(X) = \sum_{i=1}^{n} \frac{\partial F_i}{\partial x_i}.$$

Now the question of the solvability of the Dirichlet problem for system (4.95) reduces to the determination of the solvability of the equation

$$\sum_{i=1}^{n} x_i \frac{\partial \psi}{\partial x_i} - \frac{n-2}{\lambda - 2} \psi = -\frac{\lambda (n-2)}{2(\lambda - 2)} G(X). \tag{4.104}$$

This first-order equation is always solvable and has a unique solution of ψ, which is holomorphic in ball Σ if the ratio $(n-2)/(\lambda-2)$ is not an integer, i.e., if $\lambda \neq 2 + (n-2)/k$, $k = 1, 2, \ldots$, where this solution ψ is harmonic in ball Σ if G is a harmonic in Σ. Now if $\lambda = 2 + (n-2)/k$, then for Eq. (4.104) to be solvable in the class of functions holomorphic in ball Σ, it is necessary and sufficient that the expansion of function $G(X)$ in a Taylor series in a neighborhood of the origin not contain terms of degree k, i.e., that $G(X)$ be orthogonal in Σ to all homogeneous harmonic polynomials of degree k. Now the homogeneous equation corresponding to (4.104) is satisfied by a homogeneous harmonic polynomial of degree k. The orthogonality conditions imposed on function $G(X)$ are obviously orthogonality conditions imposed on the vector-function $(f_1, \ldots, f_n)$ of the boundary data of the Dirichlet problem for (4.95) in Σ. Consequently, the Dirichlet problem in ball Σ for system (4.95) for $\lambda = 2 + (n-2)/k$ is Fredholm, while the number of linearly independent solutions of the corresponding homogeneous problem is equal to the number of linearly independent homogeneous harmonic polynomials in n variables of degree k.

As $k \to \infty$ the values of $\lambda_k = 2 + (n-2)/k$ tend to 2 and the number of homogeneous harmonic polynomials of degree k increases without bound. For $\lambda = 2$ the solution of the Dirichlet problem is given by (4.90) and only the effect of the loss by a solution of smoothness in comparison with the boundary data holds. Thus, for (4.95) the value of the parameter $\lambda = 2$ is a singular value.

Analogously, introducing a parameter into system (4.61), we consider the system

$$\begin{aligned} -\Delta s + \lambda \frac{\partial}{\partial t}(s_t + u_x + v_y + w_z) &= 0, \\ -\Delta u + \lambda \frac{\partial}{\partial t}(u_t - s_x - w_y + v_z) &= 0, \\ -\Delta v + \lambda \frac{\partial}{\partial t}(v_t - w_x - s_y - u_z) &= 0, \end{aligned} \tag{4.105}$$

$$-\Delta w + \lambda \frac{\partial}{\partial t}(w_t - v_x + u_y - s_z) = 0.$$

Since the operators D and $\bar{D}$ introduced before the system (4.60) satisfy the identities $D\bar{D} = \bar{D}D = \Delta$, $D + \bar{D} = 2\partial/\partial t$, one can write system (4.105) in terms of these operators as follows:

$$\left(\frac{\lambda - 2}{2} D + \frac{\lambda}{2} \bar{D}\right) \bar{D} U = 0, \ U = (s, u, v, w). \tag{4.106}$$

System (4.106) is equivalent to system

$$\frac{\lambda - 2}{2} Du + \frac{\lambda}{2} \bar{D} u = V, \bar{D} V = 0. \tag{4.107}$$

We construct a particular solution of the first system. For $\lambda \neq 2$ we try to construct a particular solution W so that $\bar{D}W = 0$. We have $[(\lambda - 2)/2]DW = V$, but $D = 2\frac{\partial}{\partial t} + \bar{D}$, so $(\lambda - 2)\partial W/\partial t = V$. It follows from this that for a holomorphic right side of the first equation of system (4.107) there exists a holomorphic particular solution. A general solution of (4.106) for $\lambda \neq 2$ can be written as follows: $U = W + \Omega$, where Ω is the general solution of the system $[(\lambda - 2)/2]D\Omega + (\lambda/2)\bar{D}\Omega = 0$, which in detailed notation has the form

$$(\lambda - 1)s_t + u_x + v_y + w_z = 0, \ (\lambda - 1)u_t - s_x - w_y + v_z = 0,$$
$$(\lambda - 1)v_t + w_x - s_y - u_z = 0, \ (\lambda - 1)w_t - v_x + u_y - s_z = 0.$$

If $\lambda > 1$, then by the change of variable $\tau = (\lambda - 1)^{-1}t$ from this system we get $\bar{D}U = 0$, where the role of variable t is played by τ, while if $\lambda < 1$ then, by the substitution of $\tau = (1 - \lambda)^{-1}t$ we get the system $DU = 0$, where again, instead of t, we have τ. Finally, we get for solutions of system (4.105) the following representations:

$$\begin{aligned} U &= \Psi(t/(\lambda - 1), X) + \Phi(t, X), \ \bar{D}\Psi = \bar{D}\Phi = 0, \ \lambda > 1, \\ U &= \Psi(t/(1 - \lambda), X) + \Phi(t, X), \ D\Psi = \bar{D}\Phi = 0, \ \lambda < 1, \end{aligned} \tag{4.108}$$

where $X = (x, y, z)$ and U, Φ, Ψ are four-component vectors. It is easy to calculate that the characteristic determinant of system (4.105) has the form

$$\Delta = (\xi_0^2 + \xi_1^2 + \xi_2^2 + \xi_3^2)[(\lambda - 1)^2 \xi_0^2 + \xi_1^2 + \xi_2^2 + \xi_3^2],$$

so system (4.105) is elliptic for $\lambda \neq 1$ and for $\lambda = 1$ it degenerates.

With the help of representation (4.108) of solutions of (4.105) one can easily investigate the Dirichlet problem for this system in the half-space H: $\{t > 0\}$. It follows from (4.108) that for $\lambda > 1$ the homogeneous Dirichlet problem has an infinite set of linearly independent solutions $U_0 = \Phi(t/(\lambda - 1), X) - \Phi(t, X)$, where $\Phi(t, X)$ is an arbitrary solution of the system $\bar{D}\Phi = 0$ which is regular in H, i.e., Φ is an arbitrary holomorphic vector in H. Now for the solvability of the inhomogeneous problem

$$s|_{t=0}=f_1,\quad u|_{t=0}=f_2,\quad v|_{t=0}=f_3,\quad w|_{t=0}=f_4 \tag{4.109}$$

it is necessary and sufficient that the vector $Q = (f_1, f_2, f_3, f_4)$ be the trace on the hyperplane $t = 0$ of a holomorphic vector which is regular in the half-space H, since it follows from the first formula of (4.108) that the trace of a solution of system (4.105) on the hyperplane $t = 0$ coincides with the trace of a holomorphic vector $\Psi(t, X) + \Phi(t, X)$.

If $\lambda < 1$, then by the second formula of (4.108) we have $U(0, X) = \Psi(0, X) + \Phi(0, X)$, $D\Psi = 0$, $\bar{D}\Phi = 0$. Since $D\bar{D} = \Delta$, one can represent any harmonic vector $h(t, X)$ in the form of a sum $h(t, X) = \Psi(t, X) + \Phi(t, X)$ of a holomorphic and an antiholomorphic vector where, since only vectors which do not depend on t satisfy the two systems $D\Phi = 0$ and $\bar{D}\Phi = 0$ simultaneously, we get that the representation of a harmonic vector $h(t, X)$ which is regular in the half-space H as the sum of a holomorphic and an antiholomorphic vector is unique, i.e., the holomorphic vector Φ and the antiholomorphic vector Ψ are uniquely determined by the harmonic vector h. From this there follows a simple method for solving the Dirichlet problem (4.109) in the half-space H: $\{t > 0\}$ for $\lambda < 1$. From the boundary data (4.109) we construct a harmonic vector $h(t, X)$ which is regular in the half-space H, such that $h(0, X) = (f_1, f_2, f_3, f_4)$. This vector can be represented uniquely in the form

$$h(t,\ X)=\Psi(t,\ X)+\Phi(t,\ X),\quad D\Psi=0,\quad \bar{D}\Phi=0. \tag{4.110}$$

The solution of problem (4.109) for system (4.95) sought is now given by the formula $U(t, X) = \Psi(t/(1-\lambda), X) + \Phi(t, X)$, while this solution always exists and is unique.

We show how to find, from vector $h(t, X)$, the vector $\Phi(t, X)$ in (4.110). For this we act on (4.110) by operator D, and using the properties of operators D and $\bar{D}$, we get $Dh = 2 = \frac{\partial}{\partial t}\Phi(t, X)$. Let us assume that the first derivatives of the vector h are continuous up to the boundary; further, solving the Neumann problem in the half-space H, we find the vector $\Phi(t, X)$. Letting $\Phi = (\varphi_1, \varphi_2, \varphi_3, \varphi_4)$, $h = (h_1, h_2, h_3, h_4)$ in expanded form, we have the equations

$$2\frac{\partial\varphi_1}{\partial t}=\frac{\partial h_1}{\partial t}-\frac{\partial h_2}{\partial x}-\frac{\partial h_3}{\partial y}-\frac{\partial h_4}{\partial z},\quad 2\frac{\partial\varphi_2}{\partial t}=\frac{\partial h_2}{\partial t}+\frac{\partial h_1}{\partial x}+\frac{\partial h_4}{\partial y}-\frac{\partial h_3}{\partial z},\quad 2\frac{\partial\varphi_3}{\partial t}$$
$$=\frac{\partial h_3}{\partial t}-\frac{\partial h_4}{\partial x}+\frac{\partial h_1}{\partial y}+\frac{\partial h_2}{\partial z},\quad 2\frac{\partial\varphi_4}{\partial t}=\frac{\partial h_4}{\partial t}+\frac{\partial h_3}{\partial x}-\frac{\partial h_2}{\partial y}+\frac{\partial h_1}{\partial z}.$$

Solving the Neumann problem, we find

$$2\varphi_1(t,X)=h_1(t,X)-\frac{1}{2\pi^2}\int_T[(x-x_0)^2+(y-y_0)^2$$
$$+(z-z_0)^2+t^2]^{-1}\left(\frac{\partial h_2}{\partial x_0}+\frac{\partial h_3}{\partial y_0}+\frac{\partial h_4}{\partial z_0}\right)d\Omega,$$

$$2\varphi_2(t, X) = h_2(t, X) + \frac{1}{2\pi^2}\int_T [(x - x_0)^2 + (y - y_0)^2 + (z - z_0)^2 + t^2]^{-1}\left(\frac{\partial h_1}{\partial x_0} + \frac{\partial h_4}{\partial y_0} - \frac{\partial h_3}{\partial z_0}\right) d\Omega,$$

$$2\varphi_3(t, X) = h_3(t, X) + \frac{1}{2\pi^2}\int_T [(x - x_0)^2 + (y - y_0)^2 + (z - z_0)^2 + t^2]^{-1}\left(-\frac{\partial h_4}{\partial x_0} + \frac{\partial h_1}{\partial y_0} + \frac{\partial h_2}{\partial z_0}\right) d\Omega,$$

$$2\varphi_4(t, X) = h_4(t, X) + \frac{1}{2\pi^2}\int_T [(x - x_0)^2 + (y - y_0)^2 + (z - z_0)^2 + t^2]^{-1}\left(\frac{\partial h_3}{\partial x_0} - \frac{\partial h_2}{\partial y_0} + \frac{\partial h_1}{\partial z_0}\right) d\Omega,$$

where T is the hyperplane $t = 0$. Integrating by parts in these equations, we get

$$\varphi_1(t, X) = 2^{-1}h_1(t, X) - \frac{1}{2\pi^2}\int_T \frac{(x - x_0)h_2(X_0) + (y - y_0)h_3(X_0) + (z - z_0)h_4(X_0)}{[(x - x_0)^2 + (y - y_0)^2 + (z - z_0)^2 + t^2]^2} d\Omega,$$

$$\varphi_2(t, X) = 2^{-1}h_2(t, X) + \frac{1}{2\pi^2}\int_T \frac{(x - x_0)h_1(X_0) + (y - y_0)h_4(X_0) - (z - z_0)h_3(X_0)}{[(x - x_0)^2 + (y - y_0)^2 + (z - z_0)^2 + t^2]^2} d\Omega,$$

$$\varphi_3(t, X) = 2^{-1}h_3(t, X) + \frac{1}{2\pi^2}\int_T \frac{-(x - x_0)h_3(X_0) + (y - y_0)h_1(X_0) + (z - z_0)h_2(X_0)}{[(x - x_0)^2 + (y - y_0)^2 + (z - z_0)^2 + t^2]^2} d\Omega,$$

$$\varphi_4(t, X) = 2^{-1}h_4(t, X) + \frac{1}{2\pi^2}\int_T \frac{(x - x_0)h_3(X_0) - (y - y_0)h_2(X_0) + (z - z_0)h_1(X_0)}{[(x - x_0)^2 + (y - y_0)^2 + (z - z_0)^2 + t^2]^2} d\Omega$$

$X_0 = (x_0, y_0, z_0)$. In these formulas the differentiability of vector h up to the boundary is not necessarily required, and its continuity up to the boundary suffices. Knowing $\Phi(t, X)$ one can easily find the vector $\Psi(t, X)$ also, for which one gets analogous formulas, which differ only in the signs in front of the integrals.

Among elliptic systems of partial differential equations of the second order with two independent variables one can single out the class of so-called strongly connected systems [7]. This concept is introduced with the help of the representation of solutions of the system in terms of holomorphic functions of a complex variable. One can generalize the concept of strong connectedness to system (4.105) also, starting from the representation (4.108) of its solutions, where (4.105) will be the analog of a strongly connected system for $\lambda > 1$. In [38] there is given a somewhat different representation of the solutions of elliptic systems with two independent variables in terms of holomorphic functions, from which it follows that the homogeneous Dirichlet problem for a strongly connected system in a half-plane bounded by one of the coordinate axes always has an infinite set of linearly independent solutions, and for the solvability of the inhomogeneous problem, as a rule it is necessary to impose an infinite set of orthogonality conditions on the boundary data. Since one cannot always generalize the structural properties of solutions connected with their representation by harmonic functions to multidimensional elliptic systems, this property of the Dirichlet problem for the half-plane can serve as the foundation of a specific multidimensional analog of a strongly connected system [6, 7].

Definition 4.1. We shall call a multidimensional elliptic system of second-order partial differential equations with constant coefficients strongly connected if there exists a half-space H in which the homogeneous Dirichlet problem for this system has an infinite set of linearly independent solutions, or the inhomogeneous problem has an infinite set of necessary conditions for its solvability of the type of orthogonality conditions.

Obviously (4.95) is strongly connected for $\lambda = 2$, and (4.105) is strongly connected for $\lambda > 1$. The smallest terms influence the solvability of the Dirichlet problem for multidimensional strongly connected systems just as in two-dimensional systems [35]. We consider the system

$$-\Delta u_j + 2\frac{\partial}{\partial x_j}\left(\sum_{i=1}^{n}\frac{\partial u_i}{\partial x_i} + Au_n\right) = 0, \qquad j = 1, \ldots, n. \tag{4.111}$$

Its general solution has the form

$$u_j = \psi_j + \frac{\partial \omega}{\partial x_j}, \quad j = 1, \ldots, n, \; \sum_{i=1}^{n}\frac{\partial \psi_i}{\partial x_i} + A\psi_n = 0, \tag{4.112}$$

where the ψ_j are harmonic functions, and ω is a solution of the equation

$$\Delta\omega + 2A\frac{\partial \omega}{\partial x_n} = 0. \tag{4.113}$$

With the help of the present representation we investigate the Dirichlet problem in the half-space H: $\{x_n > 0\}$ for (4.111)

$$u_j|_{x_n=0} = f_j, \; j = 1, \ldots, n. \tag{4.114}$$

From (4.114) we get $[\psi_j + (\partial\omega/\partial x_j)]|_{x_n=0} = f_j$. Further, using (4.113) and the relation connecting the harmonic functions ψ_j, we find

$$F = \sum_{i=1}^{n}\frac{\partial f_i}{\partial x_i} = \left[\sum_{i=1}^{n-1}\frac{\partial \psi_i}{\partial x_i} + \left(\Delta\omega + \frac{\partial^2\omega}{\partial x_n^2}\right)\right]\Bigg|_{x_n=0} = -\left[\frac{\partial \psi_n}{\partial x_x} + A\psi_n + 2A\frac{\partial\omega}{\partial x_n} + \frac{\partial^2\omega}{\partial x_n^2}\right]\Bigg|_{x_n=0}.$$

Now to define the functions ψ_n and ω we have the following problem: Find a harmonic function ψ_n which is regular in the half-space H, and a regular solution ω of (4.113) satisfying, for $x_n = 0$, the conditions

$$\frac{\partial \psi_n}{\partial x_n} + A\psi_n + 2A\frac{\partial \omega}{\partial x_n} + \frac{\partial^2 \omega}{\partial x_n^2} = f, \quad \frac{\partial \omega}{\partial x_n} + \psi_n = g, \tag{4.115}$$

where f and g are given functions. We shall solve this problem with the help of the Fourier transform. The Fourier transforms of the functions ψ_n and ω have the form $\tilde{\psi}_n = B(\sigma)\exp(-|\sigma|x_n)$, $\tilde{\omega} = C(\sigma)\exp[-x_n(A + \sqrt{A^2 + |\sigma|^2})]$. We denote by $\alpha(\sigma)$ and $\beta(\sigma)$ the Fourier transforms of functions f and g, respectively. Passing to Fourier transforms, from (4.115) we get a system of linear algebraic equations for $B(\sigma)$ and $C(\sigma)$:

$$(A - |\sigma|)B(\sigma) + (A + \sqrt{A^2 + |\sigma|^2})(\sqrt{A^2 + |\sigma|^2} - A)C(\sigma) = \alpha(\sigma),$$
$$B(\sigma) - (A + \sqrt{A^2 + |\sigma|^2})C(\sigma) = \beta(\sigma).$$

The determinant of this system $\Delta(\sigma) = (\sqrt{A^2 + |\sigma|^2} + A)(|\sigma| - \sqrt{A^2 + |\sigma|^2})$ is nowhere zero, so (4.115) is always solvable and has a unique solution. Knowing the function ω, to determine the harmonic functions ψ_j, $j = 1, \ldots, n-1$, we get a Dirichlet problem in H, which is also solvable and has a unique solution. Although the Dirichlet problem for (4.111) in the half-space H is solvable and has a unique solution, finer analysis shows that one must require increased smoothness of the boundary data; i.e., here, as for two-dimensional systems [35], one can observe the effect of loss of smoothness.

Similarly one can consider system (4.105) as a system containing two parameters:

$$\begin{aligned}
-\Delta u + \lambda\frac{\partial}{\partial x}(u_x + v_y + w_z) + \mu\frac{\partial}{\partial t}(u_t + v_z - w_y) &= 0,\\
-\Delta v + \lambda\frac{\partial}{\partial y}(u_x + v_y + w_z) + \mu\frac{\partial}{\partial t}(v_t + w_x - u_z) &= 0,\\
-\Delta w + \lambda\frac{\partial}{\partial z}(u_x + v_y + w_z) + \mu\frac{\partial}{\partial t}(w_t - v_x + u_y) &= 0,
\end{aligned} \tag{4.116}$$

which for $\lambda = \mu = 2$ turns into the system of Shevchenko (4.69). The characteristic determinant of this system has the form

$$\Delta = (\xi_0^2 + \xi_1^2 + \xi_2^2 + \xi_3^2)\,[(\mu - 1)^2\xi_0^2 + \xi_1^2 + \xi_2^2 + \xi_3^2]\,[(\mu - 1)\,\xi_0^2 + (\lambda - 1)\,(\xi_1^2 + \xi_2^2 + \xi_3^2)]^2,$$

so the system is elliptic for all λ and μ which satisfy the condition $(\mu - 1)(\lambda - 1) > 0$. For $\lambda = \mu$ one can investigate the system just as (4.105) was investigated. For this, along with the operator considered in the construction of system (4.69), we introduce the operator which associates with the vector $U = (s, u, v, w)$ the left sides of the equations of the system

$$\begin{aligned}
&s_t + u_x + v_y + w_z = 0, \quad -u_t + s_x + v_z - w_y = 0,\\
&-v_t + s_y - u_z + w_x = 0, \quad -w_t + s_z + u_y - v_x = 0.
\end{aligned} \tag{4.117}$$

By direct calculation one verifies the identity $D_1\bar{D}_1 = \bar{D}_1 D_1 = \Delta$. Obviously the system $\bar{D}_1 D_1 U + \gamma \bar{D}_1{}^2 U = 0$ in expanded notation has the form

$$(1+\gamma)\,\Delta s = 0,$$
$$(1-\gamma)\,\Delta u + 2\gamma \frac{\partial}{\partial x}(u_x + v_y + w_z) + 2\gamma \frac{\partial}{\partial t}(u_t + v_z - w_y) = 0,$$
$$(1-\gamma)\,\Delta v + 2\gamma \frac{\partial}{\partial y}(u_x + v_y + w_z) + 2\gamma \frac{\partial}{\partial t}(v_t + w_x - u_z) = 0,$$
$$(1-\gamma)\,\Delta w + 2\gamma \frac{\partial}{\partial z}(u_x + v_y + w_z) + 2\gamma \frac{\partial}{\partial t}(w_t - v_x + u_y) = 0.$$

Consequently, setting $\gamma = \lambda/(\lambda - 2)$ one can write system (4.116) in the form

$$\bar{D}_1(D_1 + \gamma \bar{D}_1)U = 0. \tag{4.118}$$

Moreover, this system splits into Laplace's equation for s and the system (4.116) for $\lambda = \mu$ for the other components of vector U, so it suffices to take only those solutions of system (4.118) whose first components are equal to zero. One can rewrite system (4.118) as follows:

$$D_1 U + \frac{\lambda}{\lambda - 2}\bar{D}_1 U = V, \ \bar{D}_1 V = 0, \tag{4.119}$$

here one considers only those vectors U and V whose first components are equal to zero. The second system (4.119) for a vector with first component zero has the form

$$\begin{aligned} &u_x + v_y + w_z = 0, \ -u_t + w_y - v_z = 0,\\ &-v_t - w_x + u_z = 0, \ -w_t + v_x - u_y = 0. \end{aligned} \tag{4.120}$$

We try to construct a particular solution of system (4.119) in the class of vectors with first component zero, satisfying system $\bar{D}_1 Q = 0$. Let $Q = (0, p, q, r)$. In expanded form the first system (4.119) can be rewritten as follows:

$$\begin{aligned} &p_x + q_y + r_z = 0, \ -p_t + q_z - r_y = u,\\ &-q_t - p_z + r_x = v, \ -r_t + p_y - q_x = w, \end{aligned} \tag{4.121}$$

but since the functions p, q, and r satisfy (4.120), (4.121) reduces to the form

$$p_x + q_y + r_z = 0, \ 2p_t = u, \ 2q_t = -v, \ 2r_t = w,$$

and since the functions u, v, w satisfy (4.120), one can always define p, q, and r from this. Consequently, one can always represent a solution of system (4.118) in the form $U = V + W$, where V is a solution of the system $\bar{D}_1 V = 0$, and W is a solution of the system $(\lambda - 2)D_1 U + \lambda \bar{D}_1 U = 0$, where the first components of all three vectors U, V, and W are equal to zero. In expanded form, the homogeneous system corresponding to the first system (4.119) can be written as follows:

$$u_x + v_y + w_z = 0, \quad -(1+\gamma)u_t - (1-\gamma)(w_y - v_z) = 0,$$

$$-(1+\gamma)v_t - (1-\gamma)(w_x - u_z) = 0,$$

$$-(1+\gamma)w_t - (1-\gamma)(u_y - v_x) = 0,$$

$$\gamma = \lambda/(\lambda - 2).$$

By the change of variable $\tau = (1-\lambda)^{-1}t$ one can reduce this system for $\lambda < 1$ to (4.117), and for $\lambda > 1$, to (4.120). Just like (4.105), the system

$$\begin{aligned} &-\Delta u + \lambda \frac{\partial}{\partial x}(u_x + v_y + w_z) + \lambda \frac{\partial}{\partial t}(u_t + v_x - w_y) = 0, \\ &-\Delta v + \lambda \frac{\partial}{\partial y}(u_x + v_y + w_z) + \lambda \frac{\partial}{\partial t}(v_t + w_x - u_z) = 0, \\ &-\Delta w + \lambda \frac{\partial}{\partial z}(u_x + v_y + w_z) + \lambda \frac{\partial}{\partial t}(w_t - v_x + u_y) = 0 \end{aligned} \tag{4.122}$$

for $\lambda > 1$ is strongly connected, and the Dirichlet problem for it in the half-space H: $\{t > 0\}$ has an infinite set of linearly independent solutions.

It is easy to verify that one can write system (4.116) in the form

$$D_1^2 U + \alpha D_1 \overline{D}_1 U + \beta \overline{D}_1^2 U = 0. \tag{4.123}$$

Analogously one can also consider the system

$$D^2 U + \alpha D\overline{D} U + \beta \overline{D}^2 U = 0, \tag{4.124}$$

generalizing (4.105). At least when the roots of the polynomial $\chi(t) = t^2 + \alpha t + \beta$ are real, one can study system (4.124) just like (4.105), rewriting (4.124) as follows:

$$(D + \nu_1 \overline{D})(DU + \nu_2 \overline{D} U) = 0, \tag{4.125}$$

where ν_1 and ν_2 are the roots of the polynomial $\chi(t)$. Depending on the roots of the polynomial $\chi(t)$, the system will either be weakly connected or strongly connected. One can get the condition for being strongly connected from representation (4.125) of system (4.124). System (4.123) also has analogous properties.

The examples of strongly connected multidimensional elliptic systems considered in this section show that in the Dirichlet problem for such systems three new effects appear which are not observed for strongly elliptic systems: 1) the homogeneous Dirichlet problem can have an infinite set of linearly independent solutions, or it is necessary to impose an infinite set of solvability conditions of the type of orthogonality conditions on the data of the problem for it to be solvable; 2) the smallest terms influence the solvability character of the problem, i.e., in the presence in the system of smallest terms of specific structure the Dirichlet problem becomes solvable; 3) for the existence of a solution of the Dirichlet problem it is necessary to require increased smoothness of the boundary data. As is clear from the well-developed theory of elliptic systems with two independent variables [7], these phenomena are typical for second-order elliptic systems of partial differential equations. The example of system (4.95) shows that if two systems are homotopic to one another, and one of them is strongly connected, then the second will not necessarily be strongly connected, i.e., strong connectedness of a system is not necessarily preserved under homotopy. It would be interesting to investigate whether the effect of loss of smoothness of a solution of the Dirichlet problem is preserved under a homotopy of multidimensional elliptic systems. These new phenomena in the theory of the Dirichlet problem led to the consideration of new, larger classes of systems, the so-called uniformly nonelliptic systems [9] and weakly elliptic systems [36] of pseudodifferential equations, including pseudodifferential equations on the boundary of the domain, equivalent to the Dirichlet problem for systems of elliptic type, strongly connected in the domain.

We dwell further on one system generalizing (4.95):

$$-L(u_j) + \frac{\partial}{\partial x_j}\sum_{i=1}^{n} b_i(X)\frac{\partial u_i}{\partial x_i} = 0, \quad j = 1, \ldots, n, \tag{4.126}$$

$$L = \sum_{i,j=1}^{n} \frac{\partial}{\partial x_i}\left(a_{ij}(X)\frac{\partial}{\partial x_i}\right), \quad X = (x_1 \ldots, x_n).$$

Here L is an elliptic operator. System (4.126) is equivalent to the following:

$$L(u_j) = \frac{\partial \chi}{\partial x_j}, \quad j = 1, \ldots, n, \quad \chi = \sum_{i=1}^{n} b_i(X)\frac{\partial u_i}{\partial x_i}. \tag{4.127}$$

Denoting by G the Green's function of domain D for operator L, and by ψ_j, $j = 1, \ldots, n$, arbitrary solutions of the homogeneous equation $L(\psi) = 0$, from the first n equations of (4.127) we find

$$u_j(X) = \psi_j(X) + \int_D G(X, Y)\frac{\partial \chi}{\partial y_j}\, d_y\Omega, \quad j = 1, \ldots, n.$$

Substituting these expressions for U_j in the last equation of (4.127), we get the integrodifferential equation

$$\chi = \sum_{i=1}^{n} b_i(X)\frac{\partial \psi_i}{\partial x_i} + \sum_{i=1}^{n} b_i(X)\frac{\partial}{\partial x_i}\int_D G(X, Y)\frac{\partial \chi}{\partial y_i}\, d_y\Omega.$$

It reduces to a singular integral equation for χ and completely determines the solvability character of the Dirichlet problem for (4.125).

Chapter 5

OBLIQUE DERIVATIVE PROBLEM FOR HARMONIC FUNCTIONS OF TWO VARIABLES

One can investigate the oblique derivative problem for harmonic functions of two independent variables with the help of singular integral equations [29]. However, one can also study it completely, directly, bypassing the reduction to integral equations [57]. We shall recount such a method below. The solution by this method of the oblique derivative problem reduces to the solution of a Dirichlet problem and a first-order partial differential equation. For completeness of the account we first give some auxiliary information.

1. BOUNDARY PROPERTIES OF CONJUGATE HARMONIC FUNCTIONS

We shall consider harmonic functions defined on simply connected or multiconnected domains. If the domain is simply connected, then we shall consider its boundary to be a simple closed rectifiable Jordan curve, which is defined by parametric equations $x = x(s)$, $y = y(s)$, where functions $x(s)$ and $y(s)$ have continuous derivatives, except at a finite number of corners, where as parameter s one takes the arc length of the curve. Now if the domain is multiconnected, then it is assumed that its boundary consists of a finite number of such curves.

For conjugate harmonic functions $v(x, y)$ and $w(x, y)$ which are regular in the unit disk K: $\{x^2 + y^2 < 1\}$, one has the following theorem of Privalov [31]: Let $V(\theta)$ and $W(\theta)$ be the values on the circle L: $\{x^2 + y^2 = 1\}$, bounding the disk K, of the harmonic function $v(x, y)$ and its conjugate $w(x, y)$; if $V(\theta)$ satisfies a Hölder condition with exponent $h < 1$, then $W(\theta)$ satisfies a Hölder condition with the same exponent h, while if $V(\theta)$ satisfies a Lipschitz condition ($h = 1$), then $W(\theta)$ satisfies a Hölder condition with exponent $1 - \varepsilon$, where $\varepsilon > 0$ can be taken arbitrarily small, but one cannot set $\varepsilon = 0$. With the help of a conformal mapping one can generalize Privalov's theorem to simply connected domains. We consider a number of cases successively. Let us assume that boundary Γ of domain D has no corner points and the angle ψ between the tangent to Γ and the Ox axis satisfies a Hölder condition as a function of the arc length s of curve Γ. Let $z = g(\zeta)$ be a function mapping domain D conformally to disk K. Warschawski [63] proved that for such domains the ratio $dz/d\zeta$ does not vanish or become infinite in the closed domain $D \cup \Gamma$, i.e., there exist constants m and M, $0 < m < M < \infty$, such that in $D \cup \Gamma$ we have $m \leq |dz/d\zeta| \leq M$. Let arc s of curve Γ correspond on circle L to the arc σ; then we have $ds/d\sigma = |dz/d\zeta|$, from which it follows that $m \leq ds/d\sigma \leq M$. Under the conformal map $z = g(\zeta)$ the conjugate harmonic functions $v(z)$ and $w(z)$ in D go into conjugate harmonic functions $V(\zeta)$ and $W(\zeta)$ in disk K. By Warschawski's theorem, if $v(s)$ is Hölder continuous on boundary Γ of the domain, then $V(\sigma)$ is Hölder continuous with the same exponent on the unit circle, and conversely. By Privalov's theorem $W(\sigma)$ is Hölder continuous, so Warschawski's theorem guarantees the Hölder continuity with the same exponent of its preimage $w(s)$ on Γ.

One can formulate the assertion of Warschawski's theorem in a somewhat different form: let the angle ψ satisfy a Hölder condition with some exponent h on the arc PP' of boundary Γ of domain D; then the ratio $dz/d\zeta$ is bounded away from zero and infinity on any arc QQ' contained strictly inside the arc PP'. In what follows we shall call point Q of boundary Γ of domain D an ordinary point if there exists an arc PP' of curve Γ, containing point Q in its interior, such that the angle ψ on PP' satisfies a Hölder condition. Arc PP' can be arbitrarily small but not contacted to a point.

Now let us assume that boundary Γ of domain D has no corner points, but the angle ψ between the tangent to Γ and the Ox axis does not satisfy a Hölder condition, but is only continuous as a function of the arc length s of Γ. Let δs and δS be the arc lengths of Γ and of the unit circle corresponding to one another under a conformal map of D to the unit disk. Now the ratio $\delta s/\delta S$ can tend to zero or to infinity when δs and δS tend to zero; however, there are inequalities established by Lavrent'ev [56],

$$\delta S < K_1 (\delta s)^{1-\varepsilon}, \quad \delta s < K_2 (\delta S)^{1-\varepsilon_1}, \tag{5.1}$$

where $\varepsilon > 0$ and $\varepsilon_1 > 0$ are arbitrarily small numbers, and K_1 and K_2 are constants depending on ε and ε_1. Let $v(s)$ be the values on Γ of the function $v(z)$ which is harmonic in D, and let $v(s)$ satisfy a Hölder condition with exponent h. By (5.1) the function $V(S)$, which one gets from $v(s)$ by replacing s by S, satisfies a Hölder condition with exponent $h\,(1-\varepsilon)$. By Privalov's theorem the function $W(\zeta)$ conjugate to $V(\zeta)$ in D satisfies a Hölder condition with the same exponent. By (5.1) we have that the function $w(s)$, which one gets from $W(S)$ by replacing S by s, satisfies a Hölder condition with exponent $h(1-\varepsilon_1)(1-\varepsilon) = h - \eta$, where η can be taken, along with ε and ε_1, arbitrarily small.

Finally, let us assume that boundary Γ of domain D has a corner point, where the tangent rays issuing from this point make an angle $A = m\pi$, $0 \leq m \leq 2$. Taking this corner point as the origin, by the map $Z = z^{1/m}$ we carry domain D into a domain Δ with smooth boundary. We have

$$\frac{dZ}{dz} = \frac{1}{m} z^{1/m-1}, \quad \frac{dS}{ds} = \frac{1}{m} r^{1/m-1}, \quad r = |r|. \tag{5.2}$$

Consequently, the ratio dS/ds is everywhere bounded and different from zero, except at the origin. Now in a neighborhood of the origin, the coordinate s behaves like r, and S like $R = |Z|$, so S behaves like $r^{1/m}$ or like $s^{1/m}$, and s behaves like S^m.

Let $0 < m \leq 1$; by virtue of (5.2)

$$\delta S < K_1 \delta s, \quad \delta s < K_2 (\delta S)^m, \tag{5.3}$$

where K_1 and K_2 are fixed constants. Now if $v(s)$ satisfies a Hölder condition with exponent h, then $V(S)$ satisfies a Hölder condition with exponent hm, and its conjugate function $W(S)$ satisfies the given condition with exponent $hm - \eta$, where $\eta > 0$ is an arbitrarily small number. By virtue of the first inequality of (5.3), $w(s)$ satisfies a Hölder condition with the same exponent $hm - \eta$ at point $z = 0$. Let $1 < m \leq 2$; instead of (5.3) we now have

$$\delta S < K_1 (\delta s)^{1/m}, \quad \delta s < K_2 \delta S. \tag{5.4}$$

Analogously, we get that $w(s)$ satisfies a Hölder condition with exponent $h/m - \eta$, if $v(s)$ satisfies it with exponent h. Thus, if the harmonic function $v(z)$ satisfies a Hölder condition, then its dual harmonic function $w(z)$ also satisfies it, but the exponent in the Hölder condition can be smaller.

These properties of dual harmonic functions are preserved for nonsimply connected domains also. In order to prove this we generalize Privalov's theorem. Let the harmonic function $v(x, y)$ satisfy Hölder's condition on the arc AB of the unit circle. We take an arc A_1B_1, contained strictly inside the arc AB, and we represent the trace $V(\theta)$ of the function $v(x, y)$ on the unit circle in the form of a sum $V(\theta) = V_1(\theta) + V_2(\theta)$, where $V_1(\theta)$ coincides with $V(\theta)$ on the arc A_2B_2, where A_2 here is the middle of the arc AA_1, B_2 is the middle of the arc B_1B, on the complement of the arc AB, $V_1(\theta) = 0$, and on the arcs AA_2 and B_2B the function $V_1(\theta)$ varies linearly so that V_1 is Hölder continuous on the whole unit circle. Obviously $V_1(\theta)$ is Hölder continuous with the same exponent as $V(\theta)$ on A_1B_1. Consequently, by Privalov's theorem its dual harmonic function $W_1(\theta)$ is Hölder continuous with the same exponent. The function $V_2(\theta) = V(\theta) - V_1(\theta)$ is equal to zero on the arc A_1B_1; by the Schwarz symmetry principle [32] its corresponding harmonic function $v_2(x, y)$ extends analytically across the arc A_1B_1, so its dual harmonic function $w_2(x, y)$ is also analytic on the arc, and the function $W_2(\theta)$ satisfies a Lipschitz condition on A_1B_1. Thus, the following assertion is valid: If the harmonic function v, which is regular in the disk, is Hölder continuous with exponent h on the arc AB of the boundary of the disk, then on any arc A_1B_1 which is interior to AB, its conjugate harmonic function w satisfies a Hölder condition with the same exponent.

With the help of this assertion one can also prove the property of Hölder continuity of conjugate harmonic functions for multiconnected domains. Let the multiconnected domain D be bounded by closed curves $C_0, C_1, \ldots, C_n$, and let the values $V_i(s)$ on C_i of the harmonic function $v(x, y)$, which is regular in domain D, be Hölder continuous. We take an arc AB on component C_j of the boundary of D. If this arc is sufficiently small, then in D one can draw a line γ such that the arc AB and γ bound a simply connected subdomain of domain D. Now the conjugate harmonic function $w(x, y)$ to $v(x, y)$ is Hölder continuous on any arc A_1B_1 of the boundary of domain D, which is strictly interior to AB, and this means that $w(x, y)$ is Hölder continuous everywhere on the boundary of domain D if $v(x, y)$ is.

Lemma 5.1. Let $F(z) = P + iQ$ be a function which is holomorphic in the unit disk, and let Q tend to zero when point z approaches any point of the unit circle except for a finite number of points $M_1, \ldots, M_p$, in neighborhoods of which Q is unbounded. Then at each point M_i the function $F(z)$ grows no slower than the first power of the distance between z and M_i.

Proof. $F(z)$ is real everywhere on the unit circle except possibly for the points $M_1, \ldots, M_p$, so it can be extended symmetrically by $\overline{F(z)}$ to an analytic and single-valued function on the whole plane except for the points $M_1, \ldots, M_p$ [32]. Let $M_i = z_i$. In a neighborhood of the point z_i one can expand the function $F(z)$ in a Laurent series

$$F(z)=a_0+\sum_{k=1}^{\infty} a_k (z-z_i)^k + \sum_{k=1}^{\infty} b_k (z-z_i)^{-k},$$

in which at least one coefficient b_k is nonzero. This also implies the validity of the assertion of the lemma. Obviously with the help of a conformal mapping the assertion of the lemma can be extended to any domain with smooth boundary.

Lemma 5.2. Let $F(z) = P + iQ$ be a function which is holomorphic in a domain and suppose that on an arc of the boundary containing the point z_0 the functions P and Q are bounded above in absolute value by the constant K. If $F(z)$ remains bounded as z tends to z_0 along the boundary, then it will also be bounded as z tends to z_0 from within the domain.

To prove this we consider

$$\begin{aligned} F_1(z) &= P_1 + iQ_1 = [F(z) + K - iK]^{-1} \\ &= \{K+P+i(K-Q)\}[(K+P)^2+(K-Q)^2]^{-1}. \end{aligned}$$

The functions P_1 and Q_1 are positive and different from zero on an arc of the boundary containing z_0. If $F(z)$ is not bounded when z tends to z_0 from within the domain, then $F_1(z)$, and with it P_1 and Q_1, tend to zero as z tends in this way to z_0. This contradicts the properties of harmonic functions which cannot tend to zero as z approaches z_0 from within the domain if they do not tend to zero as z approaches z_0 along the boundary of the domain.

Let $V(s)$ and $W(s)$ be the values on the boundary of a harmonic function $v(x, y)$ and its conjugate $w(x, y)$. We consider the analytic function $F(z) = v(x, y) + iw(x, y)$. Let z_0 and z be points of the boundary which correspond to the values s_0 and s of the parameter. We shall set $s = s_0 + \delta s$, $\rho = |z - z_0|$, and assume the point z_0 is an ordinary point of the boundary. Without proof we cite the following two theorems.

Theorem 5.1. If a function $V(s)$ is given such that the ratio

$$[V(s)-V(s_0)]/|\delta s|^h, \quad M(s)=[V(s)-V(s_0)]\rho^{-h},$$

is bounded above in absolute value and satisfies the inequality

$$|M(s)-M(s')| < K[|s-s'|/|\delta s-\delta s'|]^{\varepsilon},$$

$$0<\varepsilon<1, \quad \delta s\delta s' \geqslant 0,$$

in a neighborhood of the point s_0, where ε is an arbitrarily small but positive number, then the ratio of $F(z) - F(z_0)$ to $(z - z_0)^h$ is bounded in absolute value for all z from a neighborhood of point z_0.

Theorem 5.2. Let $V(s)$ be a function of the arc length s given on the boundary of a domain, and let this function become infinite at the point s_0 so that the product

$$|s-s_0|^l V(s), \quad M(s)=\rho^l V(s), \quad 0<l\leqslant 1$$

remains bounded above in absolute value and $M(s)$ satisfies the inequality

$$|M(\alpha)-M(\theta)| \leqslant K|(\alpha-\theta)(\alpha+\theta)^{-1}|^{\varepsilon}. \tag{5.5}$$

Then there exists function $v(x, y)$ which is harmonic in the domain, which coincides with $V(s)$ on the boundary, while in the class of functions admitting a singularity of less than first order at point s_0 this function is unique. The function $v(x, y)$, its dual $w(x, y)$, and the analytic function $F(z) = v + iw$ are such that the products

$$\rho^l v(x, y), \quad \rho^l w(x, y) \quad (z-z_0)^l F(z)$$

are bounded above in absolute value.

2. AN AUXILIARY PROBLEM

In this section we investigate the following problem: In the simply connected domain D find two conjugate harmonic functions $A(x, y)$ and $B(x, y)$, given the ratio B/A of these functions on boundary Γ of the domain. If such harmonic functions exist, then the function $A + iB$ is a holomorphic function in domain D of the variable $z = x + iy$. Let $A + iB = R(\cos\chi + i\sin\chi)$, from which it follows that $\ln(A + iB) = \ln R + i\chi$. The functions $\ln R$ and χ are also conjugate harmonic functions. Giving the ratio B/A on boundary Γ is equivalent to giving $\tan\chi$ or, equivalently, giving the values of χ on Γ. Thus, to find the harmonic function χ it suffices to solve the Dirichlet problem in domain D. With χ known, the function $\ln R$ is defined by

$$\ln R(x, y) = \int(\chi_y\,dx - \chi_x\,dy). \tag{5.6}$$

Knowing χ and R, we find $A(x, y) = R\cos\chi$, $B(x, y) = R\sin\chi$.

We concern ourselves with the construction of function χ. Let us assume that after a circuit of the boundary Γ the values of χ given on Γ are unchanged. In this case, in domain D there exists a regular single-valued harmonic function χ, which assumes these values on Γ. The function $\ln R$, defined by (5.6), is also single-valued and bounded in domain D, so $R^2 = A^2 + B^2$ is nonzero everywhere in D. All four functions χ, $\ln R, A, B$ are now Hölder continuous in the closed domain $D \cup \Gamma$. On the right side of (5.6) one can achieve any constant, which is equivalent to multiplying A and B by a constant. Consequently, A and B are defined up to this multiplicative constant.

Suppose further that after a circuit of the boundary Γ the values of χ change by $n\pi$, $n > 0$. First we set $n = 2$. In domain D we take an arbitrary point (a, b) as pole of a system of polar coordinates r, θ. The values of χ on Γ can be represented in the form $\chi = \theta + \chi_1$. This equation defines χ_1. Obviously after a circuit of the boundary Γ the function χ_1 returns to its original value, but θ increases by 2π. From the function χ_1, just as in the first case, one constructs a single-valued harmonic function which is regular in domain D, which we shall denote by the same letter χ_1. Equation (5.6) defines a single-valued conjugate function $\ln R_1$ for χ_1 in domain D. The function θ itself is harmonic in D, but not single-valued, and $\ln r$ is its conjugate harmonic function. Consequently, we have

$$\ln (A + iB) = \ln r + i\theta + \ln R_1 + i\chi_1,$$

$$A + iB = rR_1[\cos(\theta + \chi_1) + i \sin(\theta + \chi_1)] = [z - (a + ib)]R_1 \exp(i\chi_1).$$

The quantity θ, considered as a function of the arc length on Γ, satisfies a Lipschitz condition, since $|d\theta| \le r^{-1}|ds|$, and $r \neq 0$ on Γ, so if χ satisfies a Hölder condition on Γ, then χ_1 also satisfies a Hölder condition with the same exponent. Function R_1 is positive and bounded for the same reasons as in the first case. The quantity r vanishes at the point (a, b), so A, B, $\sqrt{A^2 + B^2}$ vanish simultaneously at this point, and on the boundary A and B do not vanish simultaneously and satisfy a Hölder condition. Here, aside from an arbitrary factor, the solution A, B of the problem depends on two more real parameters a and b.

Let us assume that after a circuit of curve Γ the function χ changes by $2p\pi$, $p > 0$. In domain D, bounded by curve Γ, we choose p points $z_1, \ldots, z_p$, and we denote the polar coordinates with poles at these points by $(r_1, \theta_1), \ldots, (r_p, \theta_p)$, respectively. We set $\chi = Q + \ldots + \theta_p + \chi_1$, where χ_1 after a circuit of Γ returns to the original value. Analogously to the case $p = 1$ we have

$$A + iB = r_1 \ldots r_p \exp[i(\theta_1 + \ldots + \theta_p)] \times R_1 \exp(i\chi_1)$$
$$= (z - z_1) \ldots (z - z_p)R_1 \exp(i\chi_1),$$

where R_1 does not vanish either at zero or infinity. Now the functions A and B vanish at p points of domain D, and their expression, in addition to an arbitrary constant factor, contains $2p$ more arbitrary real parameters.

We consider odd n. Say, for example, $n = 1$. We set $\chi = \theta + \chi_1$. In order that after a circuit of Γ the quantity θ should change by π, it is necessary to take the point (a, b) on the line Γ, while this point must be an ordinary point of Γ. We have $A + iB = (z - z_0)R_1 \exp(i\chi_1)$, where $z_0 \in \Gamma$, so A and B vanish simultaneously at point z_0 of curve Γ and depend on an arbitrary real parameter, since$z_0 = a + ib$ lies on Γ. In the case of arbitrary n we set $n = 2n_1 + n_2$, $n_1 \geq 0$, $n_2 \geq 0$. Now $A + iB$ vanishes at n interior points of domain D and at n_2 points of boundary Γ. Besides a constant factor the functions A and B depend on $2n_1 + n_2 = n$ parameters. Thus we can construct functions A and B, depending on $n + 1$ arbitrary real parameters, but it is not clear whether all solutions of the problem are included among the solutions constructed. We study this question somewhat later.

Let us assume, finally, that after a circuit of Γ the quantity χ changes by $n\pi$, $n < 0$. For example, let $n = -2$, setting $\chi = -\theta + \chi_1$, where r, θ are polar coordinates with pole at the point $z_1 = (a, b)$ of domain D. Analogously to the preceding case, we find $A + iB = R_1 r^{-1}[\cos(-\theta + \chi_1) + i \sin(-\theta + \chi_1)] = (z - z_1)^{-1}R_1 \exp(i\chi_1)$, and in general $A + iB = (z - z_1)^{-1} \ldots (z - z_p)^{-1}R_1 \exp(i\chi_1)$. Now A and B become infinite simultaneously at the points $z_1, \ldots, z_p$, but do not vanish simultaneously at any point. One can also construct A and B which vanish in the domain D, but have more singular points. Such solutions are of no interest for the oblique derivative problem, so we shall not consider them. We show that in this case there are no bounded solutions A and B of the problem considered. Let us assume that $A + iB$ is a bounded solution in the case where the increment of χ on Γ is equal to $n\pi$, $n < 0$. Let $A_1 + iB_1$ be a bounded solution of the problem with boundary condition $B_1/A_1 = \tan(-\chi)$. Such a solution exists, since the increment of $-\chi$ on Γ is equal to $-n\pi$, $-n > 0$, while $A_1 + iB_1$ vanishes at least at one point of the closed domain $D \cup \Gamma$. The product $(A + iB)(A_1 + iB_1) = P + iQ$ is an analytic function in domain D, which assumes real values on boundary Γ, since the arguments of the factors are equal in absolute value and have opposite signs. Consequently, $Q \equiv 0$ in domain D, and $P \equiv$ const; since $P + iQ$ vanishes at least at one point of domain D, one has $P \equiv 0$, i.e., $A + iB = 0$.

We return to the case $n \geq 0$ and we consider the general form of the solution $A + iB$ of the problem. Let $A_0 + iB_0$ be a particular solution of the problem, all of whose zeros lie in domain D, and $A + iB$ be another solution of this same problem. The ratio $P + iQ = (A + iB)/(A_0 + iB_0)$ is a bounded real function on boundary Γ of domain D, which is analytic everywhere in the domain except for a finite number of poles. We show that the function $F(z) = P + iQ$ can be expressed simply in terms of the function $\varphi(z) = \xi + i\eta$, which maps the domain D conformally to the upper half-plane $\eta > 0$. It suffices to consider just the case $n > 0$, since for $n = 0$ the solution is determined up to a constant factor.

We start with $n = 2$. The particular solution $A_0 + iB_0$ is a holomorphic function in the domain, which has only one simple zero in domain D. Let z_0 be this zero of the functin $A_0 + iB_0$. The function $F(z) = (A + iB)/(A_0 + iB_0)$ is regular everywhere in domain D, except for the point z_0, at which it has a simple pole, so

$$F(z) = (\lambda + i\mu)(z - z_0)^{-1} + c_0 + c_1(z - z_0) + \ldots,$$

and the function $F(z) - (\lambda + i\mu)(z - z_0)^{-1}$ is holomorphic everywhere in domain D. It is also obvious that $F(z)$ is real on boundary Γ of domain D, since $B/A = B_0/A_0$ on Γ. We show that for fixed λ and μ the function $F(z)$ is determined up to a constant. Let us assume that there exist two such functions $F_1(z)$ and $F_2(z)$; then their difference

$$F_1(z) - F_2(z) = [F_1(z) - (\lambda + i\mu)(z - z_0)^{-1}]$$
$$- [F_2(z) - (\lambda + i\mu)(z - z_0)^{-1}] = P(x, y) + iQ(x, y)$$

is holomorphic in domain D and its imaginary part Q is equal to zero on F. It follows from this that $Q \equiv 0$, $P \equiv$ const in domain D, i.e., F_1 and F_2 differ by a constant.

We construct a function $F_1(z)$ of special form. Let $\varphi(z)$ be a function which maps the domain conformally to the upper half-plane so that the point z_0 goes to the point φ_0, i.e., $\varphi_0 = \varphi(z_0)$. We consider the function

$$F_1(z) = [a\varphi(z) + b]\{[\varphi(z) - \varphi_0][\varphi(z) - \bar{\varphi}_0]\}^{-1},$$

where a and b are arbitrary real constants. Since under a conformal map of the domain the boundary goes into the boundary, $\varphi(z)$ is real on Γ, and with $\varphi(z)$, $F_1(z)$ is real on Γ. By virtue of the univalence of the conformal map, $\varphi(z) - \varphi_0$ has a simple zero at the point z_0, and in the domain $D \cup \Gamma$, $\text{Im}\,[\varphi(z) - \overline{\varphi}_0] > 0$, so $F_1(z)$ has a simple pole at point z_0. At point z_0, $\varphi(z_0) - \overline{\varphi}_0 = 2i\,\text{Im}\,\varphi(z_0)$, $\varphi(z) - \varphi_0 = \varphi'(z_0) = (z - z_0) + \alpha_1(z - z_0)^2 + \ldots$, so

$$F_1(z) = \frac{a\varphi_0 + b}{\varphi'(z_0)\cdot 2i\,\text{Im}\,\varphi(z_0)}\cdot\frac{1}{z - z_0} + \beta_1(z - z_0) + \ldots$$

We express λ and μ in terms of a and b. From the equations

$$\frac{\lambda + i\mu}{z - z_0} = \frac{a\varphi_0 + b}{2i\varphi'(z_0)\,\text{Im}\,\varphi(z_0)}\cdot\frac{1}{z - z_0},$$
$$a\varphi_0 + b = 2i\varphi'(z_0)\,\text{Im}\,\varphi(z_0)(\lambda + i\mu)$$

we find, separating real and imaginary parts, that

$$a = 2\,\text{Re}\,[(\lambda + i\mu)\varphi'(z_0)],$$
$$b = 2\,\text{Im}\,\varphi(z_0)\,\text{Im}\,[\varphi'(z_0)(\lambda + i\mu)] - a\,\text{Re}\,\varphi(z_0),$$

i.e., a and b can be expressed uniquely in terms of λ and μ. Thus, $F_1(z)$ satisfies all the conditions required. Now the function $F(z)$ has the form

$$F(z) = \{a[\varphi(z)]^2 + b\varphi(z) + c\}\{[\varphi(z) - \varphi_0][\varphi(z) - \overline{\varphi}_0]\}^{-1},$$

since it differs from F_1 by a constant summand c. Finally we find $A + iB = F(z)(A_0 + iB_0)$. It follows from this that for any two solutions A, B and A_1, B_1 the ratio $(A + iB)/(A_1 + iB_1) = (a\varphi^2 + b\varphi + c)/(a_1\varphi^2 + b_1\varphi + c_1)$ is the ratio of the two quadratic trinomials with real coefficients with respect to $\varphi(z)$.

The zeros of the function $A + iB$ coincide with the zeros of the quadratic trinomail $a\varphi^2 + b\varphi + c$ with real coefficients, so $A + iB$ vanishes either at one point of domain D, which corresponds to the unique complex root of the quadratic trinomial lying in the upper half-plane, since the roots of this trinomial are complex conjugates, or vanish at two points of the boundary of domain D. From the representation

$$A + iB = (A_0 + iB_0)(a\varphi^2 + b\varphi + c)[(\varphi - \varphi_0)(\varphi - \overline{\varphi}_0)]^{-1}$$

it follows that $A + iB$ is bounded in D, since $A_0 + iB_0$ vanishes at point z_0, so $A + iB$ is holomorphic in D and has either one root inside D or two zeros on the boundary of D, i.e., our problem has no solutions other than those constructed above.

Now let $n = 2m$. We choose a particular zolutions $A_0 + iB_0$ of the problem, which vanishes at interior points $z_1, \ldots, z_m$ of domain D. For any other solution $A + iB$ the function $F(z) = (A + iB)/(A_0 + iB_0)$ is real on boundary Γ of domain D and is analytic everywhere in D except for the simple poles $z_1, \ldots, z_m$. Instead of the function $F_1(z)$ we consider the function

$$\Phi(z) = \sum_{j=1}^{m} \frac{a_j\varphi(z) + b_j}{[\varphi(z) - \varphi_j][\varphi(z) - \bar{\varphi}_j]}, \quad \varphi_j = \varphi(z_j), \ j = 1, \ldots, m.$$

Just as in the case $m = 1$, we get that for any two solutions of the problem A, B and A_1, B_1, we have

$$(A + iB)/(A_1 + iB_1) = \Phi(z) + \nu = [b_0\varphi^n + b_1\varphi^{n-1} + \ldots + b_n] \times \left\{\prod_{p=1}^{m} (\varphi - \varphi_j)(\varphi - \bar{\varphi}_j)\right\}^{-1}$$

Thus one can write

$$A + iB = (A_0 + iB_0)[b_0\varphi^n + b_1\varphi^{n-1} + \ldots + b_n] \times \left\{\prod_{p=1}^{m} (\varphi - \varphi_j)(\varphi - \bar{\varphi}_j)\right\}^{-1}, \tag{5.7}$$

where $b_0, \ldots, b_n$ are real numbers. Further, just as for $m = 1$, we show that the original problem has no solutions other than those constructed above.

We reduce the case of odd n to the preceding one by the following method. As origin we take an ordinary point of boundary Γ and we consider the product $M + iN = (A + iB)(x + iy)$, where A, B is a solution of our original problem, M, N is a solution of the analogous problem for which $\chi_1 = \chi + \theta$, so χ_1 changes by $(n + 1)\pi$ after a circuit of Γ, and $n + 1 = 2m$ is an even number. Let $M_0 + iN_0$ be a particular solution of the original problem, all of whose zeros lie inside D, while these zeros $z_1, \ldots, z_m$ are simple. We have

$$M + iN = (M_0 + iN_0)\frac{a_0\varphi^{n+1} + a_1\varphi^n + \ldots + a_{n+1}}{b_0\varphi^{n+1} + b_1\varphi^n + \ldots - b_{n+1}},$$

where $\varphi(z)$ is a function which maps domain D conformally to the upper half-plane, and the real coefficients $b_0, \ldots, b_{n+1}$ are such that the denominator has simple zeros at the point $z_1, \ldots, z_m$. In order that $A + iB$ be bounded at the point $z = 0$, $M + iN$ must vanish at this point, i.e., the polynomial $a_0\varphi^{n+1} + a_1\varphi^n + \ldots + a_{n+1}$ must be divisible by $\varphi - \varphi_0$, where $\varphi_0 = \varphi(0)$, so

$$A + iB = (M + iN)(x + iy)^{-1} = (M_0 + iN_0)(\varphi - \varphi_0)z^{-1} \times [c_0\varphi^n + c_1\varphi^{n-1} + \ldots + c_n][b_0\varphi^{n+1} + b_1\varphi^n + \ldots + b_{n+1}]^{-1}. \tag{5.8}$$

By virtue of the choice of origin and the properties of the conformal mapping function, the ratio $(\varphi - \varphi_0)/z$ has a completely definite finite limit as $z \to 0$, i.e., $A + iB$ is bounded at point $z = 0$ without any additional conditions on the coefficients $c_0, \ldots, c_n$. Thus we have found for $A + iB$ the same kind of representation for n odd as for n even, from which it follows that there are no solutions of the original problem other than those constructed at the beginning of this section.

3. OBLIQUE DERIVATIVE PROBLEM

The oblique derivative problem for harmonic functions of two independent variables is posed as follows: Find a harmonic function $v(x, y)$ which is regular in domain D, continuously differentiable in closed domain $D \cup \Gamma$, which satisfies on boundary Γ of the domain the condition

$$dv/dl \equiv av_x + bv_y = f, \tag{5.9}$$

where a, b, and f are functions given on Γ. First we shall assume that these functions are Hölder continuous and all points of Γ are ordinary, while domain D is simply connected. If in (5.9) $f \equiv 0$, then one gets the homogeneous problem corresponding to (5.9). From the homogeneous boundary condition $av_x + bv_y = 0$ we find

$$-v_x/v_y = b/a \equiv \tan\varphi.$$

Thus we arrive at the auxiliary problem considered in the preceding section by setting $v_y = A(x, y)$, $v_x = B(x, y)$, $\chi = -\varphi$.

Suppose φ after a circuit of the curve Γ returns to its original value; then χ too returns to its original value. According to the results of the preceding section, in this case there exist conjugate harmonic functions A and B, which are regular in D, such that $B/A = \tan\chi$ on Γ, while these functions are determined up to a constant factor and $A + iB \neq 0$ everywhere in D. Obviously

$$v(x, y) = \int (v_x dx + v_y dy) = \int (A dy + B dx) + \text{const},$$

where the integral is independent of the path and the constant is determined by the choice of the origin of the path. Consequently,

$$v(x, y) = C + \operatorname{Re}\{-i \int_{z_0}^{z} (A + iB)(dx + idy)\}. \tag{5.10}$$

The function v depends on two real constants, one of which is C, and the second goes into $A + iB$.

When the increment of φ on the contour Γ is equal to $n\pi$, $n > 0$, the increment of χ is equal to $-n\pi$, $-n > 0$. In this case we have $A = B = 0$ and $v \equiv$ const. There are no other solutions of the homogeneous oblique derivative problem.

If the increment of φ on contour Γ is equal to $n\pi$, $n < 0$, then the increment of χ is equal to $-n\pi$, $-n > 0$. Here there exist conjugate harmonic functions A and B which are regular in domain D and bounded in the closed domain $D \cup \Gamma$, from which, with the help of (5.10), one can reconstruct v. We set $-n = 2n_1 + n_2$; then according to the results of the preceding section we can write

$$A + iB = \psi(z)\left[a_0\varphi^{-n} + a_1\varphi^{-n-1} + \ldots + a_n\right] = a_0\psi(z) \times \prod_{k=1}^{n_1} [\varphi(z) - \bar{\varphi}_k][\varphi'(z) - \varphi_k] \times \prod_{l=1}^{n_2} [\varphi(z) - \varphi_l], \tag{5.11}$$

where $\psi(z)$ is a well-defined function, φ maps D conformally to the upper half-plane, and $a_0, \ldots, a_{-n}$ are arbitrary real parameters. Now the function v depends linearly on $1 - n$ arbitrary real constants.

We proceed to the inhomogeneous oblique derivative problem (5.9). As before, we denote by φ the angle between the direction in which one takes the derivative dv/dl on the left side of (5.9) and the Ox axis. This angle is defined up to an integral multiple of π. The derivative dv/dl is not necessarily taken in the positive direction of φ, but it can also be taken in the direction $\varphi + \pi$, but if the direction of differentiation is fixed at some point, then it varies continuously so φ also varies continuously. Let $n\pi$ be the increment in φ after a circuit of the boundary Γ. If n is even, then dv/dl is again taken after a circuit of Γ in the same direction, while if n is odd, then this derivative is taken in the opposite direction, i.e., for even n, dv/dl returns to the original value, and for odd n it changes sign.

To solve the problem for $n \geq 0$ we construct two conjugate harmonic functions A and B which are regular in domain D, satisfying the condition $B/A = \tan\varphi$ on the boundary Γ, where the direction l in which one takes the derivatives is such that $dy/dx = \tan\varphi$. We have

$$dx/A = dy/B = (A^2 + B^2)^{-1/2} dl.$$

The sign in front of the root here is chosen as follows. If the increment dl is taken in the direction of φ, then we have $dx = dl \cos\varphi$, $dy = dl \sin\varphi$, so

$$\cos\varphi/A = \sin\varphi/B = (A^2 + B^2)^{-1/2},$$

while if the increment dl is taken in the direction $\varphi + \pi$, then $dx = -dl \cos\varphi$, $dy = -dl \sin\varphi$, so

$$\cos\varphi/A = \sin\varphi/B = -(A^2 + B^2)^{-1/2}.$$

These relations uniquely determine the sign of the root. This sign can change only at those points at which A and B vanish simultaneously. Now we can write

$$\frac{dv}{dl} = \frac{\partial v}{\partial x}\frac{\partial x}{\partial l} + \frac{\partial v}{\partial y}\frac{\partial y}{\partial l}\left(A\frac{\partial v}{\partial x} + B\frac{\partial v}{\partial y}\right)(A^2 + B^2)^{-1/2}.$$

Fixing A and B, we thus fix $(A^2 + B^2)^{-1/2}$ and we get that giving dv/dl on Γ is equivalent to giving the values $dv/dl\sqrt{A^2 + B^2}$ on Γ, i.e., one can assume that on Γ the values of $P = Av_x + Bv_y$ are given. The functions A and B are conjugate harmonic functions in D, so $f(z) = A + iB$ is holomorphic in D. The gunction $g(z) = v_x - iv_y$ is also holomorphic in D, hence analytic in D, and the function $f(z)g(z) = Av_x + Bv_y + i(Bv_x - Av_y)$, i.e., $P = Av_x + Bv_y$ and $Q = Bv_x - Av_y$ are conjugate harmonic functions in D. Solving the Dirichlet problem for the values on Γ, one can reconstruct the function P everywhere in domain D, and from known P one can reconstruct the function Q conjugate to it up to a constant. Knowing P and Q in domain D, we find

$$v(x, y) = \int_{z_0}^{z} (v_x dx + v_y d_y) + C = \int_{z_0}^{z} [(AP + BQ)\, dx + (BP - AQ)\, dy] \times \quad (5.12)$$

$$\times (A^2 + B^2)^{-1} + C = C + \mathrm{Im}\left\{\int_{z_0}^{z} (P + iQ)(A + iB)^{-1} dz\right\}.$$

We note that the values of P on Γ satisfy a Hölder condition by virtue of the restirctions imposed on Γ and the coefficients of the boundary condition (5.9).

Let $n = 0$ and $A + iB$ not vanish at any point of the closed domain $D \cup \Gamma$, and let (5.12) give a solution of problem (5.9). In this formula A and B are defined up to a constant factor λ. If A and B are replaced by λA and λB, then $(A^2 + B^2)^{1/2} dv/dl$ is multiplied by λ, so $P + iQ$ is also multiplied by λ, and the ratio $(P + iQ)/(A + iB)$ is unchanged, so v, too, is unchanged. The function Q is determined by P up to an additive constant γ, so v is defined up to a function

$$v_0(x, y) = C + \mu \int_{z_0}^{z} (Bdx + Ady)(A^2 + B^2)^{-1}. \tag{5.13}$$

Let v_1 and v_2 be two solutions of (5.9); then the difference $v_1 - v_2$ is a solution of the homogeneous problem which is expressed by (5.10):

$$v_1 - v_2 = C + \int_{z_0}^{z} (A_1 dx + B_1 dy), \tag{5.14}$$

where A_1 and B_1 are conjugate harmonic functions in domain D, such that $B_1/A_1 = -\tan\varphi$ on the boundary Γ of domain D. Now the functions A and B figuring in (5.13) are such that $B/A = \tan\varphi$ on Γ, so the product $(A + iB)(A_1 + iB_1)$ is real on Γ. It follows from this that $(A + iB)(A_1 + iB_1) = K \equiv \text{const}$ in D, so

$$A_1 = KA(A^2 + B^2)^{-1}, \quad B_1 = -KB(A^2 + B^2)^{-1},$$

i.e., (5.14) becomes (5.13).

If φ varies by $n\pi$, $n > 0$, after a circuit of Γ, then the functions A and B, connected by the relation $B = A\tan\varphi$, now necessarily vanish at some points of the domain D or boundary Γ. The number n_1 of interior zeros and the number n_2 of boundary zeros satisfy the equation $n = 2n_1 + n_2$. If n is even, then one can choose A and B so that $A + iB$ vanishes only at interior points of domain D, while if n is odd, then n_2 is at least equal to one. In the case of even n the root $\sqrt{A^2 + B^2}$ is different from zero on boundary Γ and multiplication of dv/dl by this root does not introduce any extraneous solutions. The functions $P(x, y)$ and $Q(x, y)$ are now defined just as in the case $n = 0$. In (5.12) the denominator vanishes to the first order at n_1 interior points z_k, $k = 1, \ldots, n_1$, of domain D. In order that grad v and with it v be bounded in domain D, the function $P + iQ$ must vanish at the points z_k, $k = 1, \ldots, n_1$. This gives n_1 conditons on P and $n_1 - 1$ conditions on Q, since Q contains one arbitrary constant, due to which the number of conditions decreases. In all then one must impose $2n_1 - 1 = n - 1$ conditions on $P + iQ$.

As a result of the study of the homogeneous oblique derivative problem it was shown that for $n > 0$ it has a unique solution $v_0 \equiv \text{const}$. It follows from this that if in (5.12) to construct v we take another function $A_1 + iB_1$ corresponding to another solution of the auxiliary problem, the function v will differ from the function constructed from $A + iB$ only by a constant, and this means that $(P_1 + iQ_1)/(A_1 + iB_1) = (P + iQ)/(A + iB)$ or $(P_1 + iQ_1)/(P + iQ) = (A_1 + iB_1)/(A + iB)$. Since on boundary Γ of domain D we have $B_1/A_1 = B/A$ on Γ, we get

$$\frac{A_1}{A} = \frac{B_1}{B} = \frac{A_1 + iB_1}{A + iB} = \frac{\sqrt{A_1^2 + B_1^2}}{\sqrt{A^2 + B^2}} = \frac{P_1}{P} = \frac{Q_1}{Q}.$$

The function P_1 assumes on Γ the values $\sqrt{A_1{}^2 + B_1{}^2}\, dv/dl$, and P the values $\sqrt{A^2 + B^2}\, dv/dl$. Thus P_1 and Q_1 are the functions to which we would be led in deriving (5.12), replacing A and B by A_1, B_1. In the function Q_1, in general, there is an arbitrary constant, but it can be chosen so that $(P_1 + iQ_1)/(A_1 + iB_1)$ is bounded. Namely, for such a choice we have

$$(P_1 + iQ_1)/(A_1 + iB_1) = (P + iQ)/(A + iB). \tag{5.15}$$

In deriving (5.15) one can assume that some of the zeros of $A + iB$ lie on the boundary. Consequently, one can eliminate the requirement that all the zeros of $A + iB$ lie inside domain D.

In the case of odd $n > 0$ difficulties arise, connected with the fact that at least one zero of the function $A + iB$ lies on boundary Γ. Now for $(P + iQ)/(A + iB)$ to be bounded at such a boundary point it is insufficient in general that $P + iQ$ vanish. For simplicity let us assume that on Γ there lies only one zero z_0 of the function $A + iB$. As established above,

$$A + iB = (r \exp i\,Q) \prod_{k=1}^{n_1} (r_k \exp i\,Q_k)\, R_1 \exp(iX_1), \quad n_1 = (n-1)/2,$$

where R_1 is different from zero in $D \cup \Gamma$. Obviously

$$\sqrt{A^2 + B^2} = rR_1 \prod_{k=1}^{n_1} r_k.$$

The quantities r_k and R_1 are positive on Γ, while this is not so for r. In the ratios

$$\sqrt{A^2 + B^2} = A/\cos\varphi = B/\sin\varphi$$

the harmonic functions A and B change signs at the same time that the expressions $\cos\varphi$ and $\sin\varphi$ do not change signs upon passing through z_0 due to the continuity of φ, so the root must necessarily change sign. Now the change of sign of the root immediately implies change of sign of r. The values of P on the boundary Γ are given by the formula

$$P = \sqrt{A^2 + B^2} dv/dl = r\left\{\prod_{k=1}^{n_1} r_k R_1 dv/dl\right\} = rC(s),$$

where $C(s)$ denotes the function in the curly brackets. This function is Hölder continuous. The factor r at the point z_0 has derivative equal to +1 or −1, depending on whether r goes to this point from negative values to positive or from positive to negative. $P(s)$ at the point $s_0 = z_0$ has derivative

$$\left.\frac{dP}{ds}\right|_{s=s_0} = \lim_{s \to s_0} \frac{P(s)}{s - s_0} = \lim_{s \to s_0} \frac{r}{s - s_0} C(s) = \eta C(s_0).$$

Here η is the limit for $s = s_0$ of the derivative dr/ds, which is equal to +1 or −1. Since $P(s_0) = 0$, we have

$$[P(s) - P(s_0)]/(s - s_0) = C(s)r/(s - s_0).$$

This ratio satisfies a Hölder condition, since $C(s)$ and $r/(s - s_0)$ satisfy it due to the conditions imposed on Γ.

We use an assertion of [54]: Let the function $F(z) = v + iw$ be defined in domain D by the values $V(s)$ of the function v on boundary Γ of the domain; if $[V(s) - V(s_0)]/(s - s_0)$ satisfies a Hölder condition on some arc of Γ containing point s_0, function $F(z)$ at this point has completely definite bounded derivative. This assertion is proved in the same way as Theorems 5.1 and 5.2. By virtue of this assertion the function $P + iQ$ at point z_0 has a well-defined bounded derivative. Consequently, the ratio $(P + iQ - iQ_0)/(z - z_0)$, where Q_0 is the value of Q at point z_0, has a well-defined limit as z tends to z_0 from within D. The harmonic function $Q - Q_0$ vanishes at point z_0. This condition defines the function $Q - Q_0$ uniquely from the function P, and the ratio $(P + iQ)/(A + iB)$ is bounded at point z_0, if $Q = 0$ at this point. In order that the ratio considered be bounded at the $(n - 1)/2 = n_1$ zeros of $A + iB$ lying inside D also, it is necessary and sufficient that $P + iQ$ vanish at these same points, which imposes n_1 conditions on P and Q, since the arbitrary constant which occurs in Q is already determined. Thus, the number of necessary and sufficient conditions for the gradient of v to be bounded in $D \cup \Gamma$ is equal to $n - 1$ in this case, too. Further, the assumption that $A + iB$ vanishes at only one point of the boundary Γ can be lifted exactly as in the case of even n.

Theorem 5.3. A solution of the inhomogeneous oblique derivative problem (5.9) for $n > 0$ exists only if $n - 1$ conditions imposed on function f hold, and the homogeneous problem corresponding to (5.9) has only the constant solution.

The validity of this theorem follows from the arguments preceding it. In particular, for $n = 1$ the problem is unconditionally solvable.

We concern ourselves further with the case where the increment of φ on Γ is equal to $n\pi$, $n < 0$. Now there do not exist conjugate harmonic functions A and B which are bounded in the domain and satisfy the condition $B/A = \tan\varphi$ on the boundary Γ. One could investigate problem (5.9), using unbounded A and B, in the same way as in the case $n > 0$. However, it is more convenient to proceed differently. We consider two new conjugate harmonic functions U and V which, on boundary Γ of the domain, satisfy the condition $V/U = -\tan\varphi$. The pair of these harmonic functions is connected with the functions A and B by the relation $(A + iB)(U + iV) = \gamma$, where γ is a real constant, since on Γ the sum of the arguments of the factors is equal to zero, which implies the reality of the product on Γ. With the help of U and V we make the same construction as in the case $n > 0$. We have

$$P(s) = (U^2 + V^2)^{-1/2}\, dv/dl = (U^2 + V^2)^{-1}(Uv_x - Vu_y),$$

where the sign of the root $\sqrt{U^2 + V^2}$ is determined by the relations

$$\sqrt{U^2 + V^2} = U/\cos\varphi = V/-\sin\varphi.$$

The harmonic function which is regular in domain D, which is constructed with the help of $P(s)$ as a result of solving the Dirichlet problem, is the real part of the ratio $(v_x - iv_y)/(U + iV)$. From the function P we construct its conjugate harmonic function Q which is regular in domain D. We have

$$P + iQ = (v_x - iv_y)/(U + iV), \quad v_x - iv_y = (P + iQ)(U + iV),$$

from which we find

$$v(x, y) = C + \operatorname{Re}\left\{\int_{z_0}^{z} (P + iQ)(u + iV)\, dz\right\}. \tag{5.16}$$

We analyze (5.16). Let $-n = 2n_1$, i.e., n be even. Among all solutions $U + iV$ of the auxiliary problem we choose one which has n_1 zeros in D and has no zeros on the boundary Γ of domain D. Now $U^2 + V^2 \neq 0$ on Γ and the function $P(s)$ is Hölder continuous, so $P + iQ$ is bounded and holomorphic in D, and (5.16) gives a harmonic function v which is bounded in D. If $-n = 2n_1 + 1$ any bounded solution $U + iV$ of the auxiliary problem has at least one zero on boundary Γ of domain D, so at such a zero $(U^2 + V^2)^{-1/2}\, dv/dl$ becomes infinite and on the right side of the equation

$$v_x - iv_y = (P + iQ)(U + iV),$$

one gets the indeterminacy $\infty \times 0$. We get rid of this indeterminacy, assuming that $U + iV$ vanishes at one point $z_0 \in \Gamma$. As was done above, we set

$$u + iV = r\exp(i\theta)\prod_{k=1}^{n_1} r_k \exp(i\theta_k)\, R_1 \exp(i\chi_1),$$

where r_k, θ_k correspond to interior zeros of $U + iV$, and R_1 vanishes nowhere in $D \cup \Gamma$. From this we find

$$\sqrt{u^2 + V^2} = r\prod_{k=1}^{n_1} r_k R_1.$$

while r changes sign at the point z_0 according to the same rules as in the case of odd $n > 0$, and all r_k and R_1 are positive. For the values $P(s)$ of the function $P(x, y)$ on Γ we get

$$P(s) = (u^2 + V^2)^{-1/2} dv/dl = r^{-1} \left[\prod_{k=1}^{n_1} r_k R_1 \right]^{-1} dv/dl = C(s)/r,$$

where $C(s)$ is Hölder continuous. At point z_0 the function $P(s)$ becomes infinite so that $rP(s)$ has a completely definite limit at point z_0. It is easy to generalize the assertion of Theorem 5.2 to the case $l > 1$ also and to get that the product $(z - z_0)(P + iQ)$ has a well-defined limit as $z \to z_0$. Now if we write $(P + iQ)(U + iV)$ in the form

$$[(z - z_0)(P + iQ)] \left[\prod_{k=1}^{n_1} r_k \exp(i\theta_k) R_1 \exp(i\chi_1) \right],$$

since $z - z_0 = r \exp(i\theta)$, then we get the boundedness of the integrand in (5.16) for odd $n < 0$, too, everywhere in $D \cup \Gamma$.

When $n > 0$ we saw that two different solutions of the auxiliary problem led to the same solution $v(x, y)$ of (5.9). The picture is completely different for $n < 0$. First of all, one can change the function $v(x, y)$ by means of Q, choosing the constant up to which Q is defined from P, differently. According to (5.16) such changes have the form

$$\delta v(x,y) = \mathrm{Re} \left\{ i\delta C \int_{z_0}^{z} (U + iV)\, d(x + iy) \right\} = -\delta C \int_{z_0}^{z} (U dy + V dx),$$

where δC is the change in the arbitrary constant on which Q depends. The function δv has the form of the representation found above of solutions of the homogeneous problem corresponding to (5.9), for $n < 0$. One can change the function v by changing $U + iV$. If instead of U and V we take $\lambda U, \lambda V$, then $U^2 + V^2$ goes into $\lambda^2(U^2 + V^2)$, and the values of P on Γ are multiplied by λ^{-1}, so $P + iQ$ is replaced by $\lambda^{-1}(P + iQ)$, and the function $F(z) = (P + iQ)(U + iV)$ is unchanged. Thus, v does not depend on the constant factor up to which U and V are defined.

The function $v(x, y)$ changes if we replace $U + iV$ by $U_1 + iV_1$, where this latter function has different zeros in the domain $D \cup \Gamma$. Let $U_1 + iV_1$ correspond to the function $P_1 + iQ_1$, which replaces $P + iQ$. Now instead of $F(z)$ we have the function $F_1(z) = (P_1 + iQ_1)(U_1 + iV_1)$. We consider the difference

$$\delta F = F_1(z) - F(z) = (P_1 + iQ_1)(U_1 + iV_1) - (P + iQ)(U + iV)$$

and we show that it can be represented in the form $\delta F = i(U_2 + iV_2)$. This is necessary and sufficient for the function $\mathrm{Re}\,\{\int \delta F dz\}$ constructed from this difference to be a solution of the homogeneous problem corresponding to (5.9). We split δF as follows:

$$\delta F/i = U_1Q_1 - UQ + V_1P_1 - VP + i[V_1Q_1 - VQ + UP - U_1P_1].$$

One can represent the expression δF in the form $i(U_2 + iV_2)$, if on the boundary we have

$$-\tan\varphi = [V_1Q_1 - VQ + UP - U_1P_1]/[U_1Q_1 - UQ + V_1P_1 - VP].$$

However, on the boundary one has

$$\begin{aligned} &-\tan\varphi = V/U = V_1/U_1 = (VQ - V_1Q_1)/(UQ - U_1Q_1), \\ &U_1/U = V_1/V = (U_2^2 + V_1^2)^{1/2}/(U^2 + C^2)^{1/2} = P/P_1, \end{aligned} \tag{5.17}$$

since on boundary Γ one has

$$P = (Uv_x - Vv_y)(U^2 + V^2)^{-1/2};\ P_1 = (U_1v_x - V_1v_y)(U_1^2 + V_1^2)^{-1/2},$$
$$\sqrt{U^2 + V^2} = U/\cos\varphi = V/-\sin\varphi,$$
$$\sqrt{U_1^2 + V_1^2} = U_1/\cos\varphi = V_1/-\sin\varphi.$$

Comparing the two relations of (5.17), we find $UP - U_1P_1 = 0$, $VP - V_1P_1 = 0$, from which it follows that $\delta F/i = U_1Q_1 - UQ + i(V_1Q_1 - VQ)$, i.e., $\delta F/i = -\tan\varphi$ on Γ, so one can represent δF in the form $i(U_2 + iV_2)$. Hence we get the general solution of problem (5.9) if we add to a solution of the form (5.16), constructed from fixed functions U and V, the general solution of the homogeneous problem

$$v_0(x, y) = \int_{z_0}^{z} (U_1dy + V_1dx), \tag{5.18}$$

where $U_1 + iV_1$ runs through all solutions of the auxiliary problem.

Theorem 5.4. For $n \le 0$, (5.9) is always solvable, and the homogeneous problem corresponding to it has $2 - n$ linearly independent solutions.

One gets linearly independent solutions from the $1 - n$ constants which appear in $U_1 + iV_1$ in (5.18), and the constant C in (5.16). Theorems 5.3 and 5.4 completely describe the character of the solvability of problem (5.9). In the formulation of problem (5.9) additional restrictions are imposed on the boundary of the domain and the coefficients of the problem. If these restrictions are violated, the assertions of Theorems 5.3 and 5.4 may turn out to be false.

One can subject the oblique derivative problem to different transformations. One can map domain D conformally to another domain D_1. If at the point M of boundary Γ of domain D and at point M_1 of boundary Γ_1 of domain D_1 corresponding to it, there exists well-defined tangents M_1T_1 to Γ_1 and MT to Γ, then by virtue of the preservation of angles under a conformal map the direction l_1 into which the conformal map transforms the direction of differentiation l at point M, makes the same angle with M_1T_1 as l does with MT, so the given transformation does not change the number n, which characterizes the boundary condition of problem (5.9). Now the absolute value of the oblique derivative is multiplied by the absolute value of the derivative of the conformal mapping function at point M. Thus, one can always reduce the case of an arbitrary domain D to the consideration of a disk or upper half-plane.

The following transformation of problem (5.9), a special case of which was already considered for odd n, changes the number n. Let $F(z) = R\exp(i\alpha)$ be a function which is holomorphic in domain D; we introduce a harmonic function $w(x, y)$ by $w_x - iw_y = F(z)(v_x - iv_y)$. For the function w we have $w_x \cos(\varphi - \alpha) + w_y \times \sin(\varphi - \alpha) = R(v_x \cos\varphi + v_y \sin\varphi)$, i.e., the derivative of w in the direction $\varphi - \alpha$ is R times greater than the derivative of v in the direction φ. If the argument α of function F returns to the original value after a circuit of Γ, then the number n is the same for v and w, i.e., in the original and the transformed oblique derivative problems. Now if α changes after a circuit of Γ, then the number n can change. It can also happen that this type of transformation can associate an unbounded function v with a bounded function w.

4. OBLIQUE DERIVATIVE PROBLEM WITH DISCONTINUITIES IN THE BOUNDARY CONDITION

Violation of the conditions imposed on the boundary and the coefficients of the boundary condition in the formulation of problem (5.9) can lead to new effects arising. As before, we shall assume that boundary Γ of domain D in which the oblique derivative problem is considered has a continuously varying tangent, where the angle made by this tangent with a fixed direction satisfies a Hölder condition as a function of the arc length s of curve Γ. Discontinuities in the oblique derivative problem can be evoked by the fact that at certain points of the boundary, the derivative of the unknown function in the given direction has discontinuities, while at the same time the field of directions varies continuously in the neighborhood of such points. In this case the unknown function is not continuously differentiable in the closed domain. Another situation is also possible: the unknown harmonic function is continuously differentiable in the closed domain, but the field of directions in which the derivative is taken is discontinuous at a finite number of points $M_1, \ldots, M_q$ of the boundary Γ. This is precisely the situation which we consider. We denote by $\varphi_{h-\varepsilon}$ and $\varphi_{h+\varepsilon}$ the values of the angle φ, $\tan\varphi = b/a$, before passing through the point M_h and after passing through it, respectively, where the circuit of Γ is in the positive direction. The jump $\varphi_{h+\varepsilon} - \varphi_{h-\varepsilon}$ of the angle φ at point M_h is defined up to an integral multiple of π. In what follows let us agree to take this jump between the limits of zero and π, and we set $\alpha_h = \varphi_{h+\varepsilon} - \varphi_{h-\varepsilon}$, $0 < \alpha_h < \pi$, where $\varphi_{h+\varepsilon}$ is the limit of φ from one side of M_h, and $\varphi_{h-\varepsilon}$ is the limit at Γ from the other side of this point. On the rest of Γ, angle φ will be assumed to be Hölder continuous.

We denote by β the sum of the variables of the angle φ within the limits from M_h to M_{h+1} over all points M_h, assuming $M_{q+l} = M_1$, i.e., we set

$$\beta = \sum_h \int_{M_h}^{M_{h+1}} d\varphi.$$

The angle β is uniquely determined by this formula, and one has

$$\alpha_1 + \alpha_2 + \ldots + \alpha_q + \beta = n\pi, \tag{5.19}$$

where n is an integer. We show that this number plays the same role as the number n introduced in the continuous case in considering problem (5.9). We denote by r_h, θ_h polar coordinates with pole at the point M_h, where we direct the axis $\theta_h = 0$ along the inner normal to Γ at point M_h. The angle θ_h varies during a circuit of curve Γ between $-\pi/2$ and $\pi/2$, where the circuit starts from point M_h. We set

$$\varphi = -\theta_1\,\alpha_1/\pi - \theta_2\,\alpha_2/\pi - \ldots - \theta_q\,\alpha_q/\pi + \chi. \tag{5.20}$$

The angle χ, defined by (5.20), varies continuously on the whole boundary Γ, including points M_h. In fact, upon passing from the point $M_{h-\varepsilon}$ lying sufficiently close to M_h, to the point $M_{h+\varepsilon}$, also lying sufficiently close to M_h but on the other side of M_h, φ gets an increment α_h. On the right side of (5.20) all the θ_j except θ_h are continuous and $-\theta_h\alpha_h/\pi$ varies between the limits of $-\alpha_h/\pi \times \pi/2$ and $\alpha_h/\pi \times \pi/2$, i.e., the increment is exactly equal to α_h, which coincides with the increment of φ, so χ is Hölder continuous on the whole boundary. The angle χ varies continuously so its increment after a circuit of Γ is equal to an integral multiple of π. All the θ_j, $j = 1, \ldots, q$, after a circuit of Γ get no increment, so the increment of φ considering the jumps α_h upon a circuit of Γ is equal to the increment of χ, so the increment of χ is equal to $n\pi$, i.e., the quantity expressed by (5.19).

If $n = 0$, there exists a single-valued function $\chi(x, y)$ harmonic in domain D assuming, on boundary Γ, the values χ defined by (5.20). If one follows the methods of the preceding section for defining functions A and B or U and V, which are conjugate harmonic in domain D, and which on boundary Γ satisfy the conditions

$$B/A = \tan\varphi, \quad \text{or} \quad U/V = -\tan\varphi, \tag{5.21}$$

then one gets the expressions

$$\begin{aligned} A + iB &= \prod_{h=1}^{q} r_h^{-\alpha_h\pi^{-1}} R \exp\left\{i\left[-\sum_{h=1}^{q} \theta_h\alpha_h\pi^{-1} + \chi\right]\right\}, \\ U + iV &= \lambda\prod_{h=1}^{q} r_h^{\alpha_h\pi^{-1}} R^{-1} \exp\left\{i\left[\sum_{h=1}^{q} \theta_h\alpha_h\pi^{-1} - \chi\right]\right\}. \end{aligned} \tag{5.22}$$

In the second formula of (5.22) we introduce the factor λ in order to stress that one can take the pairs of functions A, B and U, V, which are defined up to an arbitrary constant factor with different factors, i.e., these constant factors do not necessarily coincide for both pairs. The function R becomes neither zero nor infinite in the closed domain $D \cup \Gamma$, so U and V are bounded in $D \cup \Gamma$ and there exists a bounded harmonic function

$$w(x, y) = \int_{z_0}^{z} (U\,dy + V\,dx) + C, \tag{5.23}$$

where C is an arbitrary constant. We show below that this formula gives the general solution of the homogeneous problem corresponding to (5.9).

We construct a solution of the inhomogeneous problem (5.9). For this it is necessary to construct a harmonic function $P(x, y)$ which is regular in domain D and on boundary Γ assumes the values

$$P(s) = \sqrt{A^2 + B^2}\, dv/dl.$$

Now the factor $\sqrt{A^2 + B^2}$ becomes infinite at all points M_h, $h = 1, \ldots, q$, but in a neighborhood of each point M_h the product $r_h{}^{\nu}P(s)$, $0 < \gamma = \alpha_h\pi^{-1} < 1$ remains bounded. According to Theorem 5.2 there exists a harmonic function $P(x, y)$ and its conjugate harmonic function $Q(x, y)$, such that $|(z - z_k)^{\gamma}(P + iQ)|$ is bounded above at point M_h. Consequently, the ratio $(P + iQ)/(A + iB)$ is defined everywhere in domain D and is bounded in the closed domain $D \cup \Gamma$, including the points M_h, $h = 1, \ldots, q$. The function

$$v(x, y) + iu(x, y) = \int_{z_0}^{z} \frac{P + iQ}{A + iB}\, dz + C \tag{5.24}$$

is holomorphic in D and differentiable in the closed domain, since $v_x - iv_y = (P + iQ)/(A + iB)$. Consequently, v is a solution of problem (5.9). Since Q is defined from P up to an additive arbitrary constant K, v is defined up to a summand

$$\mathrm{Re}\left\{i\int_{z_0}^{z} K(A + iB)^{-1} dz\right\} = \mathrm{Re}\left\{\int_{z_0}^{z} \lambda^{-1} K i\,(U + iV)\, dz\right\} = -K\lambda^{-1}\int_{z_0}^{z}(U\,dy + V\,dx),$$

i.e., up to a constant factor we have obtained a function from (5.23).

We show that with the help of (5.24) one can get the general solution of problem (5.9). We consider a solution of the second problem of (5.21), different from the solution $U + iV$, used in deriving (5.24), and we construct the product $F(z) = (A + iB)(U_1 + iV_1)$. Noting that $r^{\alpha_h\pi^{-1}} \exp(i\theta_h\alpha_h\pi^{-1}) = (z - z_h)^{\alpha_h\pi^{-1}}$, we rewrite the expression $F(z)$ as follows:

$$F(z) = \prod_{h=1}^{q} (z - z_h)^{-\alpha_h\pi^{-1}}\, [R\exp(i\chi)]\,(U_1 + iV_1).$$

The function $F(z)$ is real on boundary Γ of domain D by (5.21) and holomorphic in domain D, and at the points z_h, $h = 1, \ldots, q$, it admits a singularity of less than first order. Two cases are possible. In the first case, $U_1 + iV_1$ vanishes to order not lower than $\alpha_h\pi^{-1}$. Now function $F(z)$ is bounded in $D \cup \Gamma$ and real on Γ, so $F(z)$ is identically equal to a real constant, and $U_1 + iV_1$ differs from $U + iV$ only by a constant factor. In the second case at least one of the ratios $(U_1 + iV_1)/(z - z_h)^{\alpha_h\pi^{-1}}$ is unbounded as $z \to z_h$, but this contradicts Lemma 5.2, since $F(z)$ is real on Γ, i.e., its imaginary part is bounded in absolute value on Γ. Consequently, the second case is excluded and $U_1 + iV_1$ differs from $U + iV$ only by a constant factor, and this latter function is only defined up to a constant factor by hypothesis.

If $n > 0$, we proceed in exactly the same way as in the preceding sections of this chapter. We set $n = 2n_1 + n_2$ and we represent the χ which figures in (5.20) in the form of a sum

$$\chi = \sum_{k=1}^{n_1} \theta_k + \sum_{l=1}^{n_2} \theta_l + \chi',$$

where the θ_k which appear in the first sum correspond to polar coordinates with poles at interior points of D, and the θ_l are polar coordinates with poles at points of boundary Γ; the quantity χ', after a circuit of Γ, returns to the original value. Without loss of generality one can assume $n_2 = 0$ for even n and $n_2 = 1$ for odd n. Similarly to the case $n = 0$, assuming $n = 2n_1$ is even, we write

$$\begin{aligned} A + iB &= \prod_{h=1}^{q} r^{-\alpha_h\pi^{-1}} \prod_{k=1}^{n_1} r_k R_1 \exp\left\{i\left[-\sum_{h=1}^{q} \theta_h\alpha_h\pi^{-1} + \sum_{k=1}^{n_1} \theta_k + \chi_1\right]\right\} \\ &= \prod_{h=1}^{q} (z - z_h)^{-\alpha_h\pi^{-1}} \prod_{k=1}^{n_1} r_k R_1 \exp\left\{i\left[\prod_{k=1}^{n_1} \theta_k + \chi_1\right]\right\}, \\ U + iV &= \prod_{h=1}^{q} (z - z_h)^{\alpha_h\pi^{-1}} \prod_{k=1}^{n_1} r_k^{-1}\lambda R_2^{-1} \exp\left\{-i\left[\sum_{k=1}^{n_1} \theta_k + \chi_1\right]\right\}. \end{aligned} \tag{5.25}$$

In the second formula of (5.25) the factor λ figures for the same reasons as in (5.22). Just as for $n = 0$, the function $A + iB$ has singularities at points M_h of less than first order, but this does not hinder the construction of a pair of conjugate harmonic functions P and Q, analogous to those figuring in (5.24), and further we set $v_x - iv_y = (P + iQ)/(A + iB)$. In contrast with the case $n = 0$, now $A + iB$ vanishes at some interior points, and for the gradient of v to be bounded in D it is necessary and sufficient that $P + iQ$ vanish at these same points. As in the preceding section, this implies $n - 1$ solvability conditions for the inhomogeneous problem (5.9).

The function $U + iV$, expressed by the second formula of (5.25), becomes infinite at n_1 interior points of domain D, so it does not generate a solution of the homogeneous problem corresponding to (5.9), which is regular in domain D. Let us assume that there exists a function $U_1 + iV_1$, generating a bounded solution of the homogeneous problem corresponding to (5.9), which is not included among the functions represented by the second formula of (5.25). We consider the product

$$\begin{aligned} F(z) = (A + iB)(U_1 + iV_1) &= \prod_{h=1}^{q} (z - z_h)^{-\alpha_h\pi^{-1}} \\ &\times \prod_{k=1}^{n_1} r_k R_1 \exp\left\{i\left[\sum_{k=1}^{n_1} \theta_k + \chi_1\right]\right\}(U_1 + iV_1). \end{aligned}$$

As in the case $n = 0$, there are two possibilities: a) the function $U_1 + iV_1$ which is real on Γ vanishes at each point M_h, b) at least at one point M_h the ratio $(U_1 + iV_1)(z - z_h)^{\alpha_h\pi^{-1}}$ is unbounded. The fact that case b) is not realized is proved in exactly the same way as in the case $n = 0$. In case a) the function $F(z)$ vanishes at the same points of domain D, at which $A + iB = 0$, and there are n_1 such points, so $F(z) \equiv 0$, and this means that $U_1 + iV_1 \equiv 0$ in D. It follows from this that

$$w = \int_{z_0}^{z} (U_1 dy + V_1 dx) + C \equiv \text{const},$$

i.e., the homogeneous problem corresponding to (5.9) has only the solution identically equal to a constant. To construct a solution of (5.9) in the first formula one can take another function $A + iB$, but it is easy to show that this only changes the solution v by a constant, as in the case of continuous φ also.

In the case $n < 0$, like that of $n > 0$, it suffices to restrict oneself to even n, $-n = 2n_1$. Instead of the second formula (5.25), now we have

$$U + iV = \prod_{h=1}^{q} (z - z_h)^{\alpha_h \pi^{-1}} \prod_{k=1}^{n_1} r_k R_1^{-1} \exp\left\{i\left[\sum_{k=1}^{n_1} \theta_k - \chi_1\right]\right\}, \tag{5.26}$$

while $U + iV$ vanishes at the boundary points M_h, $h = 1, \ldots, q$, and at n_1 interior points. Further, following the usual scheme, we must construct a harmonic function $P(x, y)$ which is regular in domain D and on boundary Γ satisfies the condition

$$P(s) = (U^2 + V^2)^{-1/2}\, dv/dl,$$

and its conjugate harmonic function $Q(x, y)$. The function $P(s)$ has singularities at points M_h of less than first order, so using Theorem 5.2 we get that $P + iQ$ is holomorphic in domain D, and at points M_h of boundary Γ the functions $(z - z_h)^{\alpha_h \pi^{-1}}(p + iQ)$ are bounded. Consequently, the function $(P + iQ)(U + iV)$ is holomorphic in D and bounded in the closed domain $D \cup \Gamma$, so with the help of the formula

$$v + iu = \int (P + iQ)(U + iV)\, dz \tag{5.27}$$

one can find a solution v of (5.9) just as in the case of continuous φ.

We find the number of parameters on which the solution v of (5.9) depends. For this we take a function $U_1 + iV_1$, which is bounded in domain D and differs from the function $U + iV$ figuring in (5.27), and generating a solution of (5.9) through formula (5.27). We have

$$F(z) = (U_1 + iV_1)/(U + iV) = (U_1 + iV_1) \prod_{h=1}^{q} (z - z_h)^{-\alpha_h \pi^{-1}}$$

$$\times \prod_{k=1}^{n_1} r_k^{-1} R_1 \exp\left\{i\left[-\sum_{k=1}^{n_1} \theta_k + \chi_1\right]\right\}.$$

Function $F(z)$ is real on boundary Γ of domain D and on Γ it has singularities of less than the first order. By Lemma 5.2, $F(z)$ must also be bounded at points M_h, so exactly as in the case of a continuous angle we have

$$F(z) = [a_0\varphi^q + a_1\varphi^{q-1} + \ldots + a_q][b_0\varphi^q + b_1\varphi^{q-1} + \ldots + b_q]^{-1}, \quad q = -n > 0,$$

where $\varphi(z)$ is a function which maps domain D conformally to the upper half-plane, and all the coefficients a_j and b_j are real. The coefficients of the denominator depend on the n_1 zeros of $U + iV$ situated in the domain D and the coefficients of the numerator are arbitrary while the pair of complex conjugate roots of the polynomial in φ standing in the numerator correspond to zeros of $U_1 + iV_1$ lying in D, and the real zeros of this polynomial in φ correspond to zeros of $U_1 + iV_1$ lying on Γ. The arbitrariness in $F(z)$ is characterized by $q + 1$ coefficients of the numerator. Consequently, in this case the general solution of (5.9) depends on $2 - n$ arbitrary constants, $1 - n$ of which are contained in $F(z)$, and one appears in the integration in (5.27).

Thus, the solvability character of problem (5.9) is unchanged if one admits a finite number of discontinuities of the first kind of the angle φ, $\tan\varphi = b/a$, under the condition that the number n is determined with the help of (5.20), considering the jumps at the points of discontinuity. Let us assume that at each point of the boundary Γ the angle φ is replaced by $-\varphi$. If φ is continuous, then the characteristic n changes to $-n$. In the case of discontinuous φ this is no longer so since we have agreed to take the jump α_h between the limits of zero and π. Now when φ is replaced by $-\varphi$ one must replace α_h not by $-\alpha_h$, but by $\pi - \alpha_h$. Consequently, $\Sigma\alpha_h + \beta$ changes to $\Sigma(\pi - \alpha_h) - \beta = p\pi - \Sigma\alpha_h - \beta$, so the characteristic n changes to $p - n$.

5. VARIATION OF LEVEL LINES OF A HARMONIC FUNCTION ON A CLOSED CONTOUR

In the preceding sections it was established that the solvability character of the problem (5.9) depends essentially on the variation of the angle $\tan\varphi = b/a$. The connection between the solvability of (5.9) and the behavior of angle φ is stipulated by the behavior of the level lines of the harmonic function v, which is regular in domain D. We study such properties of the level lines of a harmonic function. We consider a closed contour Γ and a function $V(x, y)$ which is harmonic in domain D, bounded by curve Γ. We shall assume that V is continuous in the closed domain $D \cup \Gamma$ and has first derivatives V_x and V_y which are bounded and continuous in the domain $D \cup \Gamma$. Let $\varphi = \varphi(x, y)$ be the angle between the Ox axis and the tangent to the level line of function V at points (x, y). Our problem will be the calculation of the increment in angle φ, when the point (x, y) runs over boundary Γ of domain D. Obviously the angle φ is only determined up to an integral multiple of π, but this has no effect on the value of the increment of φ in a circuit of Γ, if φ varies continuously.

Angle φ is connected with the components of the gradient of V by

$$\varphi = -V_x/V_y, \tag{5.28}$$

from which we get

$$d\varphi = (V_x^2 + V_y^2)^{-1}(V_x dV_y - V_y dV_x). \quad (5.29)$$

It is easily verified by direct calculation that the expression on the right side of (5.29) is a total differential. If the coefficients in front of dx and dy in (5.29) are continuous in a closed domain Δ, bounded by a rectifiable curve C, then the integral of $d\varphi$ over C is equal to zero [32], i.e., the increment of φ on C is equal to zero. Consequently, the increment of φ on boundary Γ of domain D can differ from zero only when there are points in $D \cup \Gamma$ at which the first derivatives of function V vanish simultaneously or the second derivatives of function V are unbounded, while the integral of $d\varphi$ over Γ is equal to the sum of the integrals of $d\varphi$ over circles of sufficiently small radii with centers at these singular points.

We consider singularities lying outside D. The function V has continuous first derivatives in the closed domain $D \cup \Gamma$, so its conjugate harmonic function W exists, and the function $F(z) = V + iW$ is holomorphic in the domain D, so inside D there are no points at which the second derivatives of function V would be unbounded. All the singularities of the right side of (5.27) lying inside D are connected with zeros of the gradient of the function V, but $F'(z) = V_x + iW_x = V_x - iV_y$. In a neighborhood of zero z_0 of the derivative $F'(z)$ of function $F(z)$ one can write

$$F(z) = \gamma + i\delta + (c_p + id_p)(z - z_0)^p + (c_{p+1} + id_{p+1})(z - z_0)^{p+1} + \ldots, \quad p \geqslant 2.$$

Introducing polar coordinates with pole z_0, from this we find

$$\begin{aligned} F'(z) = V_x - iV_y &= pc_p + id_p(z - z_0)^{p-1} + \ldots \\ &= K \exp(i\alpha) r^{p-1} \exp[i(p-1)\theta] + \ldots \end{aligned}$$

For sufficiently small r one can write

$$V_x - iV_y = Kr^{p-1} \exp\{i[\alpha + (p-1)\theta]\} + O(r),$$

so the increment of the argument of $V_x - iV_y$ in a circuit of a circle $|z - z_0| = \delta$ of sufficiently small radius δ is equal to the increment of $a + (p-1)\theta$, i.e., is equal to $2(p-1)\pi$. On the other hand, denoting by ψ the argument of $V_x - iV_y$, we have

$$\tan\psi = -V_y/V_x = (-V_x/V_y)^{-1} = (\tan\varphi)^{-1} = \cot\varphi,$$

from which it follows that $\varphi = \gamma - \psi$, where γ is a constant, and this relation shows that the increment of φ on the circle $|z - z_0| = \delta$ is equal to $-2(p-1)\pi$. If in domain D there are some zeros $z_1, \ldots, z_l$ of the derivative $F'(z)$, then the increment of φ on Γ is equal to

$$-2\pi \sum_{j=1}^{j} q_{.}, \quad q_j = p_j - 1,$$

where q_j is the multiplicity of the zero. This increment is always negative.

Now we consider singular points located on boundary Γ of domain D. Let S be such a point, located on Γ; by BCA we denote an arc of a circle with center at the point S of sufficiently small radius, lying entirely in domain D, with ends at points A and B of intersection of this circle with Γ. We take the integral of $d\varphi$ over the closed contour $ASBCA$. The integral over the arc AB of boundary Γ tends to zero with the radius of the circle, since φ is assumed continuous on Γ. To calculate the integral over the arc BCA it is necessary and sufficient to calculate the increment of $\arctan(-V_x/V_y)$ on this arc. The point S can be singular owing to the second derivatives of function V being unbounded at this point, while at the same time the first derivatives V_x and V_y are continuous at this point and do not vanish simultaneously. Obviously in this case one can turn the coordinate axes so that at point S both derivatives V_x and V_y become nonzero. Now $\arctan(-V_x/V_y)$ at point S has a well-defined finite value and its increment on arc BCA tends to zero with the radius of the arc. Consequently, similar singularities do not influence the increment of φ, so it suffices to consider points on Γ at which V_x and V_y vanish simultaneously.

In the second case we preserve the figure formed by the arc ASB of boundary Γ and arc BCA of the circle, although these arcs can be replaced by an arbitrary arc of the boundary and an arbitrary arc lying in domain D such that in the closure of the figure $ASBCA$ there are no other points at which V_x and V_y vanish simultaneously. When we go around the contour $SACBS$ in the positive direction, the angle φ varies continuously, so after a circuit of the contour we get a value differing from the original one by an integral multiple of π, i.e.,

$$\int_{SBCAS} d\varphi = -k\pi, \quad k \text{ an integer.}$$

The integer k is a characteristic of point S. We take point S as the pole of a polar coordinate system r, θ and we set $V + iW = F(z)$. We consider the function $\Phi(z) = \ln[F'(z)(z - z_0)^{-k}]$, where z_0 is a point of S. We have

$$\Phi(z) = \ln[(V_x - iV_y)r^{-k}\exp(-ik\theta)] = \ln[|\text{grad } V|r^{-k}] + i(\pi/2 - \varphi - k\theta).$$

The function Φ is single-valued in domain D_1, bounded by the arc $SBCAS$, since Φ is holomorphic in D_1 and continuous in the closure of D_1, except for point S, and staying in D_1 it is impossible to avoid point S, by virtue of which $\pi/2 - \varphi - k\theta$ is uniquely defined in D_1. After a circuit of contour $SBCAS$, the quantity $\pi/2 - \varphi - k\theta$ returns to the original value, since φ gets the increment $-k\pi$ and $k\theta$ gets $k\pi$. It follows from the continuity and single-valuedness of $\pi/2 - \varphi - k\theta$ in D_1 that its conjugate harmonic function $\ln[|\text{grad } V|r^{-k}]$ is regular and continuous in the closure $\bar{D}_1$ everywhere except point S, in a neighborhood of which it remains between the limits $\varepsilon \ln r$ and $-\varepsilon \ln r$, where ε is an arbitrarily small previously given positive number. In other words, one can express this as follows: for any arbitrarily small previously given $\varepsilon > 0$ one has

$$r^{\varepsilon-k}|\text{grad } V| \to 0, \quad r^{-\varepsilon-k}|\text{grad } V| \to \infty, \tag{5.30}$$

when point z tends from within D_1 to points S. One gets these relations instead of the more precise one $c = \lim_{r\to 0} r^{-k}|\text{grad } V|$, where c is a well-defined number, due to the fact that we do not require the Hölder continuity of φ, and we do not impose any conditions on the tangent to the boundary of the domain D_1.

It follows from (5.30) that the integer k cannot be zero or negative. If this were not so, then the expression $r^{-k-\varepsilon}|\text{grad } V|$ would tend to zero at point S. Consequently, point S behaves like a multiple (saddle) point of the level lines of multiplicity $k + 1$. If we now combine the results found for interior and boundary singular points, then we arrive at the relation

$$\int_\Gamma d\varphi = -2\pi \sum_{i=1}^{q} k_i - \pi \sum_{j=1}^{m} k_j,$$

where the first sum is taken over the interior zeros of the function $V_x - iV_y$, and the second over the zeros located on the boundary of the domain. We denote by n_1 the sum, taking multiplicities into account, of the zeros lying in domain D of the function $V_x - iV_y$, and by n_2 the sum, taking multiplicities into account, of the zeros lying on boundary Γ of domain D, and we can write

$$n\pi = \int_\Gamma d\varphi = -2n_1\pi - n_2\pi, \quad n = -2n_1 - n_2.$$

We consider the case where point S at which $V_x = V_y = 0$ is a corner point of boundary Γ of domain D, while the size of the angle is equal to $m\pi$. The investigation goes exactly the same as in the preceding case, but now, after a circuit of contour $SBCAS$, function θ gets an increment $m\pi$ instead of π. Now $\Phi(z)$ must be taken in the form

$$\Phi(z) = \ln [F'(z)(z - z_0)^{-k/m}] = \ln [r^{-k/m}|\text{grad } V|] + i(\pi/2 - \varphi - k\theta/m)$$

in order that the imaginary part of $\Phi(z)$, after a circuit of the contour $SBCAS$, should return to the original value. Function $V_x - iV_y$ behaves, at point S, like $(z - z_0)^{k/m}$. However, the function $\varphi(z)$, which maps domain D conformally to the upper half-plane, behaves at point S like $(z - z_0)^{1/m}$, so one can say that the function $V_x - iV_y$ has a zero of order k at point S with respect to the difference $\varphi - \varphi_0$, $\varphi_0 = \varphi(z_0)$.

6. MULTIPLY CONNECTED DOMAINS

One can apply the method of the oblique derivative problem in a simply connected domain to the case of a multiply connected domain also. However, now all the constructions turn out to be more complicated. We shall assume that the boundary of the multiply connected domain consists of $p + 1$ closed rectifiable Jordan curves $C_0, C_1, \ldots, C_p$, all of whose points are ordinary, while the curves $C_1, \ldots, C_p$ lie in the simply connected domain, bounded by curve C_0. As the positive direction on C_0 we take the direction in which the domain remains on the left, and on $C_1, \ldots, C_p$ the direction in which the domain remains on the right. We also assume that the direction of differentiation and the values of the oblique derivative on the boundary of the domain satisfy a Hölder condition.

We denote by $\varphi_j, j = 0, \ldots, p$, the angle determining the given direction of differentiation on the curve C_j. These $p + 1$ quantities are functions of the arc length of the corresponding curve. Each function φ_j is defined up to an integral multiple of π, but the increment of φ_j after a circuit of C_j is uniquely determined and has the form $n_j\pi$, where n_j is an integer [18].

Let $v(x, y)$ be a harmonic function which is regular and single-valued in the multiply connected domain D, whose gradient exists and is bounded in closed domain $D \cup \Gamma$, where Γ is the collection of curves $C_0, C_1, \ldots, C_p$, i.e., the boundary of domain D. Assuming that v is a solution of the homogeneous oblique derivative problem, we investigate the connection between the numbers $n_j, j = 0, \ldots, p$, and the collection of critical points of function v lying in domain D. For this, with the help of slits we turn domain D into a simply connected domain D', where we make the slits so that none of them passes through critical points of v. We calculate the integral over the boundary of domain D' of the quantity $d(-v_y/v_y)$. If at some point M_h of the boundary of domain D' the gradient of v vanishes, then in the integration we replace the boundary of D' by an arc of a circle of radius ε, with center at M_h, lying in D'. For sufficiently small $\varepsilon > 0$ the integral is independent of ε. Over the slits, with the help of which domain D is turned into the simply connected domain D', the integral is taken twice, where these slits are run through in opposite directions, so the sum of the integrals over all of the slits is equal to zero. During the integration curve C_0 is traversed in the positive direction, and curves $C_1, \ldots, C_p$ in the negative one, so according to the results of the preceding section we have

$$-\left(n_0 - \sum_{j=1}^{p} n_j\right) = \nu_0 + \sum_{j=1}^{p} \nu_j + 2l, \tag{5.31}$$

where the $V_j, j = 0, \ldots, p$, are the numbers of critical points of function v, lying on curves $C_j, j = 0, \ldots, p$, and l is the number of critical points of v which lie in domain D. By virtue of the conditions imposed on the boundary of the domain and on the field of directions, the numbers $\nu_j, j = 0, \ldots, p$, and l are integral.

We denote by $-n$ the left side of (5.31). Its right side is always nonnegative [18]. Consequently, for $n > 0$ the homogeneous oblique derivative problem has no nonconstant solutions, i.e., the problem of finding a function $U + iV$ which is analytic, single-valued in domain D, and on boundary Γ of the domain satisfies the condition $-\tan\varphi = V/U$, has no solution. Conversely, the problem of finding conjugate harmonic functions A and B, which are regular in domain D and on Γ satisfy the condition $B/A = \tan\varphi$, always has bounded solutions, but they can be non-single-valued. In order that these solutions be single-valued they must satisfy p additional conditions. In this case there exists a bounded single-valued solution of our auxiliary problem only if the general solution in the class of bounded functions contains no less than p arbitrary constants. Thus, for $n > 0$ the homogeneous oblique derivative problem does not have nonconstant solutions, and a solution of the inhomogeneous problem is determined up to an additive constant, while for the existence of a solution of the inhomogeneous problem it is necessary to impose a certain number of orthogonality conditions on the values of the oblique derivative given on Γ.

We consider the case $n < 0$. We note that if $P + iQ$ is a solution of the problem $Q/P = \tan\varphi$, then $P' + iQ' = (P + iQ)^{-1} = (P + iQ)(P^2 + Q^2)^{-1}$ is a solution of the problem $Q'/P' = -\tan\varphi$, and conversely. Replacing φ by $-\varphi$ on all components of boundary Γ leads to replacing n by $-n$, so the problem of finding conjugate harmonic functions which are regular in domain D and on the boundary satisfy $B/A = \tan\varphi$ has no bounded solutions. Now the problem of finding conjugate harmonic functions which are regular in domain D and on the boundary satisfy the condition $V/U = -\tan\varphi$ has bounded solutions, but they can be multivalued. Bounded solutions of the latter problem depend linearly on $-n + 1$ arbitrary constants, and bounded and single-valued solutions contain no more than $-n + 1$ arbitrary constants, the number of which is not less than $-n + 1 - p$.

In what follows we consider in more detail only a two-connected domain, i.e., we set $p = 1$. Any two-connected domain satisfying the conditions formulated at the beginning of the section can be mapped conformally to an annulus [32], so without loss of generality we take an annulus as the domain D. In the case where one is dealing with functions which are not single-valued in the annulus, we shall consider the infinite-sheeted Riemann surface over the annulus. Under the classical map

$$z = x + iy = \pi i(\ln r_0/r_1)^{-1}[\ln r/r_1 + i\theta] = \pi(\ln r_0/r_1)^{-1}(-\theta + i\ln r/r_1),$$

where r_0 and r_1 are the radii of the outer and inner bounding circles, and the origin of the coordinates r, θ is situated at the common center of these circles, the ring is mapped to the strip bounded by the lines $y = 0$ and $y = \pi$.

For the annulus we consider the auxiliary homogeneous oblique derivative problem. We start with the case where both numbers n_0 and n_1 are equal to zero. Thus, the angles φ_0 and φ_1 return to their original values after a circuit of the circles C_0 and C_1, respectively. Here there exists a single-valued harmonic function χ which is regular in ring D and assumes the values φ_j on C_j, $j = 0, 1$. We denote by $\ln R$ the conjugate harmonic function for χ. The formula $A + iB = R\exp(i\chi)$ expresses two conjugate harmonic functions in D such that the ratio B/A coincides with $\tan\varphi$ on C_j, $j = 0, 1$. Analogously, setting $U + iV = \lambda R^{-1}\exp(i\chi)$, where λ is an arbitrary constant, we get two harmonic functions in D such that V/U assumes the values $-\tan\varphi_j$ on C_j, $j = 0, 1$. We note that the function R itself is also determined up to an arbitrary constant factor. Function R is bounded and nonvanishing, so the functions A, B, U, and V are also bounded, but generally they are not necessarily single-valued.

In investigating the single-valuedness of the functions constructed, it is necessary to distinguish two cases. In the first, let us assume that the averages of the functions φ_j over the corresponding circles are equal, i.e.,

$$\frac{1}{2\pi}\int\limits_{C_1} \varphi_1 d\theta = \frac{1}{2\pi}\int\limits_{k_0} \varphi_0 d\theta. \tag{5.32}$$

If this condition holds, the conjugate harmonic function $\ln R$ of function χ is also single-valued, and with them all the functions A, B, U, and V are also single-valued. We consider another pair A', B' of conjugate harmonic functions in D, satisfying the same boundary condition as A, B. The product $(A' + iB')(U + iV)$ is single-valued and regular in domain D together with the factors, and on the boundary of domain D this product is real, since the arguments of the factors are equal in absolute value, but opposite in sign. Consequently, the imaginary part of the product is identically equal to zero everywhere in D, and the product itself is a real constant K, i.e., $A' + iB' = K\lambda^{-1}R \exp(i\chi)$. Since R is only defined up to a constant factor, the pair of functions A', B' does not lead to a new solution of the problem.

We note that the requirement that $A' + iB'$ be single-valued is essential. If one permits multivalued solutions, then one can set

$$(A' + iB')(U + iV) = K \exp\{\pi(\ln r_0/r_1)^{-1}[-\theta + i \ln r/r_1]\}.$$

The right side of this equation is real on the circles C_j: $\{r = r_j\}$, $j = 0, 1$. From this we find the solution

$$A' + iB' = K\lambda R^{-1} \exp\{i\chi + \pi(\ln r_0/r_1)^{-1}[-\theta + i \ln r/r_1]\}.$$

This function is bounded in the annulus D, but not single-valued, since after a circuit of the origin θ gets an increment of 2π, and the function acquires the factor

$$\exp\{\pm 2\pi^2(\ln r_0/r_1)^{-1}\} = \exp(\pm a).$$

Since U and V are bounded and single-valued functions in D, the formula

$$W(x, y) = \int (U dy + V dx)$$

defines a solution of the homogeneous oblique derivative problem with bounded and single-valued gradient, since $W_x = V$, $W_y = U$. Sometimes one can restrict oneself to requiring that the gradient be bounded and single-valued, but if one requires that function W be single-valued, then besides (5.32) holding, it is necessary to impose another restriction on the function φ_0 and φ_1.

When the means of φ_0 and φ_1 are not equal to one another, i.e., when (5.32) does not hold, we introduce the angle α by the formula

$$\alpha = \frac{1}{2\pi}\int_{C_0} \varphi_0 d\theta - \frac{1}{2\pi}\int_{C_1} \varphi_1 d\theta.$$

We set $\varphi_0 = \alpha + \psi_0$. Obviously the average value of angle ψ_0 is equal to the average value of φ_1, so replacing φ_0 by ψ_0, as in the preceding case, we construct functions $A_1 + iB_1$ and $U_1 + iV_1$, corresponding to ψ_0 and φ_1. The presence of the summand α for the harmonic function χ gives increment $\delta\chi$, which is equal to zero on C_1 and equal to α on C_0. It is easy to see that $\delta\chi = \alpha(\ln r_0/r_1)^{-1}(\ln r/r_1)$. To this function corresponds the conjugate harmonic function $-\alpha(\ln r_0/r_1)^{-1}\theta$. Consequently, to the angles φ_0 and φ_1 corresponds a solution of the auxiliary problem $A + iB$, which differs from the solution $A_1 + iB_1$ of the auxiliary problem corresponding to the angles ψ_0, φ_1 by the factor

$$\exp\{\alpha[-\theta + i \ln r/r_1](\ln r_0/r_1)^{-1}\} = \exp(\alpha z/\pi).$$

From this it follows that $A + iB = (A_1 + iB_1)\exp(\alpha z/\pi)$. After each circuit of the origin the functions A, B, $(A^2 + B^2)^{1/2}$ are multiplied by the constant factor $\exp\{-2\pi(\ln r_0/r_1)^{-1}\}$. The fact that the constructed functions A and B are not single-valued does not diminish their applicability to the investigation of the oblique derivative problem, but it causes complications in the investigation and can change the number of conditions necessary for a solution.

As already noted, the angles φ_0 and φ_1 are only defined up to an integral multiple of π, i.e., to φ_j one can add $k_j\pi$, $j = 0, 1$, where k_j are arbitrary integers. Hence the angle α is also defined up to an integral multiple of π. For example, we leave φ_1 unchanged, and we increase φ_0 by π; now α is replaced by $\alpha + \pi$, and $A + iB$ acquires the factor $f = \exp z$. This factor is real for $r = r_1$ and $r = r_0$, so A and B are multiplied by factors which are identical in absolute value, positive on C_1 and negative on C_0, and the ratio B/A is unchanged on both circles. Thus, we get an infinite set of functions $A + iB$, between which there are no essential differences.

By virtue of the arbitrariness in α, this angle can always be taken between the limits of zero and π. Everything said about $A + iB$ is also valid for $U + iV$, but the factor f is replaced by the factor f^{-1}. If α is taken between the limits of $-\pi/2$ and $\pi/2$, then for $\alpha \neq 0$ the homogeneous oblique derivative problem has no solutions which are single-valued in the annulus D. The inhomogeneous problem, if it is solvable in general, has a solution which is determined up to a constant, and the gradient of this solution is determined uniquely.

To the special case considered, $n_0 = n_1 = 0$, one can reduce the more general case $n = 0$, $n_0 = n_1 \neq 0$. Suppose we have $n_0 = n_1 = 2m$, where m is an integer; we set $\varphi_0 = m\theta + \psi_0$, $\varphi_1 = m\theta + \psi_1$. When θ increases by 2π, the angles φ_0 and ψ_0 increase by $2m\pi$, so ψ_0 and ψ_1 return to the original values. Letting $A_1 + iB_1$ be a solution of the auxiliary problem corresponding to ψ_0 and ψ_1, we can write

$$A + iB = r^m(\cos m\theta + i \sin m\theta)(A_1 + iB_1).$$

The factor $r^m \exp(im\theta)$ and its inverse are bounded and single-valued in the annulus for any integer m, so the solvability character of the problem is determined by the second factor, which corresponds to $n_0 = n_1 = 0$. If n_0 and n_1 are odd, i.e., $n_0 = n_1 = 2l + 1$, then setting

$$\varphi_0 = (l + 1/2)\theta + \psi_0, \quad \varphi_1 = (l + 1/2)\theta + \psi_1,$$

we get analogously

$$A + iB = r^l \exp(il\theta)\sqrt{r} \exp(i\theta/2)(A_1 + iB_1).$$

Due to the presence of the factor $\exp(i\theta/2)$ the function $A + iB$ changes sign after each circuit of the origin. In this case the function $A + iB$ is always non-single-valued. One can show that the auxiliary problem has no other functions besides those constructed here.

Let r', θ' and r'', θ'' be coordinate systems with poles lying in the disk bounded by the circle C_1. We replace

$$\varphi_0 = m\theta + \psi_0, \quad \varphi_1 = m\theta + \psi_1 \tag{5.33}$$

by any of the relations

$$\begin{aligned} &\varphi_0 = \theta' + (m-1)\theta + \psi_{10}, \quad \varphi_1 = \theta' + (m-1)\theta + \psi_{11}, \\ &\varphi_0 = -\theta' + (m+1)\theta + \psi_{20}, \quad \varphi_1 = -\theta' + (m+1)\theta + \psi_{21}, \\ &\varphi_0 = \theta' + \theta'' + (m-2)\theta + \psi_{30}, \quad \varphi_1 = \theta' + \theta'' + (m-2)\theta + \psi_{31}. \end{aligned} \tag{5.34}$$

The following questions arise: Won't the solutions $A + iB$ and $U + iV$ obtained with the help of (5.34) differ from the solutions obtained above with the help of (5.33), and won't these new solutions be single-valued while at the same time the solutions obtained with the help of (5.33) are multivalued?

We show that the answers to these questions are negative. We take, for example, the first pair of equations (5.34). Obviously the angles ψ_{10} and ψ_{11} after a circuit of the corresponding circles return to their original values like the angles ψ_0 and ψ_1. Let θ_j, θ_j', $j = 0, 1$ be the values assumed by θ and θ' on the circles C_j, $j = 0, 1$, respectively. From (5.33) and (5.34) we have

$$\psi_0 = (\theta' - \theta) + \psi_{10}, \quad \psi_1 = (\theta' - \theta) + \psi_{11}.$$

Let χ_1 and χ be regular harmonic functions which are single-valued in the annulus D and coincide with ψ_j and ψ_{1j} on the circles C_j, $j = 0, 1$, respectively. The difference $\theta' - \theta$ is a harmonic function which is regular in the annulus D, and since the poles of both polar coordinate systems lie inside D, this difference is also single-valued in D. Consequently, we have $\chi' = \theta' - \theta + \chi$. If we add to χ' $m\theta$, or add to χ $\theta' + (m-1)\theta$, then it is clear that we get the same result, so the function $A + iB$ is unchanged.

We prove the equation

$$\frac{1}{2\pi}\left\{\int_{C_0} \psi_{10} d\theta - \int_{C_1} \psi_{11} d\theta\right\} = \frac{1}{2\pi}\left\{\int_{C_0} \psi_0 d\theta - \int_{C_1} \psi_1 d\theta\right\}.$$

For this it suffices to prove that the integral $\theta' - \theta$ is equal to zero over C_1 as well as over C_0. Obviously the integral of $\theta' - \theta$ over a circle of radius $R \geq r_1$ with center at the origin is the same for all such circles. For sufficiently large R the difference $\theta' - \theta$ tends to zero like R^{-2}, so the integral of $\theta' - \theta$ over a circle of sufficiently large radius is arbitrarily small, and this means that all such integrals are equal to zero.

We consider the case $n > 0$, i.e., $n_0 > n_1$. We take a collection of polar coordinate systems, among which we take ν_0 systems (r_{0k}, θ_{0k}) with poles at points of the circle C_0, and l systems (r_k', θ_k') with poles at interior points of the annulus D, and ν_1 systems (r_{1k}, θ_{1k}) with poles at points of the circle C_1. By (r, θ) we shall denote a polar coordinate system with pole at the center of the circles C_0 and C_1. We set

$$\varphi_0 = \left[\sum_{k=1}^{\nu_0} \theta_{0k} + \sum_{k=1}^{l} \theta_k' + \sum_{k=1}^{\nu_1} \theta_{1k} + m\theta\right]\Bigg|_{C_0} + \chi_0,$$
$$\varphi_1 = \left[\sum_{k=1}^{\nu_0} \theta_{0k} + \sum_{k=1}^{l} \theta_k' + \sum_{k=1}^{\nu_1} \theta_{1k} + m\theta\right]\Bigg|_{C_1} + \chi_1, \tag{5.35}$$

where m is any integer.

We determine what conditions the numbers ν_0, l, ν_1, and m must satisfy in order that the χ_j, defined by (5.35), after a circuit of the circles C_j, $j = 0, 1$, return to the original values. We have the relations

$$n_0 = \nu_0 + 2(l + \nu_1 + m), \quad n_1 = \nu_1 + 2m, \tag{5.36}$$

since upon a circuit of C_0 all the θ_{0k} get the increment π, and θ_k' and θ_{1k} the increment 2π, upon a circuit of C_1 all θ_{0k} and θ_k' return to the original values, and all θ_{1k} get the increment π. Subtracting the second equation of (5.36) from the first, we find

$$n_0 - n_1 = \nu_0 + 2l + \nu_1. \tag{5.37}$$

In (5.36) and (5.37) the numbers ν_0, l, ν_1 are nonnegative integers. We shall take ν_j zero or one depending on whether the number n_j, $j = 0, 1$, is even or odd. For fixed ν_0 and ν_1 (5.37) determines the number l uniquely. We construct a harmonic function χ which is regular in domain D and assumes the values χ_j on the circles C_j, $j = 0, 1$, and in the usual way we construct the function

$$A + iB = \prod_{k=1}^{\nu_0} (r_{0k} \exp i\theta_{0k}) \prod_{k=1}^{l} (r_k' \exp i\theta_k') \times \prod_{k=1}^{\nu_1} (r_{1k} \exp i\theta_{1k})\, r^m \exp im\theta \times R \exp i\chi. \tag{5.38}$$

The function $A + iB$ is bounded in the closed annulus D for any m and depends on arbitrary parameters which determine the location of the poles of the coordinate systems used in its construction. The number of these parameters is $\nu_0 + 2l + \nu_1 = n$. The function R also contains an arbitrary factor, so function (5.38) depends on $n + 1$ parameters. Function (5.38) does not give representations of all solutions of the auxiliary problem; however, one can show that all single-valued solutions of this problem are given by (5.38).

In (5.35), in place of m one can take $m' + 1/2$. Now the numbers ν_j, $j = 0, 1$, must be taken to be zeros or ones, depending on whether the number n_j, $j = 0, 1$, is odd or even. The function $A + iB$ constructed by (5.38), after a circuit of the annulus, will change sign to the opposite. This gives an example of solutions of the auxiliary problem which are obtained from (5.38) for integral m.

We investigate (5.38). We show that this formula represents a single-valued function only when the $n + 1$ arbitrary parameters appearing in it are connected by a relation, and the most general single-valued solution of the auxiliary problem can be represented so that it will contain n arbitrary parameters in a linear and homogeneous way. We consider the number

$$\alpha = \frac{1}{2\pi}\int_{C_0} \chi_0 d\theta - \frac{1}{2\pi}\int_{C_1} \chi_1 d\theta,$$

where χ_0 and χ_1 are the functions defined by (5.35). This quantity depends essentially on the position of the poles of the polar coordinate systems used in (5.35). We relocate the pole O' of the coordinate system (r_k', θ_k') at the point O'' of domain D. Replacing θ_k' by θ_k'' causes, at point M of the circle C_0, the increment $\theta_k' - \theta_k''$ of the function χ_0, but the mean value of this increment over circle C_0 is equal to zero, as was shown above. Now we take point M on circle C_1. The function $\theta_k' - \theta_k''$ is single-valued and harmonic in the disk bounded by the circle C_1. By the theorem of the mean for harmonic functions [32] this mean is equal to the value of the function $Q_k' - Q_k''$ at the center O of the circle C_1, i.e., the angle $O'OO''$, taken with the corresponding sign. If, starting from point O', point O'' completes a full revolution in D, then α gets the increment 2π, and in the course of this displacement of O'', α twice assumes values equal to an integral multiple of π, to which correspond single-valued solutions $A + iB$ of the auxiliary problem. Thus, changing one parameter, one can make the function $A + iB$ single-valued, so subjecting the parameters determining the poles of the coordinate system to one relation, in (5.38) we get a single-valued function. For this reason the general single-valued solution of the auxiliary problem contains one less arbitrary parameter than (5.38), i.e., in this solution there are n arbitrary parameters. If we displace the pole of the coordinate system (r_{jk}, θ_{jk}) on the circles C_j, $j = 0, 1$, then the calculations are analogous, but after a circuit of circle C_j, α changes by π instead of 2π.

It remains to show that (5.38) gives all single-valued solutions of the auxiliary problem, and one can choose the arbitrary parameters so that in the solution they occur linearly and homogeneously. Let us first assume that n_0 and n_1 are even, so in (5.36) we have $\nu_0 = \nu_1 = 0$. Now one can construct a particular solution of the auxiliary problem $A_0 + iB_0$, all of whose zeros lie inside D, and on the boundary of D this solution is different from zero. Let $A + iB$ be another arbitrary solution of the auxiliary problem, which is single-valued in D; then the function $F(z) = (A + iB)/(A_0 + iB_0)$ is real on the boundary of D and holomorphic everywhere in D except for $n/2$ poles, which lie at the zeros of the function $A_0 + iB_0$. Thus we have arrived at the following problem: Find a meromorphic function $F(z)$ which has $n/2$ poles in the annulus D and is real on the boundary of the annulus. It is easy to show that this function depends linearly and homogeneously on n arbitrary constants, $n - 1$ of which are factors in an elliptic function [57]. One can also consider the cases where one of the numbers n_0, n_1, or both of these numbers are odd analogously.

We have already noted that for $n > 0$ there does not exist a solution $U + iV$ of the problem $V/U = \tan\varphi$ on the boundary of D, which is not identically zero. Now one can prove this fact differently. We take any single-valued solution $A + iB$ of the auxiliary problem, whose existence for $n > 0$ is proved, and let us assume that there exists a single-valued bounded function $U + iV$, which gives a solution of the homogeneous oblique derivative problem. By the boundary conditions we have $(A + iB)(U + iV) = K \equiv$ const, but since $A + iB$ has zeros in D, $K \equiv 0$, from which it follows that $U = V = 0$, i.e., the homogeneous oblique derivative problem has no solutions which are not identically constant.

Finally, let $n_0 < n_1$, i.e., $n < 0$. It is unnecessary to consider this case specially, since it suffices to note that replacing φ_j by $-\varphi_j$, $j = 0, 1$, changes n to $-n$, so $A + iB$ and $U + iV$ change roles. From this we get the following facts: The auxiliary problem has no bounded single-valued solutions $A + iB$ which are not identically zero; for the homogeneous oblique derivative problem there exist bounded single-valued solutions $U + iV$, where $U + iV$ depends linearly and homogeneously on $n_1 - n_0$ arbitrary parameters. If as the quantity sought one takes the gradient of a solution W of the oblique derivative problem, then one can say that this problem always has bounded single-valued solutions. Now if one requires the function W to be single-valued also, then additional arguments are needed. We have

$$W(x, y) = \int (Udy + Vdx).$$

Since $U + iV$ depends linearly and homogeneously on $-n$ arbitrary parameters, then $W = \lambda_1 W_1 + \ldots + \lambda_{-n} W_{-n}$, where W_k, $k = 1, \ldots, -n$, are linearly independent particular solutions of the homogeneous oblique derivative problem.

We denote by c_k the increment undergone by W_k, $k = 1, \ldots, -n$, after the polar angle θ gets an increment 2π. Obviously some or even all the c_k may be equal to zero. We show that one can replace the system of functions by a new system in which all the functions except possibly one will be single-valued. Let c_α be the first among the numbers c_k, $k = 1, \ldots, -n$, which is different from zero, c_β be the second, and so on, and c_λ be the last number which is different from zero. We preserve the functions $W_1, \ldots, W_{\alpha-1}$, and we replace W_α by the function $Z_\alpha = c_\beta W_\alpha - c_\alpha W_\beta$; then we preserve the functions $W_{\alpha+1}, \ldots, W_{\beta-1}$, and we replace W_β by the function $Z_\beta = c_\gamma W_\beta - c_\beta W_\gamma$, where c_γ is the nonzero number among the numbers c_k, which follows c_β. Obviously after a finite number of such constructions we get a new system of linearly independent functions in which only the function W_λ will be multivalued, and all the other $-n - 1$ functions are single-valued. Thus, the general solution of the homogeneous oblique derivative problem depends on $-n - 1$ arbitrary constants. This solution can depend on $-n$ arbitrary constants, but the latter possibility is only realized when the functions φ_0 and φ_1 in the given problems satisfy an additional condition.

We arrive at the study of the inhomogeneous oblique derivative problem. Let $n_0 > n_1$, i.e., $n > 0$. We know that in this case the gradient of a solution is uniquely determined. One can get this solution using any bounded and single-valued solution $A + iB$ of the auxiliary problem (in the general case one can also use multivalued solutions). Let us assume that n_0 and n_1 are even. Now there exists a solution $A_0 + iB_0$ of the auxiliary problem which has $n/2$ zeros in the domain and vanishes nowhere on the boundary. When $A_0 + iB_0$ is single-valued, one can proceed in exactly the same way as for a simply connected domain. If on the boundary the values of dW/dl are given, then $\sqrt{A_0^2 + B_0^2}\, dW/dl$ is also known on the boundary, and this is the trace on the boundary of the function $P = A_0 W_x + B_0 W_y$, which is harmonic in the domain. From Villat's formula [61] we find the function P, and then we find its conjugate harmonic function Q, which is single-valued only when the means over the boundary circles of the function $\sqrt{A_0^2 + B_0^2}\, dW/dl$ are equal to one another. This gives one condition. The gradient of function W is defined by

$$\frac{d(G+iH)}{dz} = W_x - iW_y = \frac{P+iQ}{A_0+iB_0}.$$

In order that the gradient of W not become infinite at the zeros of the function $A_0 + iB_0$, it is necessary and sufficient that P and Q vanish at these points. Considering that Q is defined up to an additive constant, this gives $n - 1$ solvability conditions. Now together with the condition that Q be single-valued we have n solvability conditions for the inhomogeneous oblique derivative problem.

If $A_0 + iB_0$ is not single-valued, but after each circuit of the annulus D acquires a constant factor, then the function $\sqrt{A_0^2 + B_0^2}\, dW/dl$ is also multiplied by this factor. One can show that there exists a unique function $P + iQ$, which is multiplied by the same factor after each circuit and for which, on the boundary of domain, P coincides with $\sqrt{A_0^2 + B_0^2}\, dW/dl$. In order that the ratio $(P + iQ)/(A_0 + iB_0)$ be bounded in D, it is necessary to impose n conditions, i.e., in this case, too, we get n solvability conditions. In exactly the same way as for a simply connected domain one can prove that the result is independent of the choice of the solution $A + iB$ of the auxiliary problem. Analogously to the simply connected case one considers odd n_0 and n_1, but now $A_0 + iB_0$ has zeros on the boundary of the domain.

We note that if, instead of grad W, one considers W and requires it to be single-valued, then another solvability condition appears, i.e., now we have $n + 1$ solvability conditions for the inhomogeneous oblique derivative problem.

Let $n_1 < n_0$ or $n < 0$. We have already found above that in this case the homogeneous oblique derivative problem has a collection of solutions whose gradients are single-valued and depend linearly and homogeneously on $-n$ arbitrary constants. Now it suffices to construct a particular solution of the inhomogeneous problem. We find the gradient of this particular solution, defined from the equation

$$W_x - iW_y = (P + iQ)(U + iV). \tag{5.39}$$

First let us assume that n_0 and n_1 are even, and we take a particular solution $U_0 + iV_0$, which has $-n/2$ zeros in the domain and vanishes nowhere on the boundary. If $U_0 + iV_0$ is multivalued, then after each circuit it is multiplied by a constant factor, and the harmonic function P, which on the boundary assumes the values $(\sqrt{U_0^2 + V_0^2})^{-1} dW/dl$, is divided by this factor. One can also define the conjugate harmonic function Q, so that it is divided by this factor after each circuit, by choosing the arbitrary constant on which Q depends in a suitable way. The product on the right side of (5.39) will always be bounded and single-valued, i.e., grad W is defined without any conditions. One also considers the cases of odd n_0 and n_1 in the usual way.

One can represent the gradient of the general solution of the inhomogeneous oblique derivative problem in the form

$$\operatorname{grad} W = \operatorname{grad} W_0 + \sum_{k=1}^{-n} \lambda_k \operatorname{grad} W_k'.$$

where W_0 is the particular solution just constructed, and W_k', $k = 1, \ldots, -n$, are linearly independent solutions of the corresponding homogeneous problem, where $W_2', \ldots, W_{-n}'$ are assumed to be single-valued. The following three cases are possible. If W_1' is not single-valued, then for any W_0 one can always choose the constant λ_1 so that $W_0 + \lambda W_1'$ will be single-valued, i.e., in this case the general solution of the inhomogeneous problem will contain $-n - 1$ arbitrary constants. If W_1' and W_0 are single-valued, then the general solution will contain $-n$ arbitrary constants. Now if W_1' is single-valued and W_0 is multivalued, the problem will not have single-valued solutions.

It remains to consider the case $n_0 = n_1$. Here it is necessary to distinguish two situations. Let n_0 and n_1 be odd or even, but the number α introduced above not be equal to zero or to an integral multiple of π. Here there does not exist either a single-valued $A + iB$ nor a single-valued $U + iV$. Here one can repeat the arguments given for $n_0 > n_1$, and come to the following result: the gradient of a solution of the inhomogeneous oblique derivative problem is uniquely determined, and for the solution itself to be single-valued it is necessary to impose one condition on the data of the problem. Let n_0 and n_1 be even and $\alpha = 0$. Now there exist single-valued functions $A + iB$ and $U + iV$, which differ from one another by a constant factor and vanish nowhere in the domain. The homogeneous problem has one linearly independent solution which can be non-single-valued, but its gradient is single-valued. We construct a particular solution of the inhomogeneous oblique derivative problem. We form the function $\underline{(P + iQ)/(A + iB)}$. The function Q is single-valued only when the boundary values $\sqrt{A_0^2 + B_0^2}\, dW/dl$ of function P have identical means over circles C_0 and C_1. The general expression for the gradient of a solution has the form

$$\operatorname{grad} W = \operatorname{grad} W_0 + \lambda \operatorname{grad} W'. \tag{5.40}$$

If the mean values of $\sqrt{A_0^2 + B_0^2}\, dW/dl$ over circles C_0 and C_1 do not coincide, then the oblique derivative problem has no single-valued solutions nor solutions with single-valued gradients.

If the means of the function $\sqrt{A_0^2 + B_0^2}\, dW/dl$ over C_0 and C_1 coincide, then various situations are possible. When the function W' in (5.40) is not single-valued, one can always choose λ so that W will be single-valued, independent of whether W_0 is single-valued or not. If both functions W_0 and W' are single-valued, then for W to be single-valued one condition more is needed than for grad W to be single-valued, so there exists a single-valued solution depending on one constant. When W' is single-valued and W_0 is not single-valued, the problem has no single-valued solutions. Thus, the oblique derivative problem for an annulus is completely investigated.

REFERENCES

1. J. Hadamard, *Lectures on Cauchy's Problem in Linear Partial Differential Equations*, Dover, New York (1923).
2. P. S. Aleksandrov, *Combinatorial Topology* [in Russian], Gostekhizdat, Moscow (1947).
3. N. K. Bari, *Trigonometric Series* [in Russian], Fizmatgiz, Moscow (1961).
4. S. Bergman, *Integral Operators in the Theory of Linear Partial Differential Equations* [Russian translation], Mir, Moscow (1964).
5. A. Yu. Berezin, "Oblique derivative problem," *Differ. Uravn.*, **17**, No. 1, 25–30 (1981).
6. A. V. Bitsadze, *Equations of Mixed Type* [in Russian], Izd. Akad. Nauk SSSR, Moscow (1959).
7. A. V. Bitsadze, *Boundary Problems for Second-Order Elliptic Equations* [in Russian], Nauka, Moscow (1966).
8. A. V. Bitsadze, *Classes of Partial Differential Equations* [in Russian], Nauka, Moscow (1981).
9. B. V. Vainberg and V. V. Grushin, "Uniformly nonelliptic problems," *Mat. Sb.*, **72**, No. 4, 602–636; **73**, No. 1, 126–154 (1967).
10. I. N. Vekua, *Generalized Elliptic Functions* [in Russian], Fizmatgiz, Moscow (1959).
11. V. S. Vinogradov, "An elliptic system which has no Noetherian boundary problems," *Dokl. Akad. Nauk SSSR*, **199**, No. 5, 1008–1010 (1971).
12. M. I. Vishik, "Strongly elliptic systems of differential equations," *Mat. Sb.*, **29**, No. 3, 615–676 (1951).
13. F. D. Gakhov, *Boundary Problems* [in Russian], Nauka, Moscow (1977).
14. E. W. Hobson, *Spherical and Ellipsoidal Harmonics*, Chelsea, New York (1955).
15. E. Goursat, *A Course in Mathematical Analysis*, Vol. 1, Dover, New York (1904).
16. Yu. V. Egorov and V. A. Kondrat'ev, "An oblique derivative problem," *Dokl. Akad. Nauk SSSR*, **170**, No. 4, 770–772 (1966).

17. V. N. Kibirev, "Oblique derivative problem with linear coefficients for harmonic functions," *Differ. Uravn.*, **16**, No. 1, 80–85 (1980).
18. M. A. Krasnosel'skii, A. I. Perov, A. I. Povolotskii, and P. P. Zabreiko, *Vector Fields on the Plane* [in Russian], Fizmatgiz, Moscow (1963).
19. R. Courant, *Partial Differential Equations*, Wiley, New York (1962).
20. A. G. Kurosh, *A Course of Higher Algebra* [in Russian], Fizmatgiz, Moscow (1963).
21. E. E. Levi, "Linear elliptic partial differential equations," *Usp. Mat. Nauk*, **8**, 249–292 (1940).
22. S. Lefschetz, *Differential Equations. Geometric Theory*, Wiley, New York (1963).
23. V. G. Maz'ya, "Degenerate oblique derivative problem," *Mat. Sb.*, **87** (129), No. 3, 417–453 (1972).
24. *Mathematical Encyclopedia. Bitsadze Equation* [in Russian], Vol. 1, Sov. Éntsiklopediya (1977), p. 499.
25. J. Milnor, *Morse Theory*, Princeton University Press, Princeton, New Jersey (1963).
26. K. Miranda, *Partial Differential Equations of Elliptic Type* [Russian translation], IL, Moscow (1961).
27. S. G. Mikhlin, "Differentiation of series of spherical functions," *Dokl. Akad. Nauk SSSR*, **126**, No. 2, 278–279 (1959).
28. S. G. Mikhlin, *Multidimensional Singular Integrals in Integral Equations* [in Russian], Fizmatgiz, Moscow (1962).
29. N. I. Muskhelishvili, *Singular Integral Equations* [in Russian], Fizmatgiz, Moscow (1968).
30. I. G. Petrovskii, *Lectures on the Theory of Integral Equations* [in Russian], Nauka, Moscow (1965).
31. I. I. Privalov, *Cauchy Integral* [in Russian], Saratov (1919).
32. I. I. Privalov, *Introduction to the Theory of Functions of a Complex Variable* [in Russian], Nauka, Moscow (1977).
33. H. Poincaré, *Curves Defined by Differential Equations* [Russian translation], Gostekhizdat, Moscow–Leningrad (1947).
34. R. S. Saks, "Oblique derivative problem," *Soobshch. Akad. Nauk GSSR*, **63**, No. 2, 282–288 (1971).
35. R. S. Saks, *Boundary Problems for Elliptic Systems of Differential Equations* [in Russian], Izd. NGU, Novosibirsk (1975).
36. R. S. Saks, "Noetherian boundary problems for some classes of weakly elliptic systems of differential equations," in *Mathematical Analysis and Related Questions of Mathematics* [in Russian], Nauka, Novosibirsk (1978), pp. 237–253.
37. M. Z. Solomyak, "First-order linear elliptic systems," *Dokl. Akad. Nauk SSSR*, **150**, No. 1, 48–51 (1963).
38. N. E. Tovmasyan, "General boundary problem for second-order elliptic systems with constant coefficients," *Differ. Uravn.*, **2**, No. 1, 3–23 (1966).
39. T. V. Treneva, "Multidimensional analog of the system of A. V. Bitsadze," in *Analytic Methods in the Theory of Elliptic Equations* [in Russian], Nauka, Novosibirsk (1982), pp. 56–58.
40. W. Feller, "Solutions of second-order linear partial differential equations of elliptic type," *Usp. Mat. Nauk*, **8**, 232–248 (1940).

41. V. I. Shevchenko, "Boundary problem for a vector, holomorphic in a half-space," *Dokl. Akad. Nauk SSSR*, **154**, No. 2, 276–278 (1964).
42. V. I. Shevchenko, "Elliptic systems of three equations with four independent variables," *Dokl. Akad. Nauk SSSR*, **210**, No. 6, 1300–1302 (1973).
43. L. P. Eisenhart, *Riemannian Geometry*, Princeton University Press, Princeton, New Jersey (1950).
44. A. Yanushauskas, "Oblique derivative problem for harmonic functions of three independent variables," *Sib. Mat. Zh.*, **8**, No. 2, 447–462 (1967).
45. A. Yanushauskas, "Reduction of the oblique derivative problem to a Fredholm integrodifferential equation," *Differ. Uravn.*, **8**, No. 1, 179–189 (1972).
46. A. Yanushauskas, *Analytic Theory of Elliptic Equations* [in Russian], Nauka, Novosibirsk (1979).
47. A. Yanushauskas, *Analytic and Harmonic Functions of Many Variables* [in Russian], Nauka, Novosibirsk (1981).
48. A. Yanushauskas, "Generalizations of a holomorphic vector," *Differ. Uravn.*, **18**, No. 4, 699–705 (1982).
49. B. L. Borrelli, "The singular second-order oblique derivative problem," *J. Math. Mech.*, **16**, No. 1, 51–81 (1966).
50. G. Giraud, "Nouvelle méthode pour traiter certains problèmes relatifs aux équations du type elliptique," *J. Math. Pures Appl.*, **18**, 111–143 (1939).
51. G. Giraud, "Sur quelques problèmes de Dirichlet et de Neumann," *J. Math. Pures Appl.*, **11**, 389–416 (1932).
52. G. Giraud, "Equations a integrales principales. Etude suivie d'une application," *Ann. Sci. Ecole Norm. Sup.*, **51**, 251–372 (1934).
53. G. Herglotz, "Über die Bestimmung eines Linienelemente in Normalkoordinaten aus dem Riemannschen Krümmungstensor," *Math. Ann.*, **93**, 46–53 (1924).
54. O. D. Kellogg, "On the derivatives of harmonic functions on the boundary," *Trans. Am. Math. Soc.*, **33**, 486–510 (1931).
55. O. D. Kellogg, Foundations of Potential Theory, Springer, Berlin (1929).
56. M. Lavrentieff, "Sur la representation conforme," *Compt. Rend. Acad. Sci. Paris*, **184**, 1407–1409 (1927).
57. A. Liènard, "Problème plan de la dérivée oblique dans la théorie du potentiel," *J. Ecole Polytech.*, Ser. 3, No. 5, 35–158; No. 7, 177–226 (1938).
58. M. Morse and S. S. Cairns, *Critical Point Theory in Global Analysis and Differential Topology*, Academic Press, New York–London (1969).
59. W. Sternberg, "Über die elliptische Differentialgleichungen zweiter Ordnung mit drei unabhöngigen Veränderlichen," *Math. Z.*, **21**, 286–311 (1924).
60. N. Theodoresco, "La derivee areolaire," *Ann. Roum. Math. Bucarest*, No. 3, 3–62 (1936).
61. H. Villat, "Le problème de Dirichlet dans une aire annulaire," *Rend. Circ. Math. Palermo*, **33**, 134–179 (1912).
62. S. Warschawski, "Über das Randverhalten der Ableitung der Abbildungsfunktion bei konformer Abbildung," *Mat. Z.*, **35**, 321–456 (1932).